Pharmaceutical Analysis

Edited by

DAVID C. LEE
GlaxoSmithKline
Stevenage, UK

and

MICHAEL L. WEBB
GlaxoSmithKline
Stevenage, UK

Blackwell
Publishing

CRC Press

© 2003 by Blackwell Publishing Ltd,
a Blackwell Publishing Company

Editorial Offices:
9600 Garsington Road, Oxford OX4 2DQ
 Tel: +44 (0)01865 776868
108 Cowley Road, Oxford OX4 1JF, UK
 Tel: +44 (0)1865 791100
Blackwell Munksgaard, 1 Rosenørns Allé,
P.O. Box 227, DK-1502 Copenhagen V,
Denmark
 Tel: +45 77 33 33 33
Blackwell Publishing Asia Pty Ltd,
550 Swanston Street, Carlton South,
Victoria 3053, Australia
 Tel: +61 (0)3 9347 0300
Blackwell Verlag, Kurfürstendamm 57,
10707 Berlin, Germany
 Tel: +49 (0)30 32 79 060
Blackwell Publishing, 10 rue Casimir
Delavigne, 75006 Paris, France
 Tel: +33 1 53 10 33 10

Published in the USA and Canada (only) by
CRC Press LLC, 2000 Corporate Blvd., N.W.
Boca Raton, FL 33431, USA
Orders from the USA and Canada (only) to
CRC Press LLC

USA and Canada only:
ISBN 0–8493–2814–4

The right of the Author to be identified as the
Author of this Work has been asserted in
accordance with the Copyright, Designs and
Patents Act 1988.

All rights reserved. No part of this publication
may be reproduced, stored in a retrieval
system, or transmitted, in any form or by any
means, electronic, mechanical, photocopying,
recording or otherwise, except as permitted
by the UK Copyright, Designs and Patents Act
1988, without the prior permission of the
publisher.

This book contains information obtained from
authentic and highly regarded sources.
Reprinted material is quoted with permission,
and sources are indicated. Reasonable efforts
have been made to publish reliable data and
information, but the author and the publisher
cannot assume responsibility for the validity
of all materials or for the consequences of
their use.

Trademark notice: Product or corporate
names may be trademarks or registered
trademarks, and are used only for identification
and explanation, without intent to infringe.

First published 2003

British Library Cataloguing-in-Publication
Data
A catalogue record for this title
is available from the British Library

ISBN 1–84127–335–X
Originated as Sheffield Academic Press

Library of Congress
Cataloging-in-Publication Data
A catalog record for this title is available from
the Library of Congress

Pharmaceutical analysis/edited by David C. Lee
and Michael L. Webb. — 1st ed.
 p. cm.
 Includes bibliographical references and index.
 ISBN 1–84127–335–X (hb)
 1. Drugs—Analysis. 2. Pharmaceutical
chemistry. I. Lee, David.
II. Webb, Michael.

RS189.P435 2003
615′.19—dc21

2003005149

Set in 10.5/12 pt Times
by Integra Software Services Pvt. Ltd,
Pondicherry, India
Printed and bound in Great Britain by
MPG Books Ltd, Bodmin, Cornwall

For further information on
Blackwell Publishing, visit our website:
www.blackwellpublishing.com

Pharmaceutical Analysis

Analytical Chemistry Series

Series Editors: John M. Chalmers and Alan J. Handley

A series which presents the current state of the art in chosen sectors of analytical chemistry. Written at professional and reference level, it is directed at analytical chemists, environmental scientists, food scientists, pharmaceutical scientists, earth scientists, petrochemists and polymer chemists. Each volume in the series provides an accessible source of information on the essential principles, instrumentation, methodology and applications of a particular analytical technique.

Titles in the series:

Inductively Coupled Plasma Spectrometry and its Applications
Edited by S.J. Hill

Extraction Methods in Organic Analysis
Edited by A.J. Handley

Design and Analysis in Chemical Research
Edited by R.L. Tranter

Spectroscopy in Process Analysis
Edited by J.M. Chalmers

Gas Chromatographic Techniques and Applications
Edited by A.J. Handley and E.R. Adlard

Chemical Analysis of Contaminated Land
Edited by K.C. Thompson and C.P. Nathanail

Atomic Spectroscopy in Elemental Analysis
Edited by M. Cullen

Pharmaceutical Analysis
Edited by D.C. Lee and M.L. Webb

Contents

Contributors		xiii
Preface		xiv

1 Quality control and regulation — 1
C.J. MOORES

1.1 Introduction — 1
1.2 The quality of medicines — 2
    1.2.1 The meaning of quality — 2
    1.2.2 Medicines are special — 3
    1.2.3 End-product testing — 3
1.3 General quality system requirements — 4
    1.3.1 ISO 9000 — 6
    1.3.2 UKAS — 7
    1.3.3 NAMAS — 8
1.4 Good laboratory practice (GLP) — 9
    1.4.1 Organisation for economic co-operation and development (OECD) GLP guide — 9
    1.4.2 Principles of GLP — 10
1.5 Good manufacturing practice (GMP) — 11
    1.5.1 USA GMP regulations — 11
    1.5.2 EU/UK GMP requirements — 13
    1.5.3 USA/EU GMP differences — 14
    1.5.4 International GMPs — 16
1.6 International harmonisation of quality standards — 16
1.7 Quality control, quality assurance and regulatory filings — 17
    1.7.1 Pre-clinical development — 18
    1.7.2 Early phase development (Phases I/II) — 19
    1.7.3 Late phase development (Phase III) — 20
    1.7.4 Commercial manufacture — 20
1.8 Regulatory inspection key areas — 21
    1.8.1 Inspection of analytical test facilities — 21
    1.8.2 Computerised systems (21 CFR part 11) — 24
    1.8.3 Out-of-specification (OOS) test results — 26
    1.8.4 System audits — 28
1.9 Conclusions and the future of regulatory scrutiny — 29
References — 30

2 Development of achiral separation methods in pharmaceutical analysis — 31
GEORGE N. OKAFO and JOHN K. ROBERTS

- 2.1 Introduction — 31
  - 2.1.1 Historical perspective of separation methods and their uses in pharmaceutical analysis — 32
  - 2.1.2 Regulatory considerations for separation methods in pharmaceutical analysis — 34
- 2.2 General guidance for method development in separation sciences — 34
  - 2.2.1 Separation goals/objectives — 35
  - 2.2.2 Nature of the sample — 38
  - 2.2.3 Choosing the separation technique — 39
  - 2.2.4 Sample pre-treatment and detection — 41
  - 2.2.5 Developing the separation — 43
- 2.3 High performance liquid chromatography (HPLC) — 44
  - 2.3.1 Brief historical perspective of HPLC — 44
  - 2.3.2 Different modes of HPLC — 44
  - 2.3.3 Key developments in HPLC — 45
    - 2.3.3.1 Stationary phase and column technology — 45
    - 2.3.3.2 Instrumentation — 47
    - 2.3.3.3 Microcolumn liquid chromatography — 48
    - 2.3.3.4 Combined HPLC methods — 48
- 2.4 Gas chromatography (GC) — 49
  - 2.4.1 Brief historical perspective — 49
  - 2.4.2 GC in pharmaceutical analysis — 50
  - 2.4.3 Key developments in GC — 52
    - 2.4.3.1 Sensitivity enhancement with large volume injection — 52
    - 2.4.3.2 Thermally labile samples — 52
    - 2.4.3.3 Analytes in complex matrices — 52
    - 2.4.3.4 Detection systems — 53
    - 2.4.3.5 Efficiency increases in GC — 53
    - 2.4.3.6 Automation — 54
- 2.5 Capillary electrophoretic techniques — 54
  - 2.5.1 Brief historical perspective — 54
  - 2.5.2 Developments in detection modes in CE — 55
  - 2.5.3 Different modes and method development options in CE — 55
    - 2.5.3.1 Capillary zone electrophoresis — 55
    - 2.5.3.2 Micellar electrokinetic chromatography — 57
    - 2.5.3.3 Microemulsion electrokinetic chromatography — 57
    - 2.5.3.4 Capillary electrochromatography — 57

2.6		Other separation techniques	58
	2.6.1	Thin layer chromatography	58
	2.6.2	Supercritical fluid chromatography	59
2.7		Hyphenated separation techniques	59
2.8		Use of automated approaches to method development in chromatography	61
	2.8.1	Separation optimisation programmes	61
	2.8.2	Column switching devices	63
2.9		Use of chemometric approaches to method development	66
Abbreviations			67
References			68

3 Chiral analysis of pharmaceuticals   74
W. JOHN LOUGH

3.1		Significance of chirality in pharmaceutical R&D	74
3.2		Evolution of methodologies for chiral resolution	79
3.3		Recent developments in commercial CSP for LC	87
	3.3.1	Polysaccharide-based CSP	87
	3.3.2	Macrocyclic antibiotic CSP	88
	3.3.3	Synthetic multiple-interaction CSP	90
3.4		Role of *historical* CSP	92
3.5		Chiral drug bioanalysis	95
3.6		Preparative chiral separations	96
3.7		Present and future perspectives	98
	3.7.1	Alternatives to chiral LC	98
	3.7.2	Fit for intended purpose?	101
	3.7.3	The future	102
Ancillary reading			103
References			103
Commercial literature			104

4 Nuclear magnetic resonance spectroscopy in pharmaceutical analysis   105
RICHARD J. SMITH and ANDREW J. EDWARDS

4.1			Introduction	105
4.2			Structure elucidation	106
	4.2.1		Background and historical perspective	106
	4.2.2		The move to higher fields	107
	4.2.3		Modern ^{1}H NMR experiments	112
	4.2.4		Nuclei other than the proton	115
		4.2.4.1	^{13}C	117
		4.2.4.2	^{19}F	119

			4.2.4.3 ^{31}P	119
			4.2.4.4 ^{17}O	122
			4.2.4.5 ^{15}N	122
		4.2.5	Computer-assisted structure determination	125
			4.2.5.1 Computer-assisted interpretation	125
			4.2.5.2 Computer-assisted structure elucidation	126
	4.3	On-line separations		127
		4.3.1	LC/NMR	127
			4.3.1.1 Chromatographic considerations	128
			4.3.1.2 NMR mode	130
			4.3.1.3 Solvent suppression in LC/NMR	135
			4.3.1.4 Applications of LC/NMR	135
			4.3.1.5 LC/NMR/MS	137
			4.3.1.6 Capillary LC/NMR	137
		4.3.2	Hyphenation to other separation techniques	138
	4.4	Quantitation		139
		4.4.1	The basics of quantitation	139
		4.4.2	Optimising the experimental parameters	140
		4.4.3	Quantitative impurity determinations	144
		4.4.4	Summary of experimental considerations	145
		4.4.5	Method validation	145
	4.5	Solid state NMR		148
		4.5.1	Introduction	148
		4.5.2	Basic theory of solid state NMR	148
		4.5.3	Methods of assignment of solid state NMR spectra	150
			4.5.3.1 One-dimensional editing methods	150
			4.5.3.2 Two-dimensional solid state methods	151
		4.5.4	Distance measurements in the solid state – recoupling	152
			4.5.4.1 Heteronuclear recoupling	152
			4.5.4.2 Homonuclear recoupling	153
		4.5.5	Application of solid state NMR to pharmaceuticals	153
	Acknowledgements			157
	References			157
5	**Mass spectrometry in pharmaceutical analysis**			**165**
	NEVILLE HASKINS			
	5.1	Introduction		165
	5.2	Mass spectrometry		166
		5.2.1	Sample introduction systems	167
		5.2.2	Ionisation techniques	168
			5.2.2.1 Electron ionisation	168
			5.2.2.2 Chemical ionisation	169

		5.2.2.3	Electrospray	172
		5.2.2.4	Nanospray	173
		5.2.2.5	Atmospheric pressure chemical ionisation (APCI)	173
	5.2.3	Analysers		173
		5.2.3.1	Sectors	173
		5.2.3.2	Quadrupole analysers	174
		5.2.3.3	Quadrupole ion traps	175
		5.2.3.4	Time of flight analysers	176
		5.2.3.5	Ion cyclotron resonance mass spectrometers	177
		5.2.3.6	Hybrid instruments	181
	5.2.4	Ion detection systems		182
		5.2.4.1	Conversion dynode and electron multiplier	182
		5.2.4.2	Conversion dynode and photon multipliers	183
	5.2.5	Data acquisition and processing		183
5.3	Strategies for structural elucidation			183
	5.3.1	Determination of molecular weight		184
	5.3.2	Collisionally induced decomposition and MS^n		186
	5.3.3	Accurate mass measurement		191
5.4	Structural confirmation			192
	5.4.1	Library searches		193
	5.4.2	Using MS^n techniques		194
	5.4.3	Process monitoring		195
5.5	Quantitation			195
	5.5.1	Development of an assay		196
	5.5.2	The calibration process		197
	5.5.3	Use of stable isotopically labelled substance		198
	5.5.4	Use of double labelling		198
5.6	Using the data system			199
	5.6.1	Automation		199
	5.6.2	Data processing		201
	5.6.3	Data mining		201
References				202

6 Vibrational spectroscopy in pharmaceutical analysis   203
CLARE L. ANDERTON

6.1	Introduction			203
	6.1.1	Molecular motion		203
	6.1.2	Infrared spectroscopy		204
		6.1.2.1	Instrumentation for infrared spectroscopy	204
		6.1.2.2	Sample preparation for infrared spectroscopy	204

	6.1.3	Raman spectroscopy	205
		6.1.3.1 Instrumentation for Raman spectroscopy	208
		6.1.3.2 Sample preparation for Raman spectroscopy	209
	6.1.4	Applications	210
6.2	Vibrational spectroscopy to investigate molecular structure		211
	6.2.1	Introduction	211
	6.2.2	Techniques in structural elucidation and specialist applications	211
		6.2.2.1 Tautomerism	211
		6.2.2.2 Hydrogen bonding	211
		6.2.2.3 Chirality	212
		6.2.2.4 Contaminant analysis	212
		6.2.2.5 Products from combinatorial chemistry	213
		6.2.2.6 Theoretical calculations	214
	6.2.3	Routine chemical identification	216
6.3	Vibrational spectroscopy of polymorphs, hydrates and solvates		216
	6.3.1	Introduction	216
	6.3.2	Identifying structural differences between solid-state forms	218
		6.3.2.1 Polymorphs	218
		6.3.2.2 Hydrates and solvates	220
		6.3.2.3 Amorphous material	222
	6.3.3	Quantifying solid-state forms in drug substance	224
	6.3.4	Identifying the solid-state form in formulated product	226
6.4	Vibrational spectroscopy for in situ characterisation		228
	6.4.1	Investigating solid phase transformations	228
		6.4.1.1 Variable temperature spectroscopy	228
		6.4.1.2 Combining spectroscopy with other solid-state techniques	231
	6.4.2	Monitoring chemical reactions and processes	231
	6.4.3	Chromatography detection	234
6.5	Summary		236
Acknowledgements			236
References			236

7 Solid-state analysis and polymorphism   240
ULRICH J. GRIESSER and JOSEPH G. STOWELL

7.1	Introduction		240
7.2	Solid-state properties of drug compounds		242
	7.2.1	Determinant levels of solid-state properties	242

		7.2.2	Types and properties of pharmaceutical solids	244
			7.2.2.1 Crystalline solids	247
			7.2.2.2 Amorphous solids	250
			7.2.2.3 Regulatory aspects and quality control	251
	7.3	Methods of solid-state analysis		253
		7.3.1	X-ray diffraction and crystallography	254
		7.3.2	Spectroscopy	260
			7.3.2.1 Ultraviolet-visible diffuse reflectance spectroscopy	260
			7.3.2.2 Infrared spectroscopy	261
			7.3.2.3 Raman spectroscopy	266
			7.3.2.4 Solid-state NMR	269
		7.3.3	Thermal analysis and calorimetry	273
			7.3.3.1 Thermomicroscopy (hot-stage microscopy)	275
			7.3.3.2 Differential thermal analysis and differential scanning calorimetry	276
			7.3.3.3 Thermogravimetry	281
			7.3.3.4 Thermomechanical analysis	282
			7.3.3.5 Dielectric analysis	283
			7.3.3.6 Microcalorimetry	283
		7.3.4	Vapour pressure determination of solids	283
		7.3.5	Micromeritic measurements	284
			7.3.5.1 Particle-size analysis	284
			7.3.5.2 Particle shape	286
			7.3.5.3 Specific surface area	287
			7.3.5.4 Porosity	287
			7.3.5.5 Density	287
		7.3.6	Other technologies	288
			7.3.6.1 Atomic force microscopy	288
	7.4	Summary		289
	References			289
8	**Microscopy and imaging in pharmaceutical analysis**			**295**
	ROBERT A. CARLTON			
	8.1	Introduction		295
	8.2	Solid-state analysis		295
		8.2.1	Early stage solid-state analysis (polymorph discovery)	297
			8.2.1.1 Optical crystallography	298
			8.2.1.2 Solvent recrystallization experiments	298
			8.2.1.3 Thermal microscopy	300
		8.2.2	Late stage solid-state analysis (polymorph relationships)	304
		8.2.3	Summary	307

	8.3	Particle size and morphology	307
		8.3.1 Particle morphology	309
		8.3.2 Particle size analysis/image analysis	313
		8.3.3 Summary	317
	8.4	Contaminant identification	319
	8.5	Conclusion	321
		References	322

9 Process analysis in the pharmaceutical industry — 324
MARTIN WARMAN and STEVE HAMMOND

	9.1	Introduction	324
	9.2	Pharmaceutical manufacturing	325
		9.2.1 Drug substance manufacture	326
		9.2.1.1 Raw material testing	326
		9.2.1.2 Reaction mixture	328
		9.2.1.3 Reaction monitoring	328
		9.2.1.4 Crystallisation monitoring	338
		9.2.1.5 Dryer monitoring	340
		9.2.1.6 Monitoring the milling process	341
		9.2.1.7 Cleaning monitoring	343
		9.2.2 Drug product manufacturing	344
		9.2.2.1 Raw material monitoring	345
		9.2.2.2 Formulation monitoring	348
		9.2.2.3 Tablet cores	353
		9.2.2.4 Tablet coating	355
		9.2.2.5 Packaging	355
	9.3	Conclusions	355
		References	356

Index — 357

Contributors

Clare L. Anderton	GlaxoSmithKline, Gunnels Wood Road, Stevenage, Hertfordshire, SG1 2NY, UK
Robert A. Carlton	GlaxoSmithKline, 709 Swedeland Road, King of Prussia, PA, 19406, USA
Andrew J. Edwards	GlaxoSmithKline, Gunnels Wood Road, Stevenage, Hertfordshire, SG1 2NY, UK
Ulrich J. Griesser	Institute of Pharmacy, Department of Pharmaceutical Technology, Univeristy of Innsbruck, Innrain 52, A-6020 Innsbruck, Austria
Steve Hammond	Pfizer Ltd, Ramsgate Road, Sandwich, Kent, CT13 9NJ, UK
Neville Haskins	19 Anthill Close, Denmead, Hants, UK
W. John Lough	Institute of Pharmacy, Chemistry & Biomedical Sciences, University of Sunderland, Wharncliffe Street, Sunderland SR1 3SD, UK
Clive J. Moores	Bramble Cottage, Middle Road, Bells Yew Green, Tunbridge Wells, Kent, UK
George N. Okafo	GlaxoSmithKline, Gunnels Wood Road, Stevenage, Hertfordshire, SG1 2NY, UK
John K. Roberts	GlaxoSmithKline, Gunnels Wood Road, Stevenage, Hertfordshire, SG1 2NY, UK
Richard J. Smith	GlaxoSmithKline, Gunnels Wood Road, Stevenage, Hertfordshire, SG1 2NY, UK
Joseph G. Stowell	School of Pharmacy, Department of Industrial & Physical Pharmacy, Purdue University, West Lafayette, IN, 47907-2091, USA
Martin Warman	Pfizer Ltd, Ramsgate Road, Sandwich, Kent, CT13 9NJ, UK

Preface

The use of analytical sciences in the discovery, development and manufacture of pharmaceuticals is wide ranging. From the analysis of minute amounts of complex biological materials to the quality control of the final dosage form, the use of analytical technology covers an immense range of techniques and disciplines. This book concentrates on the analytical aspects of drug development and manufacture, focusing on the analysis of the active ingredient or drug substance. The book does not describe in detail the analysis of drug products – typically drug substances and excipients in a formulation. However, many of the approaches described for characterising drug substances can be applied to formulations, either by accounting for the presence of excipients in the analyte or by extracting and analysing the active ingredient.

The pharmaceutical industry is one of the most active areas for the application and development of new methods in the analytical sciences. This volume provides those joining the industry or other areas of pharmaceutical research with a source of reference to a broad range of techniques and their applications, allowing them to choose the most appropriate analytical technique for a particular purpose.

No book on the analysis of pharmaceutical materials should ignore the important area of quality and regulation. The first chapter provides an up-to-date overview of the philosophy and practicalities of working in a regulated environment, with reference to current regulations and guidance.

Subsequent chapters cover the major disciplines of separation sciences and spectroscopy. Recognising the importance and breadth of the area of separation sciences, our authors concentrate on method development in high performance liquid chromatography (HPLC or LC), capillary electrophoresis (CE), gas chromatography (GC) and thin layer chromatography (TLC), discussing traditional approaches in addition to the newer computational and chemometric methods.

One of the most important and challenging areas in the analysis of pharmaceuticals is the determination of chiral purity. It is therefore highly appropriate that a chapter is devoted to this area. The importance of chiral analysis is described, together with the development of techniques across the separation sciences and beyond.

In considering the spectroscopies, the development and widespread use of coupled techniques forms a major part of the volume in the chapters covering nuclear magnetic resonance (NMR) and mass spectrometry (MS). In the NMR chapter, extensive coverage is given to state-of-the-art coupled LC/NMR. The chapter also covers multi-nuclear NMR, computer-aided spectral interpretation, quantitative NMR and solid-state NMR – all important techniques applied in the pharmaceutical development laboratory.

Recent years have seen many important developments in MS. This book devotes a chapter to the technique, focusing on the varied instrumental capabilities, their basic principles of operation and their applicability to pharmaceutical analysis. The applications of mass spectrometry, both in structure elucidation and quantitative analysis, are considered. Quantitative analysis MS is covered in detail, to allow the reader to gain background knowledge of a technique that is becoming important in drug substance analysis.

Vibrational spectrosopy can be used to support structural elucidation by NMR and MS, but more typically it is used for identity testing, because IR and Raman spectra act as a fingerprint for molecular structure. However, both IR and Raman find their principal application in the investigation of polymorphism. Examples are described in this chapter, together with the benefits of coupling these techniques to microscopy.

Additional solid-state techniques are covered in the chapter on solid-state analysis and polymorphism. The determination and control of the solid-state form, in respect of both crystal structure and particle characteristics, are important. The physical properties of the drug substance will influence its behaviour during handling processes and formulation, and can have a dramatic effect on dissolution, solubility and therefore bioavailability.

Although microscopy and imaging are used in a number of the above disciplines, a separate chapter is devoted to the use of optical and electron imaging techniques and image analysis, which play an increasingly important role in contaminant analysis and drug-excipient distribution.

The final chapter deals with the increasingly important area of process analytical science. There is increasing interest in in-process measurements from the FDA and other regulatory bodies, and this seems certain to be a rapidly expanding area of the analytical sciences in the pharmaceutical industry.

The regulatory, environmental, technological and commercial drivers in the pharmaceutical industry have profound implications for the analytical chemist. We hope that this volume, contributed by specialists from both the industrial and the academic sectors, will prove to be a useful source of reference for all those interested in this rapidly changing field of science.

David C. Lee
Michael L. Webb

1 Quality control and regulation
C.J. Moores

1.1 Introduction

Any person working within the pharmaceutical industry or allied/support industries in a scientific capacity will be well aware that a good deal of their everyday work requires conformance to quality standards dictated by various regulatory bodies. For a scientist coming directly from university into the industry, there may well be some culture shock and bewilderment caused by the plethora of standards and procedures that require to be followed as part of their job. Additionally, it may be disturbing to discover that non-compliance with these quality standards and regulatory requirements could result in severe penalties for their employer and loss of their own job.

One of the major disciplines impacted by these requirements is the analytical sciences. The issue could arise in a variety of departments that either directly or indirectly provide data for the assessment of pharmaceutical materials or the support of various regulatory filings. These pharmaceutical materials could be drug products (the formulated dosage form), active pharmaceutical ingredients (API, also referred to as drug substance or bulk pharmaceutical chemical), raw materials, starting materials and intermediates used for the production of these pharmaceutical materials, or even materials derived from toxicology experiments. The filings could, for example, be for an investigational new drug application (IND), a new drug application (NDA) or a marketing authorisation application (MAA). This list is by no means exhaustive, but is meant to illustrate the variety of roles governed by regulatory bodies that may be encountered by an analyst within the pharmaceutical industry, whether in an R&D or a commercial environment. These terms will be explained more fully later when it will also become clear that every aspect of the work that analysts perform is influenced by the requirements of these regulatory bodies and the need for analysts to be able to defend their work.

This chapter provides an overview of the quality systems and regulations an analyst may encounter in the pharmaceutical industry and the reasons for these systems, as well as a description of the various regulatory bodies that they may have to interact with. The overview may serve as refresher training for those working in the industry at present, and as a tool to aid the understanding of the extent of regulation for those scientists thinking of entering the industry or who work outside of the pharmaceutical industry or in allied industries. The most current detail of requirements may be found in the various references quoted, which will normally direct the reader to an appropriate website.

We shall concentrate on the area of new chemical entities (NCEs) rather that new biological entities or biopharmaceuticals. These latter materials are governed by very similar regulations and quality requirements, but are somewhat outside of the experience of the author. The focus will be on regulatory requirements in the USA and Europe.

1.2 The quality of medicines

1.2.1 The meaning of quality

I could not continue to discuss quality systems, regulation and regulatory bodies that have jurisdiction over the pharmaceutical industry, without first discussing the meaning of quality in the context of medicines and why quality and hence regulation, are so important both in the commercial and R&D environments.

Most people have their own subjective view of the meaning of *quality* in everyday life. When asked to define exactly what they understand by the term, then their definitions tend to be rather woolly and it is obvious that there are different meanings depending on the environment that the term is used in. The traditional view of *quality* derives from the inspection/measurement approach used in quality control when *quality* consists of conformity with a pre-determined specification. For example, 'quality is the degree to which a specific product conforms to a design or specification' [1]. The view of *quality*, especially within the context of the pharmaceutical industry and quality assurance, has now moved away from this rather narrow view, and the definition more or less accepted at present tends to be *fitness for purpose* [2]. The International Organisation for Standardisation (ISO) has further refined this definition in the application of the ISO 9000 quality standards to mean *fitness for purpose with customer satisfaction*. This is to take into account the fact that the ISO 9000 standard covers an extremely wide range of products and services where customer satisfaction is the main determinant of quality.

When this definition of *fitness for purpose* is applied to the pharmaceutical industry, it can be seen to fit well with the concepts of quality. Personnel have to be shown to be *fit for purpose* for the job they are employed to do (training, education and experience). Equipment needs to be demonstrated as *fit for purpose* based on qualification/validation, maintenance and calibration. Manufacturing processes are deemed *fit for purpose* based on in-process testing, process validation, etc. Process materials are shown to be *fit for purpose* based on testing appropriate to their intended use in the process. This use will vary depending on the stage of processing and whether it is a primary (chemical) process or a secondary (formulation) process.

1.2.2 Medicines are special

Why are medicines different from other consumer products? Why is there so much regulation of the pharmaceutical industry? Why, in the light of modern analytical techniques, can't we rely on thorough testing of the final product (end-product testing) and do away with all of these bureaucratic systems? These are just some of the questions that are always asked of quality professionals in the pharmaceutical industry and I will try and answer them.

Medicines are special because virtually no other product is consumed by the public on such utter trust – trust that the medicine will not do them more harm than the illness it is meant to cure. Their doctor prescribes the medicine; it is dispensed by a pharmacist and is taken by the patient in the belief that it will cure their ill health and that it will not make it worse! This trust can only be assured if the medicine has been adequately tested during development. This testing should assure that side effects have been established and that the medicine is efficacious. When the medicine is given to patients it must have been appropriately manufactured, tested and packaged to assure that:

- It is the correct product.
- It is the correct strength.
- It has not degraded.
- It is free from harmful impurities and micro-organisms.
- It has not been contaminated.
- It is correctly labelled.
- It is properly sealed in a suitable container.

1.2.3 End-product testing

End-product testing (quality control) is the reliance only on appropriate analytical tests to demonstrate the quality of a medicinal material. End-product testing alone is considered as *testing quality into the product*. You only accept those materials that pass specification. This is not adequate to ensure that the medicine is free from all manufacturing faults, that test methods employed are adequate to establish its purity and that a small proportion of defective materials would be detected. I will give several examples to illustrate this point.

Example 1.1. All pharmaceutical industry regulators prohibit the manufacture of penicillins in the same facility as other medicinal materials. The reason for this is the very high potential of extremely low levels of penicillin to cause serious side effects in individuals who are sensitised to this class of compounds. For this reason, cross-contamination levels are set at zero and the only way to ensure compliance is to not have the penicillin there in the first place. Firstly, the penicillin may contaminate at such a low level that existing analytical methods would not detect the presence of it, even if it were suspected of being present. Secondly, the penicillin contamination may not be distributed evenly in the medicinal material, making detection very difficult.

Example 1.2. (This example is taken from a real incident.) A solution of a chemical intermediate used for the production of an API was stored temporarily whilst waiting further processing. The storage containers used had previously contained a potent pesticide that had not been completely removed. These pesticide residues continued through the whole process to the final drug product, without detection. Apparently, the chromophore of the pesticide was very different from that of the material under test at each stage of manufacture and the analytical methods being used could not detect these low levels of contamination.

In conclusion, end-product testing is not adequate for the following reasons:

- You only test for what you expect to find.
- You only test a small portion of the bulk of the test material. (There could be a small proportion of defective material in a batch, which only 100% testing would detect, e.g. microbiological contamination of sterile vials.)
- It is doubtful whether, in the case of medicines, the consumer would detect defects.

To assure the quality of medicinal products, quality must be built in at each stage of the manufacturing process and not merely tested in. Any factor that could have an effect on the quality of the final medicinal product must be controlled. These factors could be anything from the design of the production facility used or the environment the material is isolated in, to the analytical test methods employed at each stage of production. The philosophy of quality assurance is that batch to batch consistency should be maintained by reducing variability of all supporting processes, sub-processes and procedures. Hence, if written procedures control all of these factors and trained personnel follow these procedures, then a product consistently meeting its predetermined specification should be produced. End-product testing then becomes just a final check of the quality of the product. This testing is then used in conjunction with the written records, which demonstrate that all critical factors have been controlled, as the supporting documentation to allow the material to be released for use. This is what is meant by a quality system.

1.3 General quality system requirements

There are a number of quality systems that may be encountered by the pharmaceutical analyst but by and large the major one will be good manufacturing practice (GMP). Compliance to GMP is a requirement for the manufacture and testing of a drug product or API destined for human use, whether in the context of commercial manufacture or a clinical trial. Similarly, data supporting non-clinical safety assessment of chemicals requires to be generated to good laboratory practice (GLP). The majority of other quality systems are voluntary and are commercially driven.

Before discussing the various quality systems encountered by the pharmaceutical analyst, I would like to more clearly define some of the quality terms that are used to ensure some consistency, as these definitions may vary depending on the quality system. These definitions are taken from the GMP guide for API, ICH Q7A [3] as there is a good chance that more than one interested group has input into the definition.

Quality control (QC): Checking or testing that specifications are met.

Quality assurance (QA): The sum total of the organised arrangements made with the objective of ensuring that all materials are of the quality required for their intended use and that quality systems are maintained.

Quality unit: An organisational unit, independent of production, which fulfils both QA and QC responsibilities. This can be in the form of separate QA and QC units or a single individual or group, depending on the size and structure of the organisation.

Quality: Fitness for purpose.

Good manufacturing practice (GMP): GMP is that part of QA which ensures products are consistently produced and controlled to the quality standards appropriate to their intended use and as required by the marketing authorisation or product specification. GMP is concerned with both production and QA [4].

In previous sections, I have given a number of reasons why the whole aspect of medicine development, manufacture and testing requires to be performed in a different manner to that of any other product. There are many documented examples in the history of the pharmaceutical industry that demonstrate the necessity for regulations and regulatory bodies to oversee compliance to these regulations. There are cases of people being harmed by inadequate toxicity testing during development (e.g. thalidomide). There have been examples of the inadequacy of systems to ensure the integrity of tests performed in support of drug registration (resulting in the introduction of GLP). There have been not a few examples of the lack of assessment of the effect on the quality of drugs resulting from changes to the manufacturing environment and manufacturing route (resulting in the establishment of Good Manufacturing Practices Regulations). The establishment of these regulations has been independently mirrored outside of the pharmaceutical industry by the introduction of quality systems, such as ISO 9000, in an attempt to bring consistency of quality to other industries and services.

There is a common theme to all quality systems whether used for pharmaceutical production or not.

- The quality system must be described in written documents approved by management (policies, quality manual, etc).
- The quality system must be regularly reviewed.
- There must be senior management involvement in quality.
- All operations that can effect quality must be described in written and approved procedures (standard operating procedures, SOPs).
- Materials must be appropriately approved prior to use.
- Output (product) must be appropriately inspected prior to release.
- Equipment used must be fit for purpose (qualified/validated, calibrated and maintained).
- Personnel must be trained in the quality system and in operations they perform.
- Written records must be kept to demonstrate quality procedures have been followed.
- There must be regular internal quality audits to ensure quality is maintained.

In conclusion, all quality systems are to do with people, materials, equipment, records and procedures.

I will now give an overview of the quality systems that the analyst may come across while working in the pharmaceutical or allied industry, in either an R&D or a commercial environment.

1.3.1 ISO 9000

This quality standard is a voluntary standard operated by many industries worldwide. Compliance to this standard demonstrates to customers that defined systems have been followed for the design, manufacture and/or testing of products.

The ISO 9000 quality standard was developed from the original BS 5750 quality standards and was issued in 1987. The standard was adopted as a European standard, EN29000, in 1988, and is one set of a whole set of ISO standards. The main part of the ISO 9000 standard is made up of three separate standards:

ISO 9001 Covers design, development, production, installation and servicing.

This is the most comprehensive of the three ISO 9000 systems and is applicable to a supplier involved in original design. It is also commonly applicable to service activities where the service is being designed to meet specific requirement. This is typically found in contractors to the pharmaceutical industry where the development of a manufacturing process is performed. Similarly, it will be relevant to developers of computer software, although the ISO 9001-TickIT standard (specific to the development of software) would be more appropriate in this case.

ISO 9002 Covers production, installation and servicing.

This system is identical to ISO 9001 except it does not include the design part. This is the most common system used worldwide and can be applied to either manufacturing or service industries providing a standard product or service. A number of pharmaceutical manufacturers are certified to this standard.

ISO 9003 Covers final inspection and test.

This system has limited value and application as it encourages quality to be inspected in rather than designed in. It would be rare to find this standard applied within the pharmaceutical industry as the output of data is normally considered to be the provision of a service under ISO 9002.

These standards are operated as follows:
A national Accreditation Body, such as the United Kingdom Accreditation Service (UKAS), accredits a certification body (e.g. Lloyds, BSI, etc.) who in turn certifies individual companies that make a voluntary application to them. This certification is based on successful audits by trained ISO 9000 auditors. The company maintains its ISO standards by means of internal auditors and by annual or biannual audits and regular follow-up audits from the certification body.

These standards are a voluntary set of worldwide standards that a whole range of industries and services have adopted. Their relevance to the pharmaceutical industry has been questioned in the past, particularly by the USA Food and Drug Administration (FDA), on the basis that the system ensures consistency of development, manufacture and testing, but does not address the key issue of product quality, this being left to the customer and supplier to agree. However, these standards have been adopted by a number of pharmaceutical companies as a means of laying a quality foundation with respect to quality management and on which to base and support further quality systems such as GMP. The ISO 9000 quality systems will also be followed by the majority of material suppliers and contractors to the pharmaceutical industry where the use of these materials and services does not require compliance to a higher standard such as GLP or GMP. In fact, it would be unusual for most pharmaceutical industries to accept any of these materials or services from a company that did not follow the appropriate ISO 9000 standard. For these reasons it is important for those working in the pharmaceutical industry to have some understanding of these standards.

The latest up date of the ISO 9001 standard is ISO 9001–2000 and may be obtained via the ISO's Website [5].

1.3.2 UKAS

Analytical scientists may encounter UKAS in a number of work environments. UKAS is the sole national accreditation body recognised by government to assess, against internationally agreed standards, organisations that provide certification, testing, inspection and calibration services. UKAS is a non-profit-distributing company, limited by guarantee, and operates under a memorandum of understanding with the government through the Secretary of State for Trade and Industry [6].

The knowledge that conformity assessment organisations are accredited by UKAS gives the confidence that they have been independently evaluated for their impartiality, competence and performance capability.

One of the responsibilities of UKAS is the accreditation of laboratories to issue NAMAS (National Accreditation of Measurement and Sampling) certificates and reports.

1.3.3 NAMAS

The NAMAS designation on a report or calibration gives the assurance that the work has been performed to the highest standards and that the laboratory has been stringently assessed by independent experts. There is further assurance that the work has been performed according to agreed methods and specifications and that all measurements are traceable to national and international standards.

A laboratory may apply to UKAS for accreditation with respect to specific tests or calibrations. The laboratory is assessed by UKAS for that specific work and if it meets NAMAS requirements, then the laboratory will be accredited for those areas of work and can then issue NAMAS reports and certificates. UKAS publishes the NAMAS Concise Directory that lists all accredited laboratories and services.

As part of the accreditation process, UKAS assesses all technical aspects of the laboratory's practices and organisation and not just the quality system. Typical areas of assessment would be:

- Organisation
- Quality system
- Quality audits and review
- Personnel
- Equipment
- Measurement traceability
- Methods and procedures
- Environment
- Sample handling
- Records
- Complaints
- Sub-contractors and purchasing.

This assessment is very similar to the assessment performed for ISO 9000 certification and laboratories meeting NAMAS requirements for calibration and testing, where they are the supplier of these services, comply with the requirements of ISO 9002.

NAMAS accreditation is similar to the ISO 9000 certification process, in that it requires a thorough assessment by independently appointed industry experts. Six months following accreditation, a full follow-up visit is made by UKAS, and annual audits are made thereafter. Four years after accreditation a full re-accreditation assessment is made, although UKAS can make unannounced visits at any time. Laboratories found to be unsatisfactory on inspection will lose NAMAS accreditation until such time as that laboratory again meets the required standard.

NAMAS accreditation is an acceptable quality standard in a large number of countries both within the European Union and outside, and there exist a number of memoranda of understanding with these countries for mutual acceptance of standards.

Standards for the operation and accreditation of laboratories were originally set and published by the European Committee for Standardisation (CEN) in EN45001 and EN45002, (equating to ISO Guides 25 and 54) which were equivalent to the British standards, BS7500 series. The International Standard ISO/IEC 17025:1999 entitled *General requirements for the competence of testing and calibration laboratories* now replaces ISO/IEC Guide 25 and EN45001.

The pharmaceutical industry accepts a number of UKAS/NAMAS standards for a large variety of calibration and tests, although most companies would normally audit to confirm the acceptability of these standards in the case of very critical calibrations or tests.

1.4 Good laboratory practice (GLP)

1.4.1 Organisation for economic co-operation and development (OECD) GLP guide

The principles of GLP were originally developed under the auspices of the OECD and were first published in 1981 with later updates and guidance documents [7]. These principles are not legally binding but all OECD member countries have agreed to abide by them. However, the European Directive 87/18/EEC (amended by Commission Directive 1999/11/EC) requires that all EU member states must incorporate the OECD Principles of Good Laboratory Practice and Monitoring into national legislation. This has been accomplished in the UK as what is commonly known as The GLP Regulations [8]. In the USA, the GLP regulations have been incorporated into national law under Code of Federal Regulations (CFR) 21 part 58 and may be found on the FDA's web site [9].

1.4.2 Principles of GLP

The principles of GLP define a set of rules and criteria for a quality system concerned with the organisational process and the condition under which *non-clinical* health and environmental safety studies are planned, monitored, recorded, archived and reported. These principles have been developed to promote the quality and validity of data generated in the testing of chemicals. Apart from assuring the quality of data obtained from these studies that may have implication for human health, the other main advantage is the recognition by regulatory authorities of one country, of the data generated in other countries which hence avoids duplicative testing.

Any facility that produces data in support of these non-clinical studies is required to comply with the principles of GLP and will be audited for compliance by the appropriate regulatory authority. In the case of the UK this would be the United Kingdom GLP Compliance Monitoring Authority, which is part of the Department of Health. In the USA this would be the appropriate FDA department.

GLP must not be confused with that part of GMP that is concerned with operations in QC laboratories that assure the quality of medicines for human use (sometimes referred to as *good quality control laboratory practice*). Although there are a number of similarities in the practical aspects, GLP is in place to assure the integrity and quality of data; GMP is there to assure the quality of the product, i.e. its conformance to specification. There have been a number of statements recently from FDA sources on this subject [10].

- '(GLP)...is not the same as lab. work that tests finished drugs and active pharmaceutical ingredients. The GMP regulations have specific requirements for drug lab controls.'
- '....manufacturers should not confuse good manufacturing practices (GMPs) with good laboratory practices (GLPs). The issue occasionally arises, as some firms confuse the terms.'
- 'It's a semantic issue....but you can be sure there is no confusion at FDA. FDA does not care how a company refers to its practices. It cares about what those practices are.'
- 'While drug makers must account for many of the same issues in GMPs and GLPs, experts agreed, assessing quality and conformation to product specification is not the same as proving safety.'

One could of course argue that assurance of the integrity of data is also a requirement under any other quality system such as GMP. This assurance requires that analysts are trained, that procedures are written and approved, that analytical equipment is calibrated and maintained, that reagents and test materials are controlled and that accurate records and original raw data are kept.

1.5 Good manufacturing practice (GMP)

GMP is probably the most widespread quality system followed across the pharmaceutical industry as a whole. GMP compliance is a requirement within the R&D environment for the manufacture and testing of clinical trial materials (both drug product and API) and for commercial manufacture and testing of these materials for human and animal consumption. R&D facilities performing these operations may be subject to audit for compliance to GMP; commercial facilities *will* be audited by the appropriate regulatory authority, possibly without prior warning.

1.5.1 USA GMP regulations

The USA Food, Drugs and Cosmetics Act (FD&C Act) states that 'All drugs shall be manufactured, processed and packaged in accordance with current good manufacturing practice' [Section 501 (a)(2)(B)]. No distinction is drawn between the manufacture of drug products (secondary manufacture) and the manufacture of APIs (primary manufacture). It is also noted in the preamble to the FD&C Act that the act applies to all drugs for human use, and this therefore includes the requirement for both APIs and drug products manufactured for clinical trials, to be manufactured according to cGMP.

The requirements for compliance to cGMP are lain down in the following Code of Federal Regulations (21CFR):

Part 210 Current Good Manufacturing Practice in manufacturing, processing, packing or holding of drugs.
Part 211 Current Good Manufacturing Practice for finished pharmaceuticals.

It must be noted that the US regulations refer to *current* GMP. The regulations as detailed in 21CFR parts 210 and 211, give the pharmaceutical manufacturer plenty of scope to interpret the requirements appropriately for his specific facility and process, but in doing this the regulations require the manufacturer to adopt best *current* practice. The onus is placed upon the manufacturer to keep current with what the industry is doing (best practice), with what the current interpretation of the regulations are, and what the US FDA's expectations are.

Although the FD&C Act requires all drugs (products and APIs) to be manufactured to cGMP, the regulations 21CFR parts 210 and 211 are only mandatory for the manufacture of drug products and not APIs. In the past the onus has been on the pharmaceutical industry to interpret these requirements with respect to the manufacture of APIs. FDA has published guidelines in the form of guides for FDA investigators, to assist industry to meet compliance to cGMP and to place their interpretation on cGMP requirements for APIs and a number of other key areas such as impurities in new drugs, allowable solvent residues and stability testing. Guides issued by ICH have now supplemented most of these guidelines and these, along with other FDA guidelines, will be discussed in more detail later in this chapter.

These regulations and guidelines may not always be appropriate for the manufacture of clinical trial materials. Although most of the regulations are reasonably applicable in an R&D drug product environment they may become inappropriate where attempts are made to apply them to the early manufacture of clinical APIs within an R&D environment. It is only with the issue of the International Conference on Harmonisation (ICH) Harmonised Tripartite Guideline ICH Q7A – *Good Manufacturing Practice Guide for Active Pharmaceutical Ingredients* [3] in November 2000, that the worldwide pharmaceutical industry finally received detailed guidance for manufacture of APIs for both commercial and R&D purposes. (The scope and more detail of this guide will be discussed later.)

If one looks at the major headings of 21CFR part 211, the similarity with other quality systems becomes apparent. It is mandatory to have controls in the following areas:

Subpart A – General provisions
Subpart B – Organisation and personnel
Subpart C – Buildings and facilities
Subpart D – Equipment
Subpart E – Control of components and drug product containers and closures
Subpart F – Production and process controls
Subpart G – Packaging and labelling control
Subpart H – Holding and distribution
Subpart I – Laboratory controls
Subpart J – Records and reports
Subpart K – Returned and salvaged drug products.

The areas of these regulations that will be most important for a pharmaceutical analyst will be:

Organisation and personnel – this includes the requirement to have a QC unit having... responsibility and authority to approve and reject all components, drug product containers, closures, in-process materials, packaging materials, labelling and drug products, and the authority to review production records to assure that no errors have occurred or, if errors have occurred, that they have been fully investigated. Further requirements cover laboratory facilities and the responsibility of the quality unit for approving or rejecting all materials, specifications and procedures. The responsibilities of the quality unit must be described in written procedures.

Laboratory controls – this part covers mainly calibration of equipment, testing and release procedures, stability testing, reserve samples, laboratory animals and penicillin contamination.

Records and reports – this part describes the key records that require to be retained. These include drug product component and container/closure records, labelling records, production records, production record review, laboratory records, distribution records and complaint files.

These requirements can be further compared with the ICH guidelines for API manufacture later in this chapter when I discuss worldwide harmonisation.

In conclusion, the USA cGMP regulations apply to interstate commerce within the USA and to any facility worldwide, that exports pharmaceutical materials (drug products, APIs, or components of these products) to the USA or, wishes to perform clinical trials in the USA. These facilities are open to inspection for cGMP compliance by US FDA inspectors and for those facilities found to be in non-compliance with these requirements the material will be deemed *adulterated with respect to identity, strength, quality and purity*. Products from these facilities will be refused entry for sell or use within the USA. Data from these facilities may not be accepted in support of regulatory filings.

1.5.2 *EU/UK GMP requirements*

Two European directives lay down the principles and guidelines for GMP in the EU, one for medicinal products for human use [11] and the other for veterinary products [12]. These directives have been incorporated in the national law of member states. The European Commission has issued nine volumes of *the rules governing medicinal products in the EU*. The latest edition was issued in 1998. Volume four covers GMP for medicinal products for human and veterinary use. These are now used as a basis for inspection by the various national regulatory authorities (e.g. Medicines Control Agency (MCA) in the UK).

If one looks at the requirement of the EU GMP rule, the similarity with 21CFR part 211 is clear, as is the consistency with other quality systems. The basic requirements are detailed under the following chapter headings:

Chapter 1: Quality assurance
Chapter 2: Personnel
Chapter 3: Premises and equipment
Chapter 4: Documentation
Chapter 5: Production
Chapter 6: Quality control
Chapter 7: Contract manufacture and analysis
Chapter 8: Complaints and recall
Chapter 9: Self-inspection

There are a further 14 Annexes:

Annex 1: Manufacture of sterile medicinal products
Annex 2: Manufacture of biological medicinal products for human use

Annex 3:   Manufacture of radiopharmaceuticals
Annex 4:   Manufacture of veterinary medicinal products other than immunologicals
Annex 5:   Manufacture of immunological veterinary medicinal products
Annex 6:   Manufacture of medicinal gases
Annex 7:   Manufacture of herbal medicinal products
Annex 8:   Sampling of starting and packaging materials
Annex 9:   Manufacture of liquids, creams and ointments
Annex 10: Manufacture of pressurised metered dose aerosol preparations for inhalation
Annex 11: Computerised systems
Annex 12: Use of ionising radiation in the manufacture of medicinal products
Annex 13: Manufacture of investigational medicinal products
Annex 14: Manufacture of products derived from human blood or human plasma

1.5.3 USA/EU GMP differences

Historically there have been distinct and fundamental differences between USA regulation and EU/UK requirements for GMP. As discussed previously, the US required all *drugs* to be made to GMP requirements and performed inspections throughout the world in support of these requirements. In the UK, only drug products and biological manufacturers (not APIs, except some specified antibiotics) were inspected by the regulatory authority for compliance to GMP. Other EU countries, such as France and Italy, did require audits of API manufacturers, but the requirements and standards varied widely throughout the EU.

Although drug product manufacturers have always been audited by the UK authorities, the UK GMP guideline (The Orange Guide) was not mandatory and did not have the force of law. (Although non-compliance with GMP would result in non-approval of the facility to manufacture drug products for selling.) The original European Directive [13] defined a medicinal product as 'Any substance or combination of substances presented for treating or preventing disease in human beings or animals.' This applied to finished pharmaceutical dosage forms (drug products) only.

There are fundamental differences between a drug product and a *starting material* (API) that makes the application of many GMP drug product requirements difficult or inappropriate. An API is normally prepared by chemical processes that involve purification at each stage of manufacture, and early raw materials and processing stages may not have much influence over the quality of the final API. Impurities that are present in the final API will not be removed and will still be present in the manufacturer drug product. Similarly, if the morphic form of the API is changed through unassessed changes in the API manufacture, this could have a considerable effect on the bioavailability of the drug product. To use the API based on end-product testing, as previously discussed, is not in keeping with the principles of QA.

Historically, in the UK and Europe, there has been no legal requirement to manufacture drug products or APIs to GMP for use in clinical trials (investigational medicinal products, IMPs). This has always been a requirement under the USA FD&C Act.

The situation in the EU with respect to APIs and IMPs is now changing with the requirement for consistent standards throughout the EU and the wish to harmonise inspection standards and other regulatory requirements with other countries. The lack of GMP controls for APIs and IMPs has been seen as a major barrier to harmonisation with the USA. Harmonisation with the US through a Mutual Recognition Agreement (MRA) is seen as a big saving of inspection resources to both the EU and the USA, through mutual acceptance of facility inspection reports.

The EU *Starting Materials Directive* [14] has now been resurrected and is expected to be included as an amendment to the EU GMP legislation. The Starting Materials Directive was originally drafted in September 1997, with a second draft in 1998 after a large number of amendments. The original definition of a starting material included the API and also inactives such as excipients and the container/closure, but it is now expected to only include APIs. Because of the changes in the European Commission in 1999, the Directive was no longer seen as a priority and was never accepted by the European Parliament. With the advent of ICH Q7A, to which the EU is a signatory, the Starting Material Directive is now expected to be included as an amendment to the EU GMP legislation, with ICH Q7A used as a basis for defining GMP requirements for APIs within the EU.

The Clinical Trials Directive [15] requires clinical materials (IMPs) to be manufactured to GMP and must be incorporated into each member state's national law by 2004. The European Commission has issued the detailed requirements for IMPs as Annex 13 of volume four of the GMPs rules.

The Clinical Trials Directive:

- Mandates that all EU member states incorporate GMP for IMPs into their national legislation.
- It should provide a standard set of GMP requirements across all member states.

This Directive, through Annex 13, will require for all IMP manufacturing facilities:

- compliance of all facilities to (GMP) Directive 91/356;
- inspection by competent authority (e.g. UK MCA);

- registration of manufacturer/importer;
- licensing of site of manufacture;
- qualified person (QP) to certify/release products. (Duties of QP defined in Directive 75/319/EEC);
- all clinical trials to be conducted to good clinical practice (GCP).

1.5.4 International GMPs

Once again I must apologise for concentrating on quality systems in the USA and the EU, although I mean no disrespect to the rest of the world. There is a very good reason for this. A large number of those countries that have pharmaceutical industries, especially generic pharmaceutical manufacturers, have set up GMP QS based on USA requirements for the simple reason that the USA has the largest pharmaceutical market in the world. A large number of API manufacturers export to the USA and hence have required to comply with USA cGMP for pharmaceutical manufacture.

The World Health Organisation (WHO) [16] for a number of years has been very active in setting global GMP standards for both drug products and APIs. They have issued two chapters of guidelines for GMP. Chapter 1 concerns general considerations for GMP, more detailed guidance under the heading of *Good Practices in Production and Quality Control*, an international pharmacopoeia and a very useful glossary and definition of terms. Chapter 2 gives guidance on GMPs for APIs and excipients. It will come as no surprise that the headings of the guidance covers those already given earlier when discussing USA and EU requirements.

GMPs represent a technical standard upon which is based the 'WHO Certification scheme on the quality of pharmaceutical products moving in international commerce'.

1.6 International harmonisation of quality standards

Many countries have seen the advantages of harmonisation of standards and have formed MRA. The European Community (EC) has negotiated MRAs between the community and third countries, which include mutual recognition of standards of GMP and arrangements to ensure compliance by pharmaceutical manufacturers. At present such agreements, operating with New Zealand and Australia, are expected to be in operations soon with Canada and Switzerland, and are in a transitional evaluation phase with the USA. Negotiations are also underway with Japan. Under an MRA, the regulatory authorities accept each other's inspection reports, and routine inspection by one authority of manufacturers in the other's territory is not required. In addition, the re-testing of imported products is normally not required.

The main instrument for the harmonisation of standards between the EU, USA and Japan is through the International Conference on Harmonisation of Technical Requirements for Registration of Pharmaceuticals for Human Use (ICH). Detailed information on the guidelines issued to date can be found through their Website [17].

The following is a summary of the quality topics issued by ICH as draft or full guidelines:

Q1 Stability   Q1A Stability testing of new drugs
                   Q1B Photostability testing
                   Q1C Stability testing of new dosage forms
                   Q1D Bracketing and matrixing designs for stability testing of drug substance and drug products.
Q2 Analytical validation
Q3 Impurities
Q4 Pharmacopoeias
Q5 Biotechnological quality
Q6 Specifications   Q6A Chemical substances
                            Q6B Biotechnological substances
Q7 GMP   Q7A GMP for APIs.

1.7 Quality control, quality assurance and regulatory filings

No medicinal product may be sold without being licensed by an appropriate regulatory authority. Normally clinical trials can only be carried out under the control of the appropriate regulatory authority and a filing of some form is required in all countries.

It is very important to realise that information obtained during both the pre-clinical and clinical phases of drug development are pivotal to the success of the registration of the product. The toxicology testing is performed on early supplies of material. The health and welfare of patients on clinical trials is determined by the quality of the test material and its equivalence to the material tested in the pre-clinical programme. Eventually, all batches of the commercial material must be shown to be equivalent in quality to that used in clinical trials and the toxicology programme. No further clinical or toxicology studies are normally performed before the material is commercially manufactured.

A regulatory inspection usually concentrates on what the facility has done, rather than what it will do. Even in the latter case, as with compliance to a product licence, what the facility will do in future must be based on its past record, the systems that have been put in place and its compliance with them. At present in an R&D facility, possibility of inspection does not normally arise until an application is made for a product licence, which could be a number of years after the work has been done. For these reasons it is vital that everything that is done is accurately recorded. An FDA inspector was once quoted as stating 'If it's not written down it's only rumour.' For the pharmaceutical analyst this may mean they have to defend their work possibly five or more years after it was completed. Even worse, the analyst may have to defend the work of a peer or a subordinate. Without good records this will not be possible and failure to do so could adversely affect a regulatory submission.

There is a considerable difference in the standard and type of information required by the various national regulatory authorities throughout the world. For example, until recently it was possible to perform clinical trials within some European countries without submission of any manufacturing information; in other European countries Phase I clinical trials could be performed solely on the permission of a local ethics committee or just a doctor or dentist's approval. Requirements have now been harmonised across the EU with the Clinical Trials Directive discussed previously. Further attempts at harmonisation continue through ICH or are being driven by various MRAs. In the USA, a considerable amount of information has always been required for all phases of clinical trial from Phase I to Phase III.

In the space of this review, I cannot discuss all of the different national requirements for the various stages of development although I will give some guidance on general expectations for the pharmaceutical analyst.

1.7.1 Pre-clinical development

Materials do not have to be manufactured under GMP conditions, but it is recommended that GMP controls or controls approaching those of GMP should be in place for the following reasons:

- It is not a good practice to have dual systems (GMP and non-GMP) operating in the same facility, otherwise the standards will always drift to the lowest standard.
- It is a business risk in that undetected impurities can give a false result in a toxicology study, resulting in premature termination of the project.
- Frequently the same batch of material that is used in the pivotal toxicology study is needed for Phase I clinical trials. There is then pressure to retrospectively assign GMP status to that batch of material based on the toxicology study. This is not in accordance with QA principles where quality should be built in rather than tested in.

Having made these suggestions, *it is* a GLP requirement that the testing of these materials is performed according to GLP or an equivalent standard (see discussion previously). Facilities directly responsible for the toxicology studies or supporting those studies *will* be audited by the national or other interested GLP monitoring authority. Some national authorities will audit the test facility for the test article analysis.

It is not expected that analytical methods will be fully validated and that a specification will be in place, but the other quality requirements previously discussed such as equipment calibration, reagent control, training, etc. must be in place in the analytical facility. Some best estimate of the stability and storage conditions for the test article is required under GLP.

1.7.2 Early phase development (Phases I/II)

Clinical supplies (API and drug product) must be made and tested to an appropriate level of GMP to assure patient safety. In an analytical facility the level of GMP does not significantly change throughout development, excepting that initially analytical methods may not be fully validated, and specifications or other acceptance criteria are likely to change, as more information becomes available.

It is recognised that during early development of a manufacturing process and of the analytical methodology, there will be changes. What is vital is that all changes made to the process are assessed for their impact on the quality of the final product. Major changes to the route of manufacture normally require updating of the regulatory filing and an analytical investigation to determine if any new impurities are formed as a consequence of the change. Additionally changes at the final stage, especially for APIs, should be carefully assessed for their impact on the morphic form of the material, as changes in the form can have serious effects on bioavailability. Similarly, major changes to analytical methods must be assessed for equivalence and normally require updating in the regulatory filing. Analytical methods for the API and drug product should be stability indicating. Special attention should be paid to changes of process solvents, where the allowable levels of solvent residues may change. New stability studies on materials obtained via major route changes would normally be initiated.

At the present time, it would be unusual for a regulatory authority to audit an R&D facility that manufactures clinical supplies for Phase I and II studies, for compliance to GMP or for compliance to a clinical trials *chemistry, manufacturing, controls* (CMC) part of a submission. This is however, likely to change in Europe with the implementation of the Clinical Trials Directive, when facilities (possibly including API facilities) will require to be certified. It is still always possible that a facility named in a submission will be audited for cause. For example, if something goes wrong during a clinical trial and a patient or volunteer becomes seriously injured or dies, the appropriate regulatory authority may include in its investigation, the R&D facility that made the test material. At this time it is highly likely that all records and quality systems will be subject to close scrutiny and all personnel may have to justify decisions they have made, and defend their work. Personnel should be reminded that fraud and negligence could be severely punished by a regulatory authority.

1.7.3 Late phase development (Phase III)

Most of the GMP requirements mentioned for Phases I/II will still hold, although it is considered that the route and analytical methodology should be approaching the consistency expected of a commercial product. Specifications for the API and drug product must be in place, specifications for raw materials and intermediates should be nearing finalisation, and analytical methods should be fully validated to pharmacopoeial standards, etc.

Process validation would not be expected until commercial batches are produced but information on process boundaries and critical parameters should be obtained at this stage. It should also be pointed out that if the pivotal stability studies to be listed in the final registration (on the drug product or API) are performed on pilot batches, then this data certainly could become subject to inspection. It is becoming increasingly common in Europe and the USA for R&D facilities to be audited as well as the commercial facility before approval of a manufacturing licence, even though the R&D facility has not manufactured any commercial supplies.

1.7.4 Commercial manufacture

Throughout the world the manufacture of pharmaceutical products for sale requires to be performed under the control of an appropriate regulatory authority. A manufacturing licence of some form must be obtained and compliance of the facility with GMP is a pre-requisite. Before a licence is approved the commercial facility will need to have been subject to a recent regulatory inspection and/or have demonstrated a good compliance history via previous, recent audit. This inspection will determine that the facility complies (or should comply) with the route of manufacture, the specification specified in the licence application, and is compliant with GMP requirements. Facilities that are not in compliance will be required to give an undertaking to improve, or could be refused a licence. In some countries (e.g. the USA) where cases of fraud, extreme negligence or non-compliance are discovered, such that there is a strong possibility of risk to public health, companies and individuals can be subjected to harsh penalties including heavy fines and/or imprisonment. For the USA, companies that are not in compliance, do not respond to inspection observations (FDA 483s) adequately or promptly, or do not complete corrective action plans as agreed with the FDA, risk a warning letter that could prohibit the sell of their products in the USA. Furthermore, a Consent Decree could be issued which effectively puts production and release of each batch of the affected product under close regulatory control, at the firm's expense, until such time as there is perceived to no longer be a risk to public health.

At the present moment the European regulatory authorities tend towards the situation where a company's compliance history and the fact that the specific facility is within a two-year satisfactory inspection window, to determine

QUALITY CONTROL AND REGULATION 21

whether an audit is required. The USA authorities have in the past tended to rely on pre-approval inspections (PAI) directed specifically at the filing, along with a biennial inspection programme of all facilities. They are at present moving more towards reliance on the biennial inspection to be consistent with European authorities leading up to the MRA (and to save inspection resources). These inspections, depending on the regulatory authority, can take anything from two days to a number of weeks and can be very resource intensive for the company under inspection. They can also be unannounced although foreign inspections tend to be pre-planned for logistic reasons. The US FDA issued a useful Foreign Inspection Guide [18] that gave guidance to inspectors visiting foreign companies. This type of inspection may now change following the FDA's acceptance of system's audits (discussed later).

1.8 Regulatory inspection key areas

Some examples of key areas of concern of some regulatory authorities are given in the final part of this review, along with some actual examples taken from public files. Most of these examples are taken from the US FDA's observational findings, as these are available in the USA under the Freedom of Information Act. Many of the guidelines discussed below can be found on the web site of the appropriate regulatory authority.

1.8.1 Inspection of analytical test facilities

Problems with analytical facilities is regularly cited by the US FDA as one of the main reasons for the non-approval of a NDA or abbreviated new drug application (ANDA). The mid-Atlantic region of the US FDA issued *A Guide to Inspection of Pharmaceutical Quality Control Laboratories* [19] and although it is ten years old now, most of the information is still valid (except the more recent requirements on computerised systems which will be discussed next). In this guide to inspectors, the FDA discusses the various approaches to an inspection. The inspection might cover one or more of the following topics:

- The specific methodology which will be used to test a new product.
- The complete assessment of laboratory GMPs.
- A specific aspect of laboratory operations.

The following is a summary of some of the typical areas that an inspector would be required to scrutinise for both drug products and APIs.

Pre-approval inspections are designed to confirm that the data submitted in an application is authentic and accurate and that the procedures used in the application were actually used to produce the data contained in the application. Since not all data can be included in a submission, the inspection must determine if there is sufficient justification for certain data not being reported, especially that data pertaining to batches that fail to meet a predetermined specification. The review would include:

- A comparison of the results of analyses for those applications for which more than one batch has been produced.
- An evaluation of the methods and exceptions to the procedures or equipment actually used from those listed in the application. Confirmation that it is the same method listed in the application.
- An evaluation of raw laboratory data for tests performed on biobatches and clinical batches and comparison with the data filed in the application.

Laboratory records and logs represent a vital source of information that allows a complete overview of the technical ability of the staff and of the overall QC procedures. Laboratory data review would include:

- An evaluation of raw laboratory data worksheets and personal notebooks, laboratory procedures and methods, laboratory equipment, and method validation data to determine the overall quality of the laboratory operation and the ability to comply with GMP regulation. This will include checking that data is not missing, results have not been changed without adequate explanation and traceability and calculations are accurate and have been checked by a second person.
- Ensuring that raw data is maintained in bound, not loose-leaf, books, or on analytical sheets for which there is accountability (e.g. pre-numbered sheets).
- An examination of chromatograms and spectra for evidence of impurities, poor technique or lack of instrument calibration.
- An investigation of laboratory test failures. Evaluation of results of analyses of re-tested and failed batches. Review of decisions to release batches failing specification. Products cannot be *tested into compliance*.
- A check that data has not been transcribed without retention of the original records.
- Determining that the laboratory results are actually used to release lots of the product and ensuring that data has not been selectively used.

The review of microbiological data regarding topical and parenteral manufacture would typically include:

- A review and evaluation of the preservative effectiveness.
- A review of the bioburden from both an endotoxin and sterility perspective.
- A review of environmental monitoring methods and data.

Laboratory standard solutions must be ascertained as being suitable for use in that they are:

- stored correctly;
- have expiry dates appropriate to their stability;
- have adequate records of preparation.

Methods validation would be critically reviewed for completeness, accuracy and reliability. If an alternative to a compendial method is used, this in-house method must be compared to the compendial method to demonstrate it is equivalent or superior to the official procedure. Appropriate pharmacopoeial guidelines for methods validation should be followed.

Laboratory equipment usage, maintenance, calibration logs, repair records and maintenance procedures should be examined. Data would be suspect if it were generated from equipment known to be defective. Continuing to use and release products on the basis of such equipment would be a serious violation of GMP. The inspection would need to:

- Confirm the existence and condition of equipment specified in the method.
- Verify that equipment was present and in good working order (maintained and calibrated to pre-defined procedures) at the time the batches were analysed.
- Determine that equipment was used properly.
- Evaluate completeness of equipment and maintenance logs.

Raw materials (API) quality and purity largely determines the safety and efficacy of the drug product. The same standard of laboratory control is required for testing of APIs as for the drug product. If the API is obtained externally, the drug product manufacturer must have a system in place for evaluating the supplier with respect to quality, and assuring the accuracy of the supplier's certificate of analysis (C of A). Normally this would be performed by an audit of the API facility for compliance to GMP and qualification of the C of A through duplicate testing of a number of batches.

Stability indicating methods must be used to test the samples of the batch. Evidence that the method is stability indicating must be presented, even for compendial methods. Manufacturers may be required to carry out accelerated or forced degradation of a product to demonstrate that the test is stability indicating. The inspection should:

- Evaluate the validation report for stability testing.
- Review the raw data and the results of testing at the various stability stations to determine if the data actually reported matches the data found in site records.
- Evaluate the raw data used to generate the data filed documenting that the method is stability indicating and the level of impurities.

Other areas of inspection focus include sampling protocols, in process controls and specifications, further guidance on the quality of data, computerised systems (more details will be given in the next section), and laboratory management to determine the overall GMP compliance of the facility. This latter area would include supervisory control and review and personnel qualifications and training.

1.8.2 Computerised systems (21 CFR part 11)

Many regulatory agencies and consultants to the pharmaceutical industry throughout the world have put out some sort of guidance in the past concerning the use of computer systems. It is only within the last six years that this guidance has been given the force of law in the USA in the form of 21 CFR part 11 [20].

Guidance on the use of computerised systems within the pharmaceutical industry was given in the 1992 inspection guide discussed in the previous section. This concentrated on computerised laboratory data acquisition systems and was concerned with system security, audit trails of changes/deletions and procedures to ensure validity of the data. Since the mid-1990s it has been a US FDA expectation that major computer systems would be validated to ensure these concerns would be addressed. Occasionally the pharmaceutical newssheets would be full of serious adverse findings with respect to the computer systems of major pharmaceutical companies.

In spite of this guidance and these adverse findings, the pharmaceutical industry was originally only pressing the US FDA to come up with some more guidance on the use of electronic signatures to be used in place of traditional hand-written signatures. Following industry comment, in 1997 21 CFR part 11, Electronic Records, Electronic Signature, Final Rule was issued by FDA. The rule covered nearly 40 pages in the Federal Register (although the actual rules only covered two pages, the rest being preamble, scope, etc.) and it became effective on 20 August 1997. From this date onwards all computerised equipment records and the use of electronic signatures in support of regulatory processes were required to comply with this legislation. The US FDA saw this rule as an attempt to encourage industry to modernise with respect to the use of electronic systems and electronic submissions. The FDA stated in the preamble 'These regulations, which apply to all FDA program areas, are intended to promote the widest possible use of electronic technology, compatible with FDA's responsibility to promote and protect public health.' In many cases it has had the opposite effect, with some companies trying to revert (erroneously in most cases) to hard copy records. This rule has had far ranging effects within the pharmaceutical industry and by some estimates is likely eventually to have cost the industry considerably more than the Year 2000 preparation. One of the main reasons is that the rule applies to *all records* generated after 20 August 1997, even though virtually no equipment, at that time, could comply with the requirements. Although the US FDA gave the industry time to comply with the rule and to even put temporary dual hard copy/electronic systems in place, the expectation is that all companies will now comply with 21 CFR part 11.

Some of the main requirements of 21 CFR part 11 are as follows:

- The rule applies to electronic records that are created, modified, maintained, archived, retrieved or transmitted, under any agency regulations. It also applies to electronic records submitted to the agency under requirements of the FD&C Act, even if such records are not specifically identified in agency regulations.
- Systems must be identified as *closed* (system access controlled by people responsible for content) or *open* (system access not controlled by people with the responsibility for content).

The following requirements will concentrate only on closed systems.

- The system must be validated to ensure accuracy, reliability, consistent intended performance, and the ability to discern invalid or altered records.
- The system must have the ability to generate accurate and complete copies of records in both human readable and electronic form.
- Records must be protected to enable their accurate and ready retrieval through the record's retention period.
- The system must limit access only to authorised persons.
- The system must have equivalent levels of controls expected of a paper system such that there are computer-generated, time-stamped audit trails of all modifications, deletions, additions, etc., the system identifies the person making these changes, and the original entry is not obscured.
- Operational system checks should be in place to enforce permitted sequencing of steps and events, where appropriate.
- The system only allows authorised persons access and identifies those who are authorised to make changes to records.
- If appropriate, the system employs checks to determine the validity of the source of data input.
- All person involved with electronic records and systems have adequate education, training and experience.
- There should be adherence to written policies that hold individuals accountable and responsible for actions initiated under their electronic signature.
- There must be appropriate controls over system's documentation including distribution, revision and change control.

With respect to electronic signatures, the following are the key areas:

- Each company must certify to FDA that they intend to use electronic signatures as being equivalent to hand-written ones.

- Each electronic signature must be unique to one individual and shall not be reused by, or reassigned to, anyone else.
- The electronic signature must contain information to indicate who signed, date and time, and the meaning of the signature.
- Non-biometric signatures must employ at least two distinct identification components such as an identification code and password.
- Electronic signatures must only be used by their genuine owner.
- Systems must require that use of an electronic signature by anyone other than its genuine owner require collaboration of two or more individuals.
- Systems must be in place to ensure that each combination of identification code and password is unique to the individual.

One of the biggest misunderstandings of 21 CFR 11, is that many companies have interpreted the rule as allowing the printing of hard copy of an electronic record and signing/dating of this hard copy, as a substitute for maintaining the electronic record in compliance with the rule. Although dual systems (maintaining both hard copy and electronic raw data records) were to some extent accepted by FDA in the early days following introduction of the rule, this is not now an acceptable practice. It is certainly not acceptable to delete original electronic raw data and only retain a hard copy. It has been made clear by FDA on numerous occasions that if equipment has the ability to store an electronic record, then the electronic record is the raw data, not the hard copy print out.

This rule has wide applicability across the pharmaceutical analysis area, in that if any computerised equipment generates data that may be used in support of regulatory filings or pharmaceutical manufacture, compliance with the rule is mandatory. This includes stand-alone equipment such as HPLCs, etc., although primary target for regulatory scrutiny would be laboratory information management systems (LIMS) and data acquisition systems.

1.8.3 Out-of-specification (OOS) test results

In a ruling dated 4 February 1993, Judge Alfred Wolin of the US District Court for the District of New Jersey handed down a ruling which has had far reaching consequences as to how the pharmaceutical industry treats OOS and other suspect data. This ruling has become better known as the Barr Ruling, from the case USA vs Barr Laboratories. Although the ruling is not mandatory outside of the Court's jurisdiction, it has become the model by which the US FDA would treat laboratory failures. The US FDA considers it an expectation that every company they inspect will have a SOP that describes how these data are treated.

A summary [21] of the key points of the Barr Ruling is as follows:

- The Court rejected FDA's position that a batch failure occurs when an individual test result doesn't meet specification. OOS results can be due to laboratory errors, non-process-related (operator) errors or process-related (manufacturing) errors. Only non-process or process-related errors are properly identified as failures.
- The Court ruled that OOS results can be identified only through an investigation. The type and extent of the investigation depend on what caused the OOS result.
- Failure investigations must be documented and performed within 30 business days.
- The Court determined that laboratory errors occur when analysts make mistakes. These types of errors must be reported to and reviewed by a supervisor according to a written procedure. The review should be documented and the cause of the error identified if possible.
- Retesting is permissible if the OOS result was due to laboratory error. Other types of error require further investigation.
- Judge Wolin ruled that if a failure investigation shows that a laboratory error didn't occur, or is inconclusive, the investigation has to extend beyond the laboratory.
- The Court's opinion on outliers was that since, for chemical testing, the USP does not provide for outliers, the use of outlier testing for chemical analyses is prohibited. The Court also agreed that the USP specifically allows outlier testing for biological and antibiotic assays. (This opinion has been challenged a number of times on scientific grounds.)
- The Court ruled that re-testing is appropriate if the failure investigation has determined that re-testing is appropriate. Re-testing is not appropriate if the error was process-related, or for product failures.
- Re-testing must be done on the same sample that produced the original, failing test result. It can substitute for the original result if the error was due to an analytical mistake, and it can supplement the original result if the investigation is inconclusive. There should be predetermined testing procedures defining when re-testing ends and results should be considered in terms of overall batch and product history.
- The Court ruled that companies cannot rely on re-sampling to release a product that has failed testing and re-testing unless re-sampling is in accord with the USP standards (content uniformity and dissolution), or unless the failure investigation discloses evidence that the original sample is not representative or was improperly prepared.
- The Court ruled that although the averaging of test results can be a rational and valid approach, as a general rule this practice should be avoided because averages hide the variability among individual test results. This phenomenon is most troubling if testing generates both OOS and passing individual results which when averaged are within specification. Here,

relying on the average figure without examining and explaining the individual OOS results are highly misleading and unacceptable.

In September 1998, the US FDA issued a draft OOS guidance for industry. For further comment see [22].

It is obvious that *testing into compliance* (repeating testing until you obtain the answer you want), the averaging of failing results with passing results to give a passing result and the discarding of any data without good reason are all not acceptable practices within the pharmaceutical industry.

1.8.4 System audits

'The organisation of the future must be able to use the information it generates to improve itself.'

This comment was made by a US FDA representative at a conference in December 2001 and further was discussed in issues of the *Gold Sheet* [23, 24]. FDA is warning companies that they expect inspection findings to be acted on not just to fix a situation, but to improve the whole QS. If a large multinational pharmaceutical company receives 483 observations, then commitments for corrective and preventive actions should extend to all similar facilities throughout the company. This process is getting to the real purpose of QA and audits. It is well understood that things go wrong, equipment fails and people will always do some things wrong, but it is how each company reacts to these findings. FDA audits in future will concentrate on the QS itself. The audits will look at how the system ensures not only that corrective and preventive actions are put in place, but how management is made aware of quality issues, how the systems are improved in response to all audit findings and how management ensures that the improvements are sustained. FDA has become aware over the last few years, that as companies try to save resources, the quality unit (QA and QC) the area that becomes squeezed. Data shows that the number of employees in quality has decreased in relation to those in production and other areas such as marketing. FDA is very concerned with this situation.

System audits will always examine the whole QS in a company and additionally at least two of the other systems listed in the examples given below. These system audits may also be in conjunction with a product specific audit.

QSs would include:

- Change control
- Quality investigations
- Corrective actions

- Preventive actions
- Management review of the quality management system (QMS)
- Out-of-tolerance calibrations
- Out-of-specification results
- Equipment failures
- Training issues
- Internal and external audits.

Other systems could include:

- Facilities and equipment
- Materials
- Production
- Packaging and labelling
- Laboratory controls
- Computer Systems compliance.

FDA pre-approval inspections in the past would normally concentrate on the information submitted in the NDA. There could be (and often are) many other quality problems that would not be uncovered by this narrow focused audit. In fact dual systems could exist in a facility without auditors becoming aware of the fact.

System audits of the type discussed are more similar to the general GMP type of audits performed by the European authorities and also by the author of this work. When there is limited time, such as one or two days, to complete the audit of a large facility, then an audit of the QSs as a whole is a very powerful tool. If this is coupled with a product specific audit to challenge the operation of the systems, then this is even better for assessing the facility.

1.9 Conclusions and the future of regulatory scrutiny

In this chapter, I have attempted to give an overview of the various QSs that are operated within the pharmaceutical industry. Some of these such as ISO 9000 and NAMAS are voluntary and are followed in order to give the company a competitive advantage. Others such as GMP and GLP are mandatory for any company manufacturing materials for human or veterinary use or who supply data to regulatory authorities on the safety of chemicals, whether in an R&D or commercial environment. Hopefully, I have also tried to show that although the R&D environment has different requirements with respect to quality from the commercial manufacture of pharmaceutical materials, quality is still a vital part of most R&D's work.

With respect to where quality and regulation are going, one needs to look at the way MRAs are being introduced between regulatory authorities worldwide. In an effort to control inspection and regulatory submission overheads, there will be an emphasis on harmonisation of standards in all areas. There will be greater reliance on inspections by local inspection authorities and even further acceptance of local registration dossiers as seen within the EU at present. Inspections are certainly becoming more focused on overall QSs auditing and inspection history rather that on pre-approval type of inspections focusing on one product [23]. This approach is also described in the US FDA Compliance Guidance Manual 7356.002 [24]. Areas of scrutiny will continue to be analytical facilities but additionally with the new USA-EU MRA and the Clinical Trials Directive, R&D facilities can expect to become the focus of increased inspection activity, both for drug products and API production and testing.

References

1. Gilmore, H. *Quality progress* (June 1974).
2. Juran, J. *Quality control handbook*.
3. *Good manufacturing guide for active pharmaceutical ingredients*, ICH Q7A (November 2000).
4. UK Medicines Control Agency (MCA), www.mca.gov.uk.
5. www.iso.ch.
6. www.ukas.com.
7. www.oecd.org.
8. Statutory Instrument, no. 3106 (1999), www.hmso.gov.uk.
9. www.fda.gov.
10. Washington Drug Letter (13 August 2001).
11. Directive 91/356/EEC (13 June 1991).
12. Directive 91/412/EEC (23 July 1991).
13. Directive 65/65/EEC, Article 1(2).
14. Directive 75/319/EEC.
15. Directive 2001/20/EC.
16. www.who.int.
17. www.ifpma.org/ich1.
18. *FDA foreign inspection guide* (14 September 1992).
19. *FDA guide to inspection of pharmaceutical quality control laboratories* (22 June 1992).
20. 21 Code of Federal Regulations Part 11, Electronic Records, Electronic Signatures.
21. Madsen, Jr, R. (1994) *PDA journal of pharmaceutical science and technology*, **48**(4), 176.
22. *FDA draft guidance for industry: investigation of out of specification (OOS) test results for pharmaceutical production* (September 1998).
23. Drug GMP Report (The Gold Sheet) (March 2000, April 2001, December 2001).
24. Approach endorsed by US FDA (Nicholas Buhay), FDA CDER Compliance Office (www.fda.gov/cder/dmpg/compliance-guide) (December 2001).

2 Development of achiral separation methods in pharmaceutical analysis

George N. Okafo and John K. Roberts

2.1 Introduction

This chapter highlights the latest approaches used in the development of achiral separation methods for the analysis of impurities and degradant molecules relating to active pharmaceutical ingredients (API). This area of analytical chemistry is a diverse, wide-ranging and rapidly developing field, covering numerous analytical techniques, methods and approaches. Examples of the commonest separation techniques include high performance liquid chromatography (HPLC), gas chromatography (GC), thin layer chromatography (TLC) and capillary electrophoretic (CE) methods. Other key-related areas include sample preparation techniques, detection technology, automated approaches and tools for method optimisation. To cover fully the other aspects of each separation technique (e.g. theory and mechanisms of operation, equipment and technical requirements, general applications) is beyond the scope of this chapter, but readers can supplement their reading by referring to the following excellent publications on the individual techniques and areas. For method development in HPLC, there are many good references [1–3] that offer comprehensive reviews of practical method development and applications of HPLC, particularly in pharmaceutical analysis; GC is covered extensively in [4, 5], particularly in the areas of practical uses and applications; CE is covered comprehensively in two good publications [6] and [7]. Method development approaches and modern practical applications of TLC are covered in three review documents [8–10]. Other more recent and less commonly used separation techniques, namely supercritical fluid chromatography (SFC) is reviewed in references [11, 12].

Many method development strategies have been reported in the literature for different separation techniques [1, 4, 6]. From these, there appear to be common themes and generalised strategies that exist in all the approaches reviewed to date – this will be discussed in the early sections of this chapter in sufficient detail to set the scene for the following sections. Part of this section will also look at why particular separation methods are used, and the importance of sample preparation will be discussed. A review of the literature over the last five years has indicated that the most popular *front-line* separation techniques are HPLC, GC and CE. The next four sections will focus on each of the separation methods with particular emphasis on introducing each technique (technical and some theoretical aspects), new developments and some examples of the most up-to-date applications in API analysis. The re-emergence of analytical tools such as TLC and SFC will also be mentioned.

Detection and visualisation of analytes, impurities and degradants is a critical component of any separation technique and will be covered by highlighting the latest approaches. Additionally, the enormous utility of combining existing separation techniques with other more specialised systems (e.g. mass spectrometers and other structural characterisation tools) will be discussed briefly.

Other new strategies in modern method development include the use of automation, computer software to optimise separations, and statistical and chemometric tools to enhance existing approaches. The application of these method development tools will be discussed in the final sections of this chapter.

Method development approaches for biopharmaceutical analysis are not going to be covered in this chapter, but there are some recent publications which may be referred to for a comprehensive overview of this application area [13–15].

2.1.1 Historical perspective of separation methods and their uses in pharmaceutical analysis

The historical development of separation sciences can be traced as far back as the early 1900s, when Michael Tswett first coined the term chromatography [15]. Since then, a number of major advances in the developmental path of separation sciences have progressed through instrument technology [16–18], mechanistic understanding [16, 19], column technology [20, 21], the advent of computers [22, 23] and robotic automation [24, 25]. These major developments have established separation sciences as a critical component in modern analytical laboratories. Another key driver in the advancements of separation sciences has been the increasingly demanding requirements of the petrochemical [26], agrochemical [27] and pharmaceutical industries [28]. In these application areas, there is a continuous and relentless need to generate reliable, high quality analytical data usually in ever decreasing time frames [29].

In pharmaceutical laboratories, the use and importance of separation sciences extends throughout almost all aspects of the development process [28]. Separation methodology is used in the very early stages of discovery research, where numerous biologically active chemical entities must be separated to provide information on relative purities and impurities levels. In the pre-clinical and clinical phases of the drug development process, separation methods play a pivotal role in fully characterising chemical entities prior to use in human clinical studies. Throughout the process, batches of API are prepared via a synthetic route that is continually being improved and optimised. Hence, characterisation of drug substance batches can yield very important information about the synthetic process. A good example of this is that for the antihistamine drug Pafenolol (1) [30]. Liquid chromatographic separation of a crude batch of (1) reveals the presence of several important intermediates, by-products from side reactions.

Within the last decade, timelines for developing pharmaceuticals from pre-clinical to clinical phases have become extremely short [29], typically between 36 and 60 months. A consequence of this aggressive time frame is the requirement for analytical data in the early development phases to be used to make critical decisions. By ensuring that the right decisions are made in these early development phases and backed up by sound reliable data, the likelihood of the successful progression of a drug candidate through the clinical phases and beyond is increased [29].

Separation methods have been applied in pharmaceutical analysis in numerous ways for drug substances and for formulated drug products in areas such as the assessment of the levels of degradants in batches of drug substance and products – a key regulatory requirement [28]. This information is often used to determine the shelf life of the compounds, to recommended storage conditions and transportation conditions. Degradation can sometimes occur during storage of the drug substance/product caused by interaction between the compound, the storage container, excipients and packaging materials.

2.1.2 *Regulatory considerations for separation methods in pharmaceutical analysis*

It is important to recognise that all drug substances contain impurities that arise from many different sources. Impurities can originate from raw materials, reagents, as reaction by-products and through degradation during manufacture and storage. The fact that impurities can have implications for safety and efficacy, both the API manufacturer [28] and the regulatory agencies [31, 32] pay particular attention to this subject. Nowhere is this more important than in the development of separation methods to resolve, identify and quantify these impurities. The current guidelines [33] indicate that identification of all recurring impurities at or above the 0.10% level is expected in all batches manufactured by the proposed commercial process. Similarly, degradation products observed in stability studies should also be identified.

The current FDA guidelines [33] for the classification of impurities divide them into two categories: organic and inorganic. The organic category covers starting materials, by-products, intermediates, degradation products, reagents, ligands and catalysts. Inorganic materials will not be covered in this chapter. For drug products, the analytical challenges for identifying impurities are even greater because of the presence of excipient components present in the formulation. Here, impurities can arise through interactions between the API and excipients, packaging materials and through degradation. The registration of a new drug product requires significant documentation for submission to the regulatory agencies [28] including justification and rationale for the methods developed.

With a threshold of 0.10% for the impurities level in most APIs, the implication is that a limit of quantitation (LOQ) of approximately 0.05% will be required for most separation methods [33]. Hence, to ensure that key impurities are resolved and that main peak homogeneity is maintained, it is becoming essential to increase selectivity, improve sensitivity and enhance the consistency of separation methodology. Current approaches include the use of alternative detection technologies, derivatisation strategies, orthogonal coupling of chromatographic techniques, e.g. HPLC/TLC and HPLC/CE, or the coupling of chromatographic separations with information-rich spectroscopic methods such as HPLC/MS or HPLC/NMR. These topic will be covered in later sections.

2.2 General guidance for method development in separation sciences

For most separation techniques (HPLC and GC), technological advances, ease-of-use and greater commercial availability have led to separation-scientists having a good practical working knowledge and understanding of chromatographic separation principles. Moreover, knowledge of the effects of making changes to the experimental conditions has significantly improved. This knowledge usually forms the basis of any systematic approach to method development. With this in mind, Snyder *et al*. [1] summarised method development strategy in the following way – *the length of time taken to achieve the desired separation varies from performing relatively few experiments in a short time to exhaustive experimentation*. The strategy should be one that requires as many experimental runs to achieve the desired final result. In this case, each experimental run should contribute towards the desired result.

Nowadays, in the pharmaceutical industry, the need to obtain accurate and reliable analytical data faster and more cost-effectively, has meant that approaches to method development are becoming simpler and more straightforward. These simpler approaches have the benefits of more consistent methods, lower method diversity, reduced inventory costs for separations consumables and more easily transferable methods from R&D laboratories to manufacturing quality control (QC) laboratories. Many pharmaceutical companies have adopted this approach.

A review of method development in separation sciences indicates that individual approaches exist which are based on previous experience of compounds or previous knowledge; method development often follows the same sequence of events. These are summarised in the flowchart shown in Fig. 2.1. In the following sections, each step is discussed in some detail.

2.2.1 Separation goals/objectives

Before the method development commences, it is important to define as clearly and in as much details as possible what the objectives and goals are for the intended separation. There are many publications [1, 34–36] that detail the types of questions that should be asked and in many cases, these questions are specific for HPLC [35, 36]. However, by considering these questions in a broader sense, they can apply equally to most, if not all, common separation techniques. The following questions should be asked:

1. Primary purpose

    - Qualitative analysis (i.e. for identification purposes)
    - Quantitative analysis (i.e. determine impurity levels or purity of the main compound)
    - Rugged separation
    - Recovery or isolation of a sample fraction or purified material.

Defining the aims and objectives of the separation helps in deciding the choice of separation technique. Moreover, if the aim is to isolate purified material, then separation techniques such as GC and CE are not always easy to use.

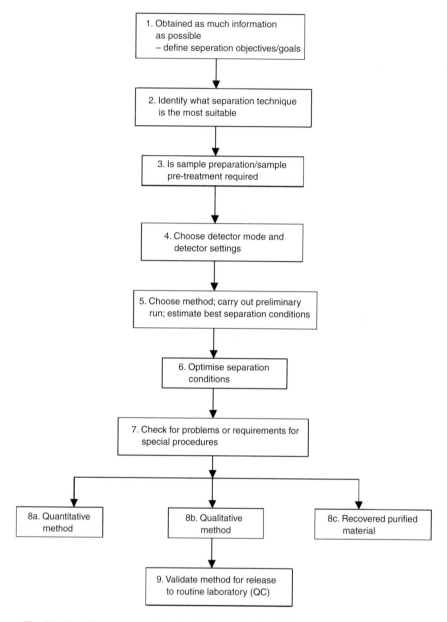

Fig. 2.1 Flow diagram summarising the basic steps involved in developing separation methods.

2. Impurities

- Should all the impurities and degradant be resolved from each other and the main component?
- Is it only necessary to separate the impurities and degradant from the main component?
- Are all the known impurities, degradants and solutes available as standards?

For this question, it may be necessary to separate all impurities and degradants from the main component for a purity assay. However, if the objective is to obtain an impurity profile, then all components will need to be separated from each other. In some cases, it is also possible to obtain both impurity profile data and a purity assessment of the main component from a single method.

3. Quantitative analysis

- What level of accuracy and precision is required?
- What range of concentrations for the relevant components will be required?
- What is the desired analysis time?

For pharmaceutical analysis, ICH guidelines [32] would normally recommend accuracy value ranges of 95–105% and precision of values less than 10%. Typically, the choice of concentration range would be dependent on the spectroscopic properties of the compound(s), the chosen detection mode and the volume (amount) of sample to be analysed. Deciding the desired analysis or run time is important because certain applications need fast methods for a high sample turnover.

4. Sample matrices

- How many different sample matrices should the method be designed for?
- Will the sample require extraction from the matrix?
- Can the separation be achieved with untreated samples?
- Is sample pre-concentration required?

This question is an important one because the component in the matrix can significantly influence the separation characteristics of the desired components. This is particularly relevant when dealing with biological fluids (where proteinaceous materials can interfere), environmental samples (interfering compounds may be present at very high levels) and in formulations (where excipients can interact with the separation column).

The analysis of untreated samples may also be chosen, however, the risks here are that the separation column can become blocked or damaged with prolonged use. Not surprisingly, this approach is not commonly used. In cases where the levels of matrix component far exceed the compounds of interest, it is important to extract the compounds and concentrate the compound.

5. Number of samples

- How many samples will be analysed at any one time?
- Speed vs Resolution?

When a large number of samples require analysis, then run time becomes very important. This is particularly relevant when the separation method is being used to monitor multiple chemical reactions, and samples are analysed at different timepoints. Quite often, it is desirable to trade in a decrease in sample resolution for a shorter run time. In situations where the sample numbers are fewer and the information required is quality critical in nature, then longer run times are sometimes required to achieve the desired resolution of critical impurities.

6. End-user of final method

- Will the separation method be used on different instruments makes and models?
- Will the separation method be used by different analysts with varying skill levels?
- Will the separation method be used in different laboratories?

In most pharmaceutical analyses, the answer to these questions is quite simple. The separation method will have to satisfy all the above criteria, because by meeting this criteria, the ruggedness of the method is tested. The end-user is most likely the QC laboratory in a manufacturing facility, and the ease with which a separation method can be transferred in most cases, R&D laboratories to a QC laboratory will be highly dependent on the method development strategy used [37].

2.2.2 Nature of the sample

If some information on the nature of the sample is known before beginning method development, it often provides valuable clues to the choice of separation technique, detection mode, extraction procedure (if required) and initial starting conditions for the separation. Table 2.1 summarises the type of sample-related information that would be required. Quite often, very little is known about the sample composition at the beginning of method development. This is particularly true for compounds presented at the discovery or pre-clinical phases of drug

Table 2.1 Summary of information relating to the physical and chemical properties of the test sample

	Physico-chemical property
1	Physico-chemical property of the sample – b.p., m.p.
2	Number of compounds/components present.
3	Chemical structure or functional groups present – are there anionic, cationic, neutral or zwitterionic moieties present?
4	Molecular weight(s) – are components high (MW >1000 Da) or low MW?
5	pKa values of compound(s) – is the compound acidic (pKa <4), basic (pKa >7) or neutral (pKa = 7)?
6	UV spectra – does the compound end-absorb (l < 200 nm) or is it coloured (l > 400 nm)?
7	Does the compound(s) fluoresce?
8	Is the compound electrochemically active?
9	Sample solubility in organic and aqueous solvent systems.
10	Concentration ranges of compounds in samples of interest.
11	Is the compound volatile – i.e. b.p. <90°C?

development and it is usually at this stage that separation scientists begin to start collecting critical analytical information on potential drug compounds.

There appear to be two practical approaches utilised for method development by most separation scientists. The first approach is to rely on past experience of the compound or previous separations of a compound with a similar structure and then supplement this information with some information from the literature. This can usually lead to the development of some initial starting conditions. The alternative approach is usually to not consider the sample information and proceed directly to some generic starting conditions. Both of these approaches have been characterised by Snyder *et al.* [1] as theoretical for the former and empirical for the latter. Although both approaches have been shown to work, the optimum strategy for method development is quite often a combination of both approaches.

2.2.3 Choosing the separation technique

One of the first steps in any method development process is to decide which separation technique to use. The choices may be as diverse as GC, LC in its many modes, CE in its many modes, TLC, SFC or some other technique. Within HPLC, choices may include size-exclusion, normal-phase, reversed-phase (RP-HPLC), ion-exchange and/or ion-pairing. Within RP-HPLC, the silica stationary phase can be modified with a range of different functional groups each imparting its own unique chromatographic properties e.g. octadecyl, octyl, phenyl, amino or cyano groups. Newer stationary phases such as mixed mode, porous graphitic carbon, fluorinated and monolithic phases should also be considered. Figure 2.2 summarises the current choices of HPLC modes available. A further factor to consider is the scale of the separation; if there is very little sample available then techniques such as capillary HPLC are appropriate [66]. If the goal of separation is to isolate material then preparative scale chromatography [198] or at a larger scale, simulated moving bed (SMB) techniques are more

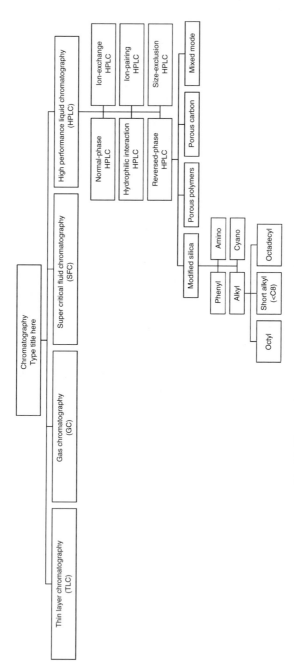

Fig. 2.2 Diagram showing the different modes of chromatography.

appropriate choices [199]. For GC, TLC, CE and SFC, there are fewer, but still significant choices.

To select a separation method for a suitable application, the separation scientist usually relies on a number of factors:

- the intended end-user of the separation method;
- any phyisco-chemical information on the sample;
- personal insights and experiences;
- information from the literature and the manufacturer;
- availability of equipment and consumables;
- ease of use;
- reliability of technique.

There may be other factors to add to the list above, but the seven factors listed are considered to be the most common factors. The first factor relates to who and where the method is likely to end up. The availability of sample information will to a certain extent dictate the most appropriate separation method to use. The use of personal insights and experience is likely to be a dominant one if similar problems have been solved before using a particular separation technique. This is the case for compounds with similar structures. If the separation scientist encounters a completely novel area or analytical solution, then the use of information from published work or from instrument manufacturer is usually of more importance. The availability of instrumentation and consumables in the long and short term is quite often an important requirement particularly if the separation technology is not well established, expensive or is sourced from relatively fewer manufacturers. Quite often, if the separation techniques require very specialised knowledge or experience, then the choice of method would usually be one that is simple, reliable and easy to use.

In pharmaceutical analysis, separation methods are usually developed with a view to ensuring that the technique chosen can:

1. Solve the analytical problem and provide reliable data.
2. The method is robust and reliable enough to be transferred to other locations and laboratories.
3. Spare parts and consumables are available for repairs and maintenance.
4. The technique is relatively easy to use by operators with different skill levels.

2.2.4 Sample pre-treatment and detection

Sample preparation is an essential part of the separation process and is one of the key considerations at the start of any method development strategy. Numerous publications are available on this subject [38, 39]. In brief, the primary aim is to provide a reproducible and homogenous solution that is ready for injection into the separation system. Sample preparation will produce a sample solution that is relatively free of interfering compounds that can damage the separation column and is compatible with the intended separation method. By satisfying these conditions, the separation characteristics of the method are unlikely to be affected. For most separation techniques, sample preparation is predominantly a manual process and as a result, can sometimes be a time-consuming component of the whole method development process. Despite this, it is still a very important and sometimes challenging area and requires sufficient planning and effort because key method validation parameters such as precision and accuracy are frequently dependent on a reproducible and robust sample preparation strategy for example in weighing, dilution and dissolution. Figure 2.3 summarises the sample preparation choices available.

Samples can be presented in various forms as solids, liquids, suspensions and even gases. In the solution form, these samples can either be ready for direct injection into the separation system or require further manipulation, i.e. dilution, buffering or addition of internal standards. When samples are in the solid form, they must first be either dissolved or extracted using a suitable solvent. For samples that consist of complex matrices either chemically based (e.g. formulated drug products) or biological in nature (e.g. blood, urine or plasma), some form of sample pretreatment to remove the interfering components is usually required. Direct injection of these matrix samples will quite often damage the separation column or instrument and should be avoided. In circumstances where the sample is relatively soluble, direct dissolution in a compatible solvent is the preferred mode of sample preparation because it is convenient and offers greater precision.

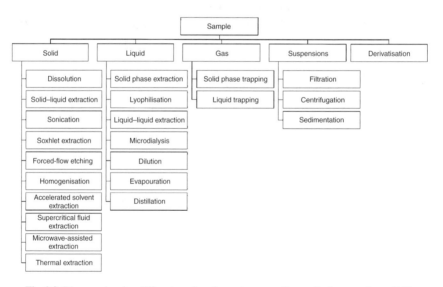

Fig. 2.3 Diagram showing different modes of sample preparation methods currently available.

For these sample types, the choice of solvent is important because choosing a solvent system that closely matches the separation solvent minimises baseline problems and other unwanted separation effects.

There are many sample preparation procedures published in the scientific literature, and within the scope of this chapter, only the most current and popular methods will be discussed. By far, the commonest and most popular method used for pretreatment of liquid samples is solid phase extraction (SPE) [40, 41]. For solid samples, several techniques are available including supercritical fluid extraction (SFE) [42, 43], microwave-assisted solvent extraction (MASE) [44, 45] and accelerated solvent extraction (ASE) [46, 47]. Solvent extraction methods have long been established as the standard approach to sample preparation, but the increasingly demanding needs of industries like the pharmaceutical, agrochemical and petrochemical for greater productivity, faster assays, and increased automation have led to the development of newer ways of sample preparation summarised in Fig. 2.3.

Derivatisation offers another means of preparing the sample ready for analysis. Derivatisation involves chemical reactions between the analyte molecule and a reagent to change the phyisco-chemical properties of the analyte molecule. By using this approach to sample preparation, the sample's properties can be modified to improve detectability, change its overall molecular structure and polarity, change the matrix for better separation, stabilise an unstable analyte and change its physical properties. By improving the detectability of the compound(s), poorly detected compounds can be derivatised to form derivatives which can have very strong UV absorbence, be highly fluorescent or possess electrochemical activity. Additionally, derivatisation can also be used to convert compounds that are not volatile into analytes that can be volatilised for GC analysis. Additional information on derivatisation strategies can be found in a recent publication by Blau and Halket [48].

During method development, it is critical that the derivatisation reaction is rapid, quantitative and produces minimal by-products or side reactions. These criteria are not always easy to achieve, hence, derivatisation as a sample preparation method is quite often chosen as a last resort. In pharmaceutical analysis, the use of derivatisation as a means of sample preparation is not usually the method of choice because the primary objective of any method development is to detect all major impurities and degradants in batches of the drug substance. Quite often, many of these compounds will not have derivatisable functional groups or have impurities that are present at such low levels that the derivatisation reaction is not optimised.

2.2.5 *Developing the separation*

The final stage in the simple step-by-step approach to method development is to develop the separation. When the separation technique has been chosen and using the information obtained from the earlier stages, a series of exploratory experimental runs are usually performed. The conditions chosen for these initial runs will depend on the separation method used. The following sections will provide some guidance on some of the new approaches adopted in method development for small molecules in HPLC, GC, CE and other less common separation techniques.

2.3 High performance liquid chromatography (HPLC)

2.3.1 Brief historical perspective of HPLC

The historical development of liquid chromatography has been extensively reviewed and can be traced as far back as the early 1900s [15], where the Russian botanist Zwett used a variant of liquid chromatography to separate some coloured plant substances. A much more detailed historical assessment of liquid chromatography can be found elsewhere [16, 17]. Here, the focus will be on modern developments in HPLC, a term that was coined in the late 1960s with the advent of more sophisticated instrumentation, better engineered separation columns, and reliable and highly efficient stationary phases and packing materials [18]. These technological advances have been, in part, fuelled, by the need to separate an increasingly larger variety of differing compound classes encountered as APIs, e.g. antibiotic, sulphonamides, nucleotides and nucleosides, fat-soluble vitamins, neutral and non-polar compounds. Additional challenges include developing faster and more consistent HPLC methods requiring higher flow rates, while maintaining peak-to-peak resolution, retention characteristics, peak shape, peak symmetry and efficiencies. Another important analytical challenge is the desire to detect and accurately quantify low levels of impurities at the 0.10% level present in API materials.

2.3.2 Different modes of HPLC

Nowadays, there are many different modes of HPLC available to equip the separation-scientist with the necessary tools to separate compounds that possess either acidic, basic, zwitterionic, oligomeric or aromatic functional groups (Fig. 2.2). Detailed texts are available for each mode of HPLC, along with the descriptions of their operation and applications but in this section, the focus will be primarily on the use and application of RP-HPLC. Although the other modes are routinely used in the pharmaceutical industry, RP-HPLC is the mode of choice. This is because many of the molecules of interest are weak organic bases and the separation is dramatically influenced by factors such as mobile phase pH, ionic strength, organic modifier and possible use of ion-pair reagents.

In RP-HPLC, the stationary phase is less polar than the mobile phase and is usually comprised of spherical silica particles (typically, 3–5 μm in diameter). The acidic functionalities on the silica material have been modified by derivatisation with alkyl (C2 to C18), phenyl, cyano and amino groups. Typical mobile phases used in RP-HPLC consist of mixtures of aqueous buffers mixed with water-miscible organic solvents, such as methanol and acetonitrile. In addition to modified silica stationary phases, other new developments in RP-HPLC are now available, e.g. porous polymeric, carbon and mixed modal phases.

2.3.3 Key developments in HPLC

Over the last decade, the most significant developments in liquid chromatography, specifically in HPLC, have been in the areas of instrumentation and equipment [49–52], advances in stationary phase chemistry and their design [51, 52], improvements in column design and dimensions [53, 54], use of different additives in mobile phases [55] and sample preparation techniques [38, 39]. All of these factors have helped to significantly improve the way separation-scientists develop their methods.

2.3.3.1 Stationary phase and column technology

The column is often called the *heart* of the HPLC separation process, and the availability of stable, high performance stationary phases and columns is critical to the development of rugged, reproducible and robust methods. Modern commercial columns can differ widely among suppliers and these differences can sometimes affect the development process of the desired HPLC method. Specifically, different columns can vary in terms of plate numbers (N), retention characteristics (k') and resolution (Rs). For these reasons, column and stationary phase manufactures have developed technologies to help ensure that these separation materials are produced in a more consistent and reproducible manner. An excellent reference by Snyder *et al.* [1] provides a comprehensive overview of modern stationary phases and column technology.

Most HPLC column packing materials use silica particles as the basis of the support. Silica is known for its reliable strength and rigidity, relative inertness and ability to be modified chemically. Significant scientific effort has been devoted to developing and improving the morphological and physico-chemical characteristics of these silica materials (Fig. 2.4).

Some of the more modern silicas include micropellicular particles [55], which consist of solid silica core with a thin outer skin of interactive stationary phase, particles range typically from 1.5 to 2.5 μm. Columns made of these microparticulates generate very sharp, low-volume peaks. Perfusion particles [56] have very large pores (typically 4000–8000 Å) throughout the support and also includes smaller interconnecting pores. These types of highly porous silica particles are

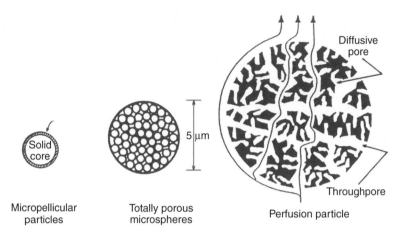

Fig. 2.4 Particle types used in modern stationary phases in HPLC. Figures represent approximate relative sizes of particle used.

more suited for separations of biopolymer and not for routine analytical separation of small molecules. Fully porous silica microspheres are the most commonly used [57, 58] because they offer many important benefits in HPLC columns such as good efficiency, high sample loading, durability and wide commercial availability. A recent development is the availability of monolithic silica columns where the column consists of a very porous silica rod with a large surface area for effective separations but with significantly reduced back pressure over a conventional particle-packed column allowing much higher flow rates. Over the next few years as this technology improves and the column types and chemistry becomes more diverse, these columns may well become the most commonly used in the pharmaceutical industry [58].

One unfavourable characteristic of silica is its solubility at extreme pH values and at elevated temperatures, thus leading to reduced lifetime and deterioration of peak shape and resolution. With the advent of particles formed by the aggregation of silica sols, so-called sol-gel silicas, this property of silica has been significantly reduced. The recent development of hybrid silica particles also address this problem [59].

Many compounds developed as biologically active drugs within the pharmaceutical industry are amine-containing compounds. In the past, the separation of amino compounds using earlier HPLC columns has resulted in significant peak tailing and poor peak shapes. This effect is now known to be due to the presence of metal ions (e.g. Fe, Al, Ni and Zn) that contaminate and interact with the silica surface and form complexes with silanol groups [60]. To circumvent the problem associated with analysing amine compounds, a range of highly purified silicas have been developed by stationary phase and HPLC column manufacturers. Nowadays, the column of choice for performing pharmaceutical analysis will most likely be made using ultra-high pure silica.

Similarly, manufacturers have put significant effort in recent years in improving the range of stationary phases which are available. Apart from the classic reverse phase C18 and C8 phases, other column chemistries are available including amine, phenyl, diol, nitrile and many others which impart selectivity into separations. Phases also containing various polar-embedded groups which modify selectivity and shield analytes from silanol interactions now give much improved peak shapes for polar compounds and prevent phase collapse in high aqueous content buffers. Phases with hydrophilic end capping are also available which permit separations of very polar compounds to be carried out in 100% aqueous mobile phases. The pH range that separations can be carried out at on silica-based columns has also been significantly extended in recent years with improvements in the silica and bonding chemistries. Separations can now be carried out at pH values less than 2 and up to 11 on certain phases. The extension of the basic pH range now permits separation of basic pharmaceuticals under reverse-phase conditions at pHs above the pKa for the molecule.

HPLC column configurations have also undergone important developments over the past few years [61, 62]. Most columns are now made from lengths of stainless steel with highly polished interior walls because of its mechanical strength and chemical inertness. However, in very few cases, some corrosive chloride-containing mobile phases can slowly erode the steel surface [62]. Some of the newer column configurations include the use of plastic material [63], glass-lined inner surfaces [63] and outer column surfaces made from the rigid polymer, polyether ether ketone (abbreviated as *PEEK*) [64]. Another novel approach is to use a soft polymeric material for the outer surface, and by compressing the column radially, the efficiency of the separation is enhanced as the stationary phase is pressurised [65].

2.3.3.2 Instrumentation

The manufacturers of modern HPLC systems have adopted a modular approach to the design and utility of the instrumentation [49–51]. This modular approach is an efficient and flexible means of performing HPLC because, different modules (e.g. different detectors) can be configured easily. Additionally, this set-up allows for easy stepwise tracking of problems that may arise during operation of the HPLC system (e.g. a high back pressure due to a blockage in the system). The modules consist of a pumping system, an autosampler, a column oven and the detector, all controlled by a computer and appropriate software. Although there are small variations in this set-up, all commercially available HPLC systems subscribe to this format. In later sections, the key technological developments that have occurred in HPLC instrumentation will be discussed in more detail.

Some important technological advances in HPLC have been associated with instrumentation. Specifically in the areas of mobile phase pump design, sampling introduction systems, detection systems, connections and detection systems. For supplemental reading, references [49, 50] are recommended. Two areas are worthy of specific mention, namely column ovens and eluent mixing systems. The introduction of column ovens has significantly improved method development by ensuring methods are reliable and more reproducible. This is of particular importance in a pharmaceutical environment. The development of eluent mixing systems in HPLC has meant that efficient mixing of solvents is now possible for isocratic and gradient elution.

2.3.3.3 Microcolumn liquid chromatography
Any discussion on new developments in HPLC is not complete without mentioning microcolumn liquid chromatography. In microcolumn LC, columns with internal diameters of less than 1 mm are used with HPLC systems equipped with microtubing, a micro injection device and a specialised detector. The most important advantages of microcolumn LC are the ability to work with small sample sizes, lower solvent consumption, small volumetric flow rates and the enhanced detection performance with the use of concentration sensitive detection. A recent review on microcolumn HPLC is recommended for further reading [66]. The main application areas of this technique are bioanalysis, neuroscience and protein/peptide research. However, in small molecule analysis, microcolumn LC is not used as routinely as conventional HPLC. This is most probably due to current instrumentation being less reliable and not as robust as conventional systems. However, this area of liquid chromatography is progressing rapidly, and in the future, the status of this technique in applications such as routine API analysis will change.

2.3.3.4 Combined HPLC methods
To meet the speed and high efficiencies in separations demanded by the pharmaceutical industry, combined HPLC methods have been frequently used to simultaneously determine combination products [67]. A stability-indicating method for the simultaneous determination of aspirin and warfarin in warfarin sodium/aspirin combination tablets has been recently developed and validated [68]. In another example [69], the simultaneous determination of enalapril (2) and its two degradants, enapril-DKP (3) and enalapril-diacid (4), and felodipine (5) and its degradant, named H152/37 (6) was achieved using combined method approach.

The chromatogram in Fig. 2.5 demonstrates that the retention of (3), (5) and (6) is governed by reversed-phase partitioning processes, whereas for (2) and (4), both reverse-phase and cation-exchange mechanism are operating. Again, this method was validated to be precise, accurate, specific and robust.

2.4 Gas chromatography (GC)

2.4.1 Brief historical perspective

Gas chromatography (GC) was originally developed in the 1950s [70] and commercialised in the late 1950s. The principle as with all chromatography techniques depends on the partitioning of the analyte between a stationary phase and a mobile phase. In the case of GC, the mobile phase is a gas which means that the analyte must be volatile to progress through the column and most

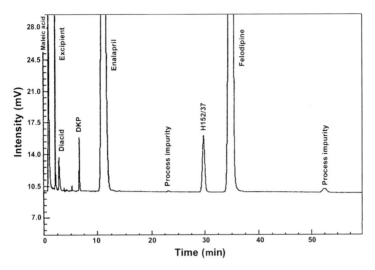

Fig. 2.5 Typical HPLC chromatograms showing the separation of compounds (2) to (6) and a UV-absorbing excipient using the HPLC conditions listed in ref [70].

analytes have little interaction with the mobile phase. Essentially, there are two modes of operation, gas solid chromatography (GSC) and gas liquid chromatography (GLC). In these early days, the technique used packed columns where the analyte partitioned either with the solid stationary phase (GSC) or with a liquid coating on the solid stationary phase (GLC). Detection was by total conductivity detection (TCD) and later with flame ionisation detection (FID). GC was rejuvenated in the 1980s with the advent of chemically bonded fused capillary columns which offered high efficiencies and thermal stability. GC is used widely for analysis in the pharmaceutical industry and although it is a mature technique, improvements in columns, injection and detection systems, and ovens have permitted the expansion of the technique to analyses previously beyond the scope of GC. Some of these technical improvements are discussed below and are illustrated by analytical examples. In line with most other separation techniques, technical improvements have given rise to the possibility of increasing the efficiency of the technique by reducing cycle time, often with an associated improvement in limits of detection for analytes, and some of these technical innovations will also be discussed.

2.4.2 GC in pharmaceutical analysis

GC is used widely for analysis in the pharmaceutical industry, and applications for assays and impurity profiles for raw materials, intermediates and APIs are commonplace. GC is often chosen as the method of analysis when compounds have poor or widely different UV chromophores. A common case is when the

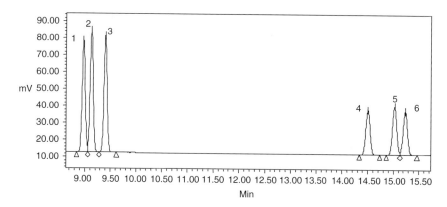

Fig. 2.6 Gas chromatographic separation of 4-nitrobenzoyl chloride (4) and 4-chlorobenzoyl chloride (2) and their regiosiomers.

reduction of a nitrated aromatic species to the corresponding amino species gives rise to large shifts in the UV maxima and hence significant UV response factor differences between starting materials, products and impurities. Another common use is illustrated by the example below, where an impurity profile method was required to analyse (7) and to specifically quantify the levels of the corresponding regioisomers in the samples (Fig. 2.6). The large number of theoretical plates available on a standard GC column makes the technique particularly suited to this kind of separation, especially here in the case of acid chlorides which are difficult to analyse by HPLC.

(7)

GC is the most commonly used technique for residual solvent and organic volatiles analysis in API and has also found application where trace level detection of analytes is required. An airborne monitoring method for SB-202026-A (8) was required and due to the low occupational exposure limit (OEL) for this molecule it was necessary to be able to quantify (8) in solutions at 100 ng ml^{-1}. This was achieved readily using a 1 microlitre splitless injection, and detection with a nitrogen phosphorus detector without the need for sample pre-concentration.

(8)

2.4.3 Key developments in GC

2.4.3.1 Sensitivity enhancement with large volume injection

The programmable temperature vapouriser (PTV) inlet allows the enrichment of analytes in the PTV insert rather than relying on extraction procedures and facilitates the use of large volume injection to obtain analysis on low-level analytes. It also permits the coupling of GC with other sample preparation techniques such as solid phase extraction (SPE), liquid–liquid extraction (LLE) or SFE and a method for the coupling of LC to GC. Many applications are recorded in the literature predominantly in the field of environmental analysis. Engewald et al. have reviewed PTV techniques and applications [71]. For analytes eluting close to the solvent front, the sensitivity benefits of large volume injections can be obtained using cool-on-column injection with solvent vapour exit (COC-SVE) [72].

2.4.3.2 Thermally labile samples

The requirement for volatility often precluded the use of GC for analysis of thermally labile samples. Injection of the sample directly into the column using cool-on-column injection (COC) with volatilisation occurring on the column has meant such samples can be readily analysed [73a,b].

2.4.3.3 Analytes in complex matrices

Often interference effects from either solvents [74] or other components in sample matrices can cause significant problems especially with direct injection of such solutions. Headspace analysis has been shown to be of great value for residual solvent analysis in drug substance [75] and drug product [76] because the drug itself is not introduced into the system. Similarly, residual solvent analysis in pharmaceuticals using thermal desorption [77] and solid phase microextraction (SPME) [78] has been shown to be of benefit. For more complex matrices such as blood or environmental samples, problems with interferences can be more acute and previously required significant sample preparation prior to analysis. SPME has proved to be very useful for these samples [79] especially as the technique can be combined with PTV [80]. SPME can also be used for in situ derivatisation [81] or can be molecularly imprinted to facilitate specific analyte detection [82].

2.4.3.4 Detection systems

One of the main advantages of GC analysis is that it can be readily combined with a large number of detection techniques and hence a wide range of applications are available. Interfaces connecting GC with the mass spectroscopy [83] to allow structure elucidation have now become commonplace. More recently, the coupling of GC to Fourier transform infra red (FTIR) spectroscopy and computer modelling of spectral features has proved a reliable and fast screening technique [84]. Element-specific detectors such as the nitrogen–phosphorus detector (NPD) have been used to identify low-level components of interest in complex samples especially when combined with GC/MS [85]. The atomic emission detector (AED) has further expanded the detection possibilities to almost any element and provided the ability to give element ratios and hence empirical formulae for analytes of interest [86]. The electron capture detector (ECD) has historically been used in the pharmaceutical industry for the analysis of low-level halogenated compounds but suffers from the disadvantage of the requirement for a radioactive source. For many analyses, the non-radioactive pulsed discharge helium ionisation detector (PDHID) in electron capture mode can replace the ECD with no loss of sensitivity [87]. The PDHID has also proved to be very valuable in normal ionisation mode with excellent sensitivity for a wide range of analytes [88]. An application of GC that is becoming increasingly important in the pharmaceutical industry is that on inverse gas chromatography (IGC). In this technique, the retention of volatile probes is used to determine surface energy information. Changes in surface energy may be used to detect batch to batch differences due to recrystallisation from different solvents and following milling operations. Grimsey *et al.* have reviewed the use of IGC in the analysis of surface energies of pharmaceutical powders [89].

2.4.3.5 Efficiency increases in GC

In common with most other separation techniques, technological advances have not only increased the applications of the technique but have also allowed results to be obtained with much shorter analysis times with no loss of sensitivity. In line with HPLC, improved column technology has allowed capillary columns to become commercially available with internal diameters less than 100 µm. These columns combined with the ability for GC ovens to heat at rates of greater than 100°C/min have enabled analysis times to be reduced by factors of up to 10 times with no loss of resolution [90 A.M. Godwin & J.K. Roberts, unpublished work]. Conversion of existing methods to fast GC methods is also a simple process with the availability of online software to translate the methods [91]. Most modern GC instrumentation is now available with electronic pressure control which will allow flow rate programming to optimise separation efficiency throughout the run [92]. Another technique which can increase the speed of separations is a variant of backflushing [93] where late eluting peaks (e.g. make up solvent in residual solvents analysis) can be backflushed from the column, saving on analysis and re-equilibration time. A relatively recent development has been in the use of resistive column heating in GC separations. In this application, the column is in close contact with a restive heating element and hence very high reproducible heating rates of around $1000°C\,min^{-1}$ are possible. Coupled with this, cooling rates are much faster than with a conventional GC oven and limits of detection are significantly lower due to the narrow peak widths [94]. Cycle time reductions for analysis of the order of 5–50 times have been reported [95].

2.4.3.6 Automation

In today's pharmaceutical industry, the developments discussed above in improving GC instrumentation to reduce cycle times also need to be implemented in an automated and a manner compliant with good manufacturing practise (GMP) to achieve maximum benefit. Most instrument manufacturers now offer data systems which are capable of controlling instruments, storing the data in a compliant manner and allow the use of unattended operation with built-in acceptance criteria. Automated sample preparation [96] is also becoming more important, with the goal being to achieve *round-the-clock analysis* with automated results reporting for samples with minimal human intervention. For some analyses where generic methodology can be used (e.g. residual solvents analysis), most of the above goals are now achievable and systems are in place for total analysis from a solid drug substance [97]. Similarly, systems are also being developed to automate the procedure of method development using GC [98].

2.5 Capillary electrophoretic techniques

2.5.1 Brief historical perspective

The development of capillary electrophoretic (CE) techniques dates as far back as the late 1960s [99] where, Hjerten separated some UV-absorbing compounds using a 300 μm capillary tube [99]. Since then, the technique has developed further by Virtanen [100], Mikkers, Everaerts and Verheggen [101], who use small diameter Pyrex and Teflon tubing to separate inorganic and organic ions. In the late 1980s, Jorgenson and Lukacs [102] used narrower silica capillary tubes (typically <100 μm); this is most likely where the term capillary electrophoresis was popularised. It was not until the late 1980s that interest in CE really began to take off [103]. This era saw the advent of commercial capillary electrophoresis instrumentation and the transformation of a research technique into a routine technology present in many industrial settings [6, 7].

In modern pharmaceutical laboratories, CE is still an important technique, but its application in pharmaceutical analysis is less widespread. The rapid developments in LC have to a greater extent overtaken CE techniques, hence the role of CE is either in niche applications such as chiral analysis or as an orthogonal analytical technique to HPLC.

2.5.2 Developments in detection modes in CE

Instrumental developments in CE have largely been focused in two areas: expanding the range of detectors available and the hyphenation to mass spectrometry (this is covered in later sections). UV absorbance is by far the commonest detection mode available in CE and offers sensitive analysis, particularly for the separation of low-level impurities. The low path length of the detection area in the capillary, which is effectively the internal diameter (ID) of the capillary, often limits the minimum amounts that can be detected when compared to other techniques. The use of larger ID capillaries, pre-concentration and focusing effects and the use of low-ionic strength sample solvents has helped to circumvent this disadvantage. Diode array detectors offering wavelength selection between 190 and 700 nm are also available providing good quality spectral analysis. Other sensitive but less common detection systems such as fluorescence [104, 105], mass spectrometry [106, 107], Raman [108], amperometry [109], conductivity [110], radiochemical [111] and NMR [112] have also been used with CE.

2.5.3 Different modes and method development options in CE

Several modes of CE have been described in the literature over past decade [6, 7]. The most common are open tubular or capillary zone electrophoresis (CZE), micellar electrokinetic chromatography (MEKC), microemulsion electrokinetic chromatography (MEEKC), capillary electrochromatograpy (CEC), capillary gel electrophoresis, capillary isoelectric focusing and capillary isotachophoresis. Three recent reviews by Watzig [113], Tagliaro et al. [114] and Riekkola et al. [115] summarise the current method development options available to manipulate selectivity. In pharmaceutical analysis, CZE, MEKC, MEEKC and CEC are commonly used.

2.5.3.1 Capillary zone electrophoresis
Method development in CE involves optimising the experimental parameters such as pH, organic modifiers, surfactant additives, ion-pair reagents, cyclodextrins, polymer additives, complexation agents and combinations of these additives. The use of pH is a very powerful tool for manipulating and influencing

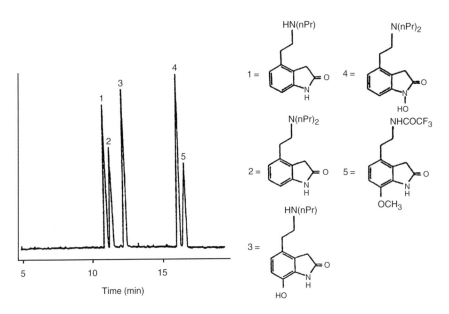

Fig. 2.7 Capillary electrophoresis of the drug ropinirole (peak 2) and a number of related molecules.

selectivity, particularly for analytes that have small differences in their pKa values. A good example [6] of this is in the separation of the anti-Parkinson drug ropinirole (9) from a number of related molecules shown in Fig. 2.7. Here the pKa differences and overall charge densities between (9) and its impurities are small and by using a phosphate buffer at pH 7.8, all five compounds are baseline resolved. Altering the pH value also changes the magnitude of the electroosmotic flow (EOF), which can affect migration times and the ionisation of the analytes.

(9)

Other experimental parameters to be considered include the nature of the column and its dimensions. The recommended columns are fused silica materials with typical IDs of 50–75 μm and column lengths of between 25 and

75 cm. Buffers types consisting of phosphates, borates and formates are usually recommended with ionic strengths of 25 and 150 mM. Where the CZE system uses a mass spectrometer as the detector, volatile buffers such as ammonium formates, acetates and bicarbonates are used.

The use of deuterated solvents such as deuterated water has been used to improve the resolution of a number of analytes including pharmaceuticals [116, 117]. This approach has some very unique benefits such as different ionisation equilibria of polar compounds, higher thermal conductivity and higher density, which can give rise to some dramatic improvements in separation.

2.5.3.2 Micellar electrokinetic chromatography
MEKC is used to separate compounds that are neutral and uncharged and cannot be resolved using simple buffer systems. The separation mechanism closely resembles chromatography. In MEKC, different surfactants can be used giving rise to significant differences in selectivity e.g. bile salts [118], cationic surfactants such as cetyltrimethylammonium bromide [119] and neutral surfactants such as Tween [120]. Further manipulation of selectivity can be achieved by varying surfactant concentration, the addition of urea, cyclodextrins, ion-pair reagents, temperature, pH, buffer type and concentration. A more recent development in MEKC has been to perform MEKC in the absence of EOF [121]. This is achieved using either coated capillaries or at low pH. This is particularly useful for acidic species which would be ionised at higher pH values and would not interact with the micellar phase.

2.5.3.3 Microemulsion electrokinetic chromatography
MEEKC is a relatively new technique in CE and in many ways is very similar to MEKC except that the solute partitions between the aqueous phase and oil droplets which are moving through the capillary [122]. The microemulsion composition usually consists of a high pH buffer (typically borate at pH 9–10), octane, butan-1-ol and a surfactant (usually sodium dodecyl sulphate (SDS)). The water-immiscible octane forms minute oil droplets that are coated with butan-1-ol and SDS, which reduce the surface tension of the solution allowing a stable microemulsion to form. The same parameters used in MEKC to modulate selectivity can also apply to MEEKC, but it is generally accepted that MEEKC is more applicable to a broader range of solutes and most closely matches the retention and partitioning mechanism described in reverse-phase chromatography. MEEKC has been used to resolve a range of neutral compounds and pharmaceuticals [123].

2.5.3.4 Capillary electrochromatography
CEC is a relatively new development in CE that strives to combine the best features of HPLC and CE. Over the last decade, publications in the use of CEC have increased significantly [124–127] signalling the importance of

mode of CE and its applications in the area of pharmaceutical analysis. Other modes of CEC have been explored including coupling to mass spectrometry and the use of pressurised flow capillary electrochromatography (PEC) to further extend the applicability of CEC. PEC has been reported to address difficult applications such as the separations of highly basic pharmaceutical compounds [127]. Method development considerations in CEC are the same as those described in Section 2.2 for HPLC and in CE. These include the choice of stationary phase, choice of organic modifier and concentration, applied voltage, pH, temperature, electrolyte concentration and the use of gradient elution.

2.6 Other separation techniques

Two other chromatographic techniques that are worthy of mention are TLC and SFC.

2.6.1 Thin layer chromatography

TLC is still used in many pharmaceutical laboratories in either manual or semi-automatic operation on conventional, high performance or modified stationary phases [9, 10, 126]. TLC offers a quick, inexpensive, flexible and portable technique that has been the subject of some recent new developments. Despite these advancements, modern TLC has largely served as a complementary technique to other column-based liquid chromatographic methods such as HPLC.

Advances in stationary phase technology have led to commercial availability of adsorbents such as high performance silicas, aluminas, polyamides, celluloses and derivatised silicas [9, 10]. The development of automated method development (AMD) systems [127] now allow multi-step gradients of different elution strengths to be achieved in a relatively short time compared to earlier manual approaches. AMD systems are ideally suited for separation of complex mixtures with a wide range of polarities. Further improvements in sample resolution and reduced method development times in TLC include the use of two-dimensional development approaches [128] and forced-flow development by over-pressure liquid chromatography (OPLC) [129].

The main improvements in qualitative and quantitative detection systems in TLC are centred around the introduction of densitometry. These involve the use of slit scanning densitometers or video or CCD camera (image processing) [130].

Other method development approaches used in TLC include unidimensional multiple development [127] and multi-modal separation techniques [127], where TLC, in normal phase mode is used in conjunction with reversed-phase liquid column chromatography [131] or GC [131] to provide additional information in separations. This complementary strategy can prove very important even for well characterised mixtures where new and unexpected peaks can sometimes be revealed. An example of where this complementary strategy has proved successful is the analysis of polar aromatic flavour compounds in vanilla extracts [132]. In this case, the RP-HPLC method was able to separate and quantify most of the principal aromatic compounds in the vanilla extracts. However, by using normal phase TLC, an additional peak was observed, isolated and later identified as 5-(hydroxymethyl)-2-furfural, a known compound found in vanilla extracts.

In the above example, the new impurities were relatively harmless, but in pharmaceutical analysis, the potential dangers are much higher and so there is a need to ensure that all identified and unidentified impurities are accounted for. Other multi-modal approaches in TLC include GC/TLC [133] which has largely been superceded by GC/MS today [131]. Others include SFC/TLC, combining the chromatographic technique to a flame ionisation or flame-thermionic ionisation detector (TLC/FID/FTID) or non-dispersive infrared (TLC/NDIR) detection [131].

2.6.2 Supercritical fluid chromatography

SFC has been undergoing a renaissance in its use in modern analytical laboratories over the last decade, particularly in chiral and achiral analysis of pharmaceutical compounds. Many of the problems associated with implementing SFC have come from the technical and operational difficulties in earlier instruments [134, 135]. New developments in SFC include improved back-pressure-regulation, more consistent SFC flow rates, more reliable sample injection systems and improved flow cell designs [136, 137].

2.7 Hyphenated separation techniques

The need to obtain more information about samples and increased sensitivity for low-level impurities has lead to the coupling of existing separation techniques with specialised detection methods. Nowhere has this been more prominent than in the hyphenation of chromatographic methods to mass spectrometric detectors. A mass spectrometer can offer many significant benefits during method development compared to other detection modes (e.g. UV and diode array). Here, the key analytical challenges are sensitivity, selectivity and speed. The combination to MS systems can avoid common analytical problems by tracking and identifying individual peaks in the chromatogram between experiments. In systems linked to MS detectors, one can distinguish compounds of interest from minor compounds or interferences. LC/MS and GC/MS represent the most important of these techniques, and this is reflected in the speed of their development and the spread and acceptance of these instruments in analytical laboratories. In the pharmaceutical industry, LC/MS has become the method-of-choice for analytical support in many stages of drug development. Other separation techniques such as GC, CE, TLC and SFC have been hyphenated to mass spectrometers, NMR, Raman spectrometers and FTIR systems. Table 2.2 summarises a review of the key hyphenated techniques currently being used in pharmaceutical laboratories for impurity and degradant analysis. The developments and applications of separation techniques combined to MS and NMR systems will be covered more extensively in other chapters in this book.

Another new dimension in the area of hyphenated techniques that offers some very significant benefits in pharmaceutical analysis is that of multi-dimensional chromatography. Various set-ups involving coupling GC, HPLC and CE systems together in different configurations have been studied for analysing many different sample types. Examples include coupling RP-HPLC with CZE for separating ionic analytes, ion exchange chromatography with RP-HPLC [138], size exclusion chromatography coupled to with RP-HPLC and CE [139] and GC coupled to LC [140]. Since RP-HPLC and CE techniques are capable of high resolution separation with orthogonal separation mechanisms, combining both techniques in a two-dimensional mode can produce very high peak capacities and extremely high resolving power, particularly useful for complex mixtures. LC coupled to CE has been successful in a number of applications such as determination of impurities in the antifungal agent fenticonazole

Table 2.2 Summary of hyphenated separation techniques used in pharmaceutical analysis

Separation technique	Hyphenated mode	References
Liquid chromatography	Liquid chromatography-mass spectrometry (LC/MS)	174, 175
	Liquid chromatography-Fourier-transform infrared spectrometry (LC-FTIR)	176, 177
	Liquid chromatography-nuclear magnetic resonance spectroscopy (LC/NMR)	178, 179
	Liquid chromatography-inductively coupled plasma mass spectrometry (LC-ICPMS)	180
Gas chromatography	Gas chromatography-mass spectrometry (GC/MS)	181
	Gas chromatography-Fourier-transform infrared (GC-FTIR)	182, 183
	Gas chromatography-FTIR-MS (GC-FTIR-MS)	184, 185
Capillary electrophoresis	Capillary electrophoresis-mass spectrometry (CE/MS)	186, 187
	Capillary electrophoresis-nuclear magnetic resonance spectrometry (CE/NMR)	188, 189
	Capillary electrophoresis-surface enhanced Raman spectrometry (CE-SERS)	190, 191
Thin layer chromatography (TLC)	Thin layer chromatography-mass spectrometry (TLC/MS)	192, 193
	Thin layer chromatography-surface enhanced Raman spectrometry (TLC-SERS)	194, 195
Supercritical fluid chromatography/ extraction (SFC/SFE)	Supercritical fluid extraction-capillary gas chromatography-mass spectrometry (SFE-CGC-MS)	196
	Supercritical fluid-Fourier-transform infrared (SFC-FTIR)	197

[141] and the use of capillary-HPLC-CE to resolve hundreds of neutral compounds present in traditional Chinese medicines [142].

2.8 Use of automated approaches to method development in chromatography

Despite the considerable advances in chromatography, the basic approaches to GC and HPLC method development are still based mainly on a trail-and-error experimentation and can be very time-consuming and very inefficient. Within the last few years, there have been many reported examples of strategies [1, 143, 144] that improve method development efficiency, reduce time, produce more consistent methods and ultimately reduce overall costs. These strategies all use automation to varying degrees, ranging from automated method development software programs [144–146] through to column-switching systems and automated multi-column [145, 147, 148], multi-eluent approaches and the use of robotic sample preparation systems [154]. Many of these approaches are now available as commercial product that can be retrofitted to existing HPLC systems.

In developing separations methods in the pharmaceutical industry, there exists two contradictory objectives: the first objective is to achieve optimum resolution of critical peaks for the API and related impurities, the second objective is to achieve this separation in as short an analysis time as possible. By using these automated approaches, a balance between resolution and run time should be achievable.

2.8.1 Separation optimisation programmes

There are a number of commercially available computer software packages that have been used recently by many separation scientists in pharmaceutical laboratories. These computer programs are designed to simulate the actions of a practising scientist developing a method. It plans, experiments, collects data, evaluates the results, makes decisions and automatically searches for the optimum conditions. Furthermore, some software systems can learn from previous experiments and observations. To achieve a fully automated system, these software programs are integrated with computers that control the operational parameters of the instrumentation. Conversely, they can also be used as off-line computer assisted HPLC method development. A review of the current literature reveals that the currently available method development software systems can be classified according to its capabilities or function and has been summarised in Table 2.3. All the software packages listed in the table are commercially available. Many of these expert systems have focused on HPLC because this separation technique is modular e.g. sample preparation, method development and method validation, in which the basic rule is set, and is relatively well defined.

Table 2.3 Summary of method optimisation software currently available

Function/capability	Example	References
Changes one variable at a time and predicts separation as function of that variable	Drylab	165, 166, 170
Changes one or more variables at a time and predicts separation as a function of those variables	ICOS, DIAMOND	167, 168, 171
Changes column conditions (column dimensions, particle size and flow rate) and predicts separation for any column condition	Drylab, ENHANCER	165
Changes gradient conditions and predicts separation for any gradient condition	Drylab	153, 165, 166
Changes one or more conditions and examines experimental chromatograms for best separation	PESOS	169, 170
Expert systems to predict best initial separation conditions on the basis of sample component molecular structures	ELUEX, CHROMDREAM, HPLC-METABOLEXPERT, ProDigest-LC, LABEL	165, 166, 172, 173

Many of these software systems work (particularly for Drylab™) by using mathematical algorithms to predict separations for a number of other conditions after a few experimental runs have been performed. Typical predictions are made for changes made to the mobile phase conditions, temperature, isocratic or gradient separations, or changes to the column conditions (e.g. column dimensions, particle size and flow rate). The references listed in Table 2.3 provide more specific descriptions of each type of software system.

Presently, Drylab is probably the most widely used method development software package for HPLC [149, 150] and GC [151, 152] in the pharmaceutical industry and has been used to optimise the separation methodology for numerous examples [153]. A good example of the utility of Drylab is in the development of the related substances method for SB-243213 (10) [200].

An initial method was developed to separate SB-243213 from 11 related impurities and degradants which ran for 30 min and utilised a multi-segmented gradient to achieve adequate resolution. Four trial experimental runs permitted the construction of a 3D contour plot displaying critical resolution as a function of gradient time and temperature (Fig. 2.8). Based on these data, eluent starting conditions, gradient slope and temperature were defined to predict an optimised separation in terms of resolution and run time. The predicted and actual separations using the conditions are shown in Fig. 2.9.

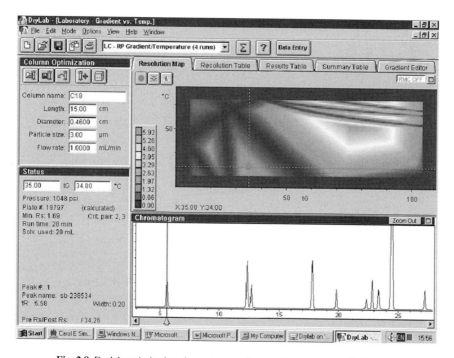

Fig. 2.8 Drylab optimisation chromatograms for the pharmaceutical, SB-243213.

Actual and predicted chromatograms agree very well with an overall decrease in method run time and a reduction in gradient complexity which will make the method much better suited for transfer to the QC environment.

2.8.2 Column switching devices

Column switching devices fitted to existing HPLC systems can offer significant improvements in method development time in LC [147, 148]. They offer

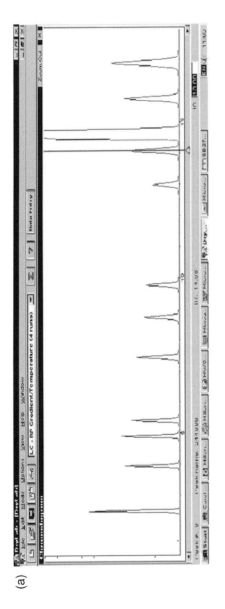

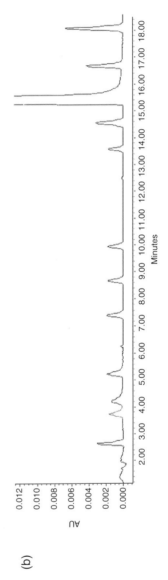

Fig. 2.9 Diagram showing (a) Drylab predicted, and (b) Actual separation for the pharmaceutical, SB-243213 and its related substances.

the ability to screen numerous columns and conditions with minimal manual intervention. They are easy to automate, especially with modern HPLC systems and can be configured to perform different functions (e.g. backflush, heart-cutting and on-column concentration). When column switching is used for method development, decreased analysis times and higher sample throughput can be achieved. This approach has significant benefits in the development of a quantitative method for the X-ray enhancement agent, 3,5-bis (acetylamino)-2,4-6-triiodobenzoate (11) [148]. Here, six HPLC columns (C-1, C-6, cyano, C-18, amino and phenyl) were screened rapidly (Fig. 2.10). The HPLC set-up is shown in the diagrammatic representation in Fig. 2.3.

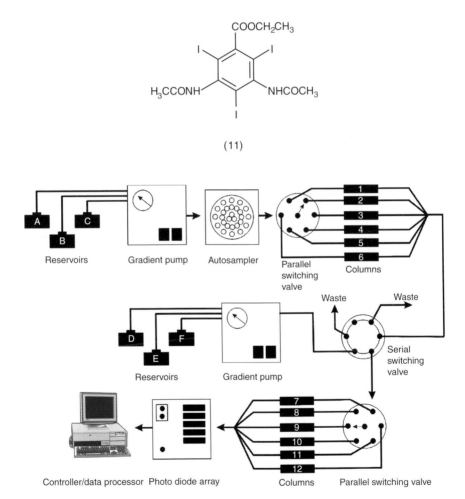

Fig. 2.10 Diagram of instrument layout for column switching screening.

Other applications where the use of column switching devices have offered significant benefits include the direct analysis of the drug indomethacin in rat plasma [154].

2.9 Use of chemometric approaches to method development

Chemometrics has been defined in some texts [155] as the entire process whereby data are transformed into information used for decision-making. It is this definition that is the most applicable to separation sciences, more specifically in method development and optimisation in liquid chromatography. In this example, chemometrics has been used to predict optimum separation conditions based on empirical data and other separation information. Chemometric approaches to method development can be based on either sequential simplex models [156] or simultaneous fixed factorial designs [157] or interactive mixture designs [158] which combine the advantages of simultaneous and simplex models.

In HPLC, the combination of wide utilisation of the technique, its high degree of automation and computer control, and the high degree of method development processes have made this area particularly attractive to chemometricians. Chemometrics are particularly applicable in mobile phase optimisation, method development and post-separation data processing. In mobile phase optimisation, the sequential simplex model is most often used because it is easily understood and requires no prior knowledge of the sample and chromatographic system. However, it is limited by its inability to distinguish between local and global optima, hence it is best suited to fine-tuning when global optimum regions have been pre-defined. Another common chemometric approach is the use of factorial or mixture designs, which are regression methods that aim to model one or more of the chromatographic parameters. Examples of the use of a factorial design in method development are that of the separation of metronidazole and tinidazole [159] and for the method development of a complex mixture of monoamine neurotransmitters [160]. Further use of chemometric approaches has been demonstrated by Djordjevic *et al.* [161] who by combining a factorial experimental design with a sequential strategy and computer simulation software was able to determine retention characteristics for five pharmaceutical compounds.

Other recently developed chemometric approaches include Heuristic evolving latent projections (HELP) [162] which is primarily used for the analysis of qualitative and quantitative data from 2-dimensional data (e.g. HPLC DAD) used for complex mixtures, and HELP lessens the demand on developing complicated separation methods; mixed level orthogonal array design (ML-OAD) has been used to optimise the parameters involved in HPLC detection [163]. The LC methods for determining the purity of tetracycline HCl and resolution of its impurities were optimised using two new chemometric approaches, namely othoganol projection approach (OPA) and fixed size moving window evolving factor analysis approach (FSW-EFA) [164].

Abbreviations

ACE	Accelerated solvent extraction
AED	Atomic emission detector
AMD	Automated method development
API	Active pharmaceutical ingredient
CCD	Charge coupled device
CE	Capillary electrophoresis
CEC	Capillary electrochromatography
COC-SVE	Cool-on-column with solvent vapour exit
CZE	Capillary zone electrophoresis
EOF	Electroosmotic flow
FDA	Food and drug administration
FID	Flame ionisation detection
FSW-EFA	Fixed size moving window-evolving factor analysis
FTID	Flame thermionic ionisation detector
GC	Gas chromatography
HELP	Heuristic evolving latent projections
HPLC	High performance liquid chromatography
ICH	International conference on harmonisation
IGC	Inverse gas chromatography
MASE	Microwave-assisted solvent extraction
MEEKC	Microemulsion electrokinetic chromatography
MEKC	Micellar electrokinetic chromatography
ML-OAD	Mixed level-orthoganol array design
NACE	Non-aqueous capillary electrophoresis
NDIR	Non-dispersive infra red
NPD	Nitrogen phosphorus detector
OPA	Orthoganol projection approach
OPLC	Over pressure liquid chromatography
PDHID	Pulsed discharge helium ionisation detector
PEC	Pressurised flow capillary electrochromatography
PTV	Programmable temperature vapouriser
QC	Quality control
RP-HPLC	Reversed phase-high performance liquid chromatography
SDS	Sodium dodecyl sulphate
SFC	Supercritical fluid chromatography
SFE	Supercritical fluid extraction

SPE   Solid phase extraction
SPME Solid phase micro-extraction
TLC   Thin layer chromatography

References

1. Snyder, L.R., Kirkland, J.J. & Glajch, J.L. (1997) *Practical HPLC Method Development*, John Wiley & Sons Inc., New York.
2. Snyder, L.R. & Stadalius, M.A. (1986) *High Performance Liquid Chromatography, Advances in Perspectives*, Academic Press, San Diego.
3. Weber, S.G. & Carr, P.W. (1989) *High Performance Liquid Chromatography*, Wiley-Interscience, New York.
4. Grob, R.L. (1995) *Modern Practice of Gas Chromatography*, Wiley-Interscience, New York.
5. Rao, G.R., Murthy, S.S.N. & Khadagapathi, P. (1987) *East. Pharm.*, **30**(353), 35.
6. Camilleri, P. (1997) *Capillary Electrophoresis, Theory and Practice*, 2nd edn, CRC Press, Boca Raton.
7. Weinberger, R. (1993) *Practical Capillary Electrophoresis*, Academic Press, San Diego.
8. Hahn-Deinstrop, E. (2000) *Applied Thin-Layer Chromatography – Best Practice and Avoidance of Mistakes*, Wiley-VCH, Weinheim.
9. Dalls, F.A.A., Read, H., Ruane, R.J. & Wilson, I.D. (1987) *Recent Advances in Thin Layer Chromatography*, Plenum Press, New York.
10. Zlatkis, A. & Kaiser, R.E. (1977) *HPTLC – High Performance Thin-Layer Chromatography*, Elsevier, Amsterdam.
11. Salvador, A., Angeles, J.M., Becarra, G., De La Guardia & Fresenius, M. (1996) *J. Anal. Chem.*, **356**, 109.
12. Lee, M.L. & Markides, K.E. (1990) *Analytical Supercritical Fluid Chromatography and Extraction*, Chromatography Conferences Inc., Provo.
13. Hage, D.S. (2002) *J. Chromatogr. B*, **768**, 3.
14. Ricker, R.D., Sandoval, L.A., Permar, B.J. & Boyes, B.E. (1979) *Biological/Biomedical Applications of Liquid Chromatography* (ed. G.L. Hawk) Marcel Dekker, New York, pp. 93–105.
15. Berezkin, V.G. & Horwood, E. (1990) *Chromatographic Adsorption Analysis. Selected works of Mikhail Semenovich Tswett*, New York.
16. Desty, D.H. (1989) *LC-GC*, **4**(5), 32, 40.
17. Wixom, R.L., Gehrke, G.W. & Bayer, E. (2002) *Chromatography – A Century of Discovery*, Abstracts of Papers, 223rd ACS National Meeting, Orlando, FL, 7–11 April.
18. Adlard, E.R. (2002) *LC-GC*, **11**, 12.
19. Ettre, L.S. (2000) *Chromatographia*, **51**(1/2), 7.
20. Dandeneau, R.D. & Zerenner, E.H. (1979) *J. High Resoln. Chromatogr.*, **2**, 351.
21. Jorgenson, J.W. & Lucas, K.D. (1981) *Anal. Chem.*, **53**, 1298.
22. Isenhour, T.L. (1985) *J. Chem. Information and Computing*, **25**, 292.
23. Stockwell, P.B. & Corns, W.T. (1996) *Automatic Chemical Analysis*, Taylor & Francis, London.
24. Dessy, R. (1983) *Anal. Chem.*, **55**, 1100A.
25. Hawk, G.L. & Kingston, H.M. (1988) *Laboratory Robotics and Trace Analysis in Quantitative Trace Analysis of Biological Materials*, Elsevier, Amsterdam.
26. Lynch, T. (1999) *CAST, Chromatography and Separation Technology*, **8**, 4.
27. Vinall, M. (1992) *Anal. Proc.*, **29**(11), 466.
28. Cavalla, D., Flack, J. & Jennings, R. (1997) *Modern Strategy for Preclinical Pharmaceutical R & D – Towards the Virtual Research Company*, John Wiley & Sons, New York.
29. Lendrem, D. (1995) *Script Magazine*, 22 December.

30. Jannsson, S.O. & Johannsson, S. (1982) *J. Chromatogr.*, **242**, 41.
31. Berridge, J.C. (1995) *J. Pharm. Biomed. Anal.*, **14**, 7.
32. D'Arcy, P.F. & Harron, D.W.G. (1992) *Proceedings from the First International Conference on Harmonisation*, Queen's University of Belfast.
33. FDA CDER Guidance for Industry document, 'ANDa's: Impurities in Drug substances', November 1999.
34. Catalano, T., Madsen, G. & Demarest, C. (2000) *Am. Pharm. Rev.*, **3**(3), 62.
35. Schoenmakers, P.J. & Mulholland, M. (1988) *Chromatographia*, **25**(8), 737.
36. Whiting, R. (1994) *Chem. N.Z.*, **58**(6), 16.
37. Saeed, S.A., Shaikh, K.F. & Salam Khan, M.A. (1983) *J. Pharm. Uni. Kar.*, **2**(1), 57.
38. Yoshihiro, S., Maki, K., Motohiro, I., *et al.* (2002) *Anal. Sci.*, **18**, 1.
39. David, N. (2002) Drugs and the pharmaceutical sciences, Vol. **117**, *Handbook of Pharmaceutical Analysis*, pp. 59–86.
40. Gilar, M., Bouvier, E.S. & Compton, B.J. (2001) *J. Chromatogr. A*, **909**(2), 111.
41. Drummer, O.H. (1999) *J. Chromatogr. B*, **733**(1–2), 27.
42. Kaiser, C.S., Rompp, H. & Schmidt, P.C. (2001) *Pharmazie*, **56**(12), 907.
43. Karlsson, L., Torstensson, A. & Taylor, L.T. (1997) *J. Pharm. Biomed. Anal.*, **15**(5), 601.
44. Lopez-Avila, V., Young, R. & Beckert, W.F. (1994) *Anal. Chem.*, **66**, 1097.
45. Hasty, E. & Reves, Z. (1995) *Am. Lab.*, **27**(4), 66.
46. ASE Application Note 321 (1996) Dioncx, Salt Lake City, UT.
47. ASE Application Note 322 (1996) Dionex, Salt Lake City, UT.
48. Blau, K. & Halket, J. (eds) (1994) *Handbook of Derivatives for chromatography*, John Wiley & Sons, New York.
49. McGrath, A. & Britton, V.J. (1984) *Liq. Chromatogr. HPLC Mag.*, **2**(1), 44.
50. Glick, M.R., Tanabe, K., Berthod, A. & Winefordner, J.D. (1988) *Anal. Instrum. (NY)*, **17**(3), 277.
51. Dorsey, J.G., Cooper, W.T., Siles, B.A., Foley, J.P. & Barth, H.G. (1998) *Anal. Chem.*, **70**(12), 591R.
52. Toshihiko, H. (2000) *Advances in Chromatography (New York)*, **40**, 315.
53. Noctor, T. (1996) *High Perform. Liq. Chromatogr.*, 97.
54. Kirkland, J.J. (1994) *Am. Lab. (Shelton, Conn.)*, **26**(9), 28K.
55. Huber, C.G. (1998) *J. Chromatogr. A*, **806**(1), 3.
56. Gordon, N., McCoy, M., Mu, N., Nadler, T., Londo, T. & Whitney, D. (1997) *Chromatography*, **18**(4), 236.
57. Kirkland, J.J., Truszkowski, F.A., Dilks, C.H. & Engel, G.S. (2000) *J. Chromatogr. A*, **890**(1), 3.
58. Kirkland, J.J., Dilks, C.H. & DeStefano, J.J. (1993) *J. Chromatogr.*, **635**(1), 19.
59. Stella, C., Rudaz, S., Veuthey, J.-L. & Tchapla, A. (2001) *Chromatographia*, **53**(Suppl.) S113.
60. Lubda, D., Cabrera, K., Kraas, W., Schaefer, C., Cunningham, D. & Majors, R.E. (2001) *LC-GC Europe*, **14**(12), 730.
61. Snyder, L.R. & Antle, P.E. (1985) *LC Mag.*, **3**(2), 98–100, 102, 104, 108–109.
62. Dorsey, J.G., Cooper, W.T., Siles, B.A., Foley, J.P. & Barth, H.G. (1998) *Anal. Chem.*, **70**(12), 591R.
63. Neue, U.D. (1997) *HPLC Columns: Theory, Technology, and Practice*, Wiley-VCH, New York, 393.
64. Slingsby, R.W., Bordunov, A. & Grimes, M. (2001) *J. Chromatogr. A*, **913**(1–2), 159.
65. Dunn, D.L. & Thompson, R.E. (1983) *J. Chromatogr.*, **264**(2), 264.
66. Vissers, J.P.C. (1999) *J. Chromatogr. A*, **856**, 117.
67. Qin, X.-Z. (1997) *Process Control and Quality*, **10**, 1.
68. Montgomery, E.R., Taylor, S., Segretario, J., Engler, E. & Sebastian, D. (1996) *J. Pharm. Biomed. Anal.*, **15**, 73.
69. Qin, X.-Z., DeMarco, J. & Ip, D.P. (1995) *J. Chromatogr. A*, **707**, 245.
70. Martin, A.J.P. & James, A.T. (1952) *J. Biochem.*, **50**, 679.

71. Engewald, W., Teske, J. & Efer, J. (1999) *J. Chromatogr. A*, **856**(1/2), 259.
72. Klee, M.S., Nixon, D.D. & Wylie, P.L. (1998) *Am. Lab. (Shelton, Conn.)*, **30**(5), 104H.
73a. Klick, S.J. (1995) *J. Chromatogr. A*, **689**(1), 69.
73b. Klick, S.J. (1995) *J. Pharm. Biomed. Anal.*, **13**(4/5), 563.
74. Urakami, K., Kobayashi, C., Miyazaki, Y., *et al*. (2000) *Chem. Pharm. Bull.*, **48**(9), 1299.
75. Natishan, T.K. & Wu, Y. (1998) *J. Chromatogr. A*, **800**(2), 275.
76. Markovich, R.J., Ong, S. & Rosen, J. (1997) *J. Chromatogr. Sci.*, **35**(12), 584.
77. Hashimoto, K., Urakami, K., Fujiwara, Y., Terada, S. & Watanabe, C. (2001) *Anal. Sci.*, **17**(5), 645.
78. Camarasu, C.C., Mezei-Szuts, M. & Varga, G.B. (1998) *J. Pharm. Biomed. Anal.*, **18**(4/5), 623.
79. Namera, A., Watanabe, T., Yashiki, M., Iwasaki, Y. & Kojima, T. (1998) *J. Anal. Toxicol.*, **22**(5), 396.
80. van Hout, M.W.J., van Egmond, W.M.A., Franke, J.P., de Zeeuw, R.A. & de Jong, G.J. (2002) *J. Chromatogr. B*, **766**(1), 37.
81. Staerk, U. & Kulpmann, W.R. (2000) *J. Chromatogr. B: Biomed. Sci. Appl.*, **745**(2), 399.
82. Cresecenzi, C., den Hoedt, W., Koster, E.H.M., de Jong, G.J. & Ensing, K. (2001) *Abstracts of Papers*, 222nd ACS National Meeting, Chicago, IL, United States, 26.
83. Lauko, A. (2000) *Prog. Pharm. Biomed. Anal.*, **4**, 183.
84. Praisler, M., Dirinck, I., Van Bocxlaer, J.F., De Leenheer, A.P. & Massart, D.L. (2001) *J. Anal. Tox.*, **25**(1), 45.
85. Aebi, B. & Bernhard, W. (1999) *Forensic Sci. Int.*, **102**(2/3), 91.
86. Bos, R. & Barnett, N.J. (1997) *J. Anal. At. Spectrom.*, **12**(7), 733.
87. Wentworth, W.E., Cai, H., Madabushi, J. & Qin, Y. (1993) *Process Control Qual.*, **5**(2/3), 193.
88. Hunter, M.C., Bartle, K.D., Seakins, P.W. & Lewis, A.C. (1999) *Anal. Commun.*, **36**(3), 101.
89. Grimsey, I.M., Feeley, J.C. & York, P. (2002) *J. Pharm. Sci.*, **91**(2), 571.
90. Chen, T.K., Phillips, J.G. & Durr, W. (1998) *J. Chromatogr. A*, **811**(1/2), 145.
91. Method Translation Software available from http://www.agilent.com.
92. Hermann, B.W., Freed, L.M., Thompson, M.Q., Phillips, R.J., Kleinand, K.J. & Snyder, W.D. (1990) *J. High Resoln. Chromatogr.*, **13**(5), 361.
93. Eppert, G.J. (1968) *Gas Chromatogr.*, **6**(7), 361.
94. Dalluge, J., Ou-Aissa, R., Vreuls, J.J. & Brinkman, U.A.Th. (1999) *J. High Resoln. Chromatogr.*, **22**(8), 459.
95. Surve, M., Knight, D. & Warren, N. (2001) *LaborPraxis*, **25**(9), 40.
96. Penton, Z. (1992) *J. High Resoln. Chromatogr.*, **15**(5), 329.
97. Squicciarini, C. (2001) *Eur. Pat. Appl.*, CODEN: EPXXDW EP 1136808 A1 20010926 13.
98. Jayatilaka, A. & Poole, C.F. (1993) *J. Chromatogr. Biomed. Appl.*, **617**(1), 19.
99. Hjerten, S. (1967) *Chromatogr. Rev.*, **9**, 122.
100. Virtanen, R. (1974) *Acta Polytechnica Scandinavia*, **123**, 67.
101. Mikkers, F.E.P., Everaerts, F.M. & Verheggen, T.P.E.M. (1979) *Electrophoresis*, **169**, 11.
102. Jorgenson, J. & Lukacs, K.D. (1983) *Science*, **222**, 266.
103. Perret, D., Database on Capillary Electrophoresis, with entries to December 1995.
104. Kuhr, W.G. & Yeung, E.S. (1988) *Anal. Chem.*, **60**, 1832.
105. Bergquist, J., Douglass, G.S., Ewning, A.G. & Ekman, R. (1994) *Anal. Chem.*, **66**, 3512.
106. Reinhoud, N.J., Tinke, A.P., Tjaden, U.R., Niessen, W.M.A. & van der Greef, J. (1992) *J. Chromatogr.*, **627**(1/2), 263.
107. Okafo, G., Tolson, D., Monte, S. & Marchbank, J. (2000) *Rapid Comm. in Mass Spect.*, **14**(23), 2320.
108. Ruddick, A., Batchelder, D.N., Bartle, K.D., Gilby, A.C. & Pitt, G.D. (2000) *Applied Spectroscopy*, **54**(12), 1857.
109. Zhang, S.S., Yuan, Z.B., Liu, H.X., Zou, H. & Wu, Y.J. (2000) *J. Chromatogr. A*, **872**(1/2), 259.
110. Zemann, A.J. (2001) *Trends in Anal. Chem.*, **20**(6/7), 346.

111. Klunder, G.L., Andrews, J.E., Grant, P.M., Andresen, B.D. & Russo, R.E. (1997) *Anal. Chem.*, **69**(15), 2988.
112. Olson, D.L., Peck, T.L., Webb, A.G. & Sweedler, J.V. (1996) *Proc. 14th Am. Pept. Symp., Pept. Chem., Struct. Biol.*, 730.
113. Watzig, D.C. (1994) *Pharmazie*, **49**(2/3), 83.
114. Tagliaro, F., Deyl, Z., Miksik, I. & Ulfelder, K.J. (1998) *Methods Biochem. Anal.*, **38**, 41.
115. Riekkola, M.L., Wiedmer, S.K., Valko, I.E. & Siren, H. (1997) *J. Chromatogr. A*, **792**(1/2), 13.
116. Greenaway, M., Okafo, G., Manallack, D. & Camilleri, P. (1994) *Electrophoresis*, **15**(10), 1284.
117. Camilleri, P. & Okafo, G. (1991) *J. Chromatogr.*, **541**(1/2), 489–495.
118. Neubert, R.H.H., Mrestani, Y., Schwarz, M. & Colin, B. (1998) *J. Pharm. Biomed. Anal.*, **16**(5), 893.
119. Hilhorst, M.J., Somsen, G.W. & De Jong, G.J. (1998) *J. Pharm. Biomed. Anal.*, **16**(7), 1251.
120. Terabe, S. (1992) *J. Pharm. Biomed. Anal.*, **10**(10/12), 705.
121. Shah, R.S., Wang, Q., Lee, M.L. & Milton, L. (2002) *J. Chromatogr. A*, **952**(1/2), 267.
122. Altria, K.D. (2002) *J. Capillary Electrophoresis Microchip Tech.*, **7**(1/2), 11.
123. Altria, K.D. (2000) *J. Chromatogr. A*, **892**(1/2), 171.
124. Tzovolou, D., Mitropoulou, T., Antonopoulos, A., Goulas, S. & Karamanos, N.K. (2001) *Pharmakeutike*, **14**(1), 27.
125. Idei, M. & Hajos, G. (2001) *Magyar Kemikusok Lapja*, **56**(11), 398.
126. Hilhorst, M.J., Somsen, G.W. & De Jong, G.J. (2001) *Electrophoresis*, **22**(12), 2542.
127. Krull, I.S., Stevenson, R.L., Mistry, K. & Swartz, M.E. (2000) *Capillary Electrochromatography and Pressurised Flow Capillary Electrochromatography – an Introduction*, HNB Publishing, New York.
128. Touchstone, J.C. (1993) *LC-GC*, **11**(6), 404, 406, 409.
129. Birkinshaw, F.L. & Waters, D.G. (1995) *J. Planar Chromatogr. – Mod. TLC*, **8**(4), 319.
130. Prosek, M., Drusany, I. & Golc-Wondra, A. (1991) *J. Chromatogr.*, **553**(1/2), 477.
131. Mincsovics, E. & Tyihak, E. (1998) Recent Adv. Thin-Layer Chromatogr., *Proc. Chromatogr. Soc. Int. Symp.* **57**.
132. Mustoe, S.P. & McCrossen, S.D. (2001) *Chromatographia*, **53**(Suppl.), S474–S477.
133. Cserhati, T. & Forgacs, E. (1998) *J. AOAC Int.*, **81**(2), 329–332.
134. Sherma, J. (2000) *J. Chromatogr.*, **880**(1–2), 129–147.
135. Poole, C.F. (1999) *J. Chromatogr.*, **856**(1/2), 399.
136. Chester, T.L. & Pinkston, J.D. (2000) *Anal. Chem.*, **72**(12), 129.
137. Chester, T.L., Pinkston, J.D. & Raynie, D.E. (1998) *Anal. Chem.*, **70**(12), 301R.
138. Nishino, I., Fujitomo, H. & Umeda, T. (2000) *J. Chromatogr. B: Biomedical Sciences and Applications*, **749**(1), 101.
139. Williams, R.A., Macrae, R. & Shepherd, M.J. (1989) *J. Chromatogr.*, **477**(2), 315.
140. Quigley, W.W.C., Fraga, C.G. & Synovec, R.E. (2000) *J. Microcolumn Separations*, **12**(3), 160.
141. Quaglia, M.G., Donati, E., Bossu, E., Desideri, N. & Campana, F. (2001) *J. Sep. Sci.*, **24**(5), 392.
142. Huang, S., Xu, S. & Zhang, X. (2000) *Fenxi Huaxue*, **28**(12), 1467.
143. Berridge, J.C. (1985) *Techniques for the automate optimization of HPLC Separations*, John Wiley & Sons, New York.
144. Stockwell, P.B. & Corns, W.T. (1996) *Automatic Chemical Analysis*, Taylor & Francis, London; Pfeffer, M. & Windt, H. (2001) *J. Anal. Chem.*, **369**(1), 36.
145. Drouen, A., Dolan, J.W., Snyder, L.R., Poile, A. & Schoenmakers, P.J. (1991) *LC-GC*, **9**(10), 714.
146. Snyder, L.R. (1996) *Methods Enzymol.*, **270**, 151.
147. Gmeiner, G., Geisendorfer, T., Kainzbauer, J., Nikolajevic, M. & Tausch, H. (2002) *J. Chromatogr. B*, **768**(2), 215.
148. Wu, D., Berna, M., Maier, G. & Johnson, J. (1997) *J. Pharm. Biomed. Anal.*, **16**, 57.
149. Bonfichi, R. (1994) *J. Chromatogr. A*, **678**(2), 213.

150. Bye, C.A., Larmann, J.P., Clarke, H. & Norris, K.J. (2001) *Abstracts of Papers*, 222nd ACS National Meeting, Chicago, IL, United States.
151. Molnar, I. (1993) *LaborPraxis*, **17**(12), 40, 44.
152. Abbay, G.N., Barry, E.F., Leepipatpiboon, S., *et al.* (1991) *LC-GC*, **9**(2), 100, 104, 106, 110, 112, 114.
153. Schmidt, A.H. & Molnar, I. (2002) *J. Chromatogr. A*, **948**(1/2), 51.
154. Lui, S., Kamijo, M., Takayasu, T. & Takayama, S. (2002) *J. Chromatogr. B*, **767**(1), 53.
155. Beebe, K.R., Pell, R.J. & Seasholtz, M.B. (1998) *Chemometrics A Practical Guide*, John Wiley & Sons, New York.
156. Marengo, E. & Gennaro, M.C. (2000) *Chemom. Intell., Lab. Syst.*, **53**(1/2), 57.
157. Dimov, N. (1997) *Anal. Lab.*, **6**(3), 163.
158. Berridge, J.C. (1989) *Chemom. Intell. Lab. Syst.*, **5**(3), 195.
159. Kettaneh-Wold, N. (1991) *J. Pharm. Biomed. Anal.*, **9**(8), 605.
160. Pullan, L.M. (1988) *J. Liq. Chromatogr.*, **11**(13), 2697.
161. Djordjevic, N.M., Erni, F., Schreiber, B., Lankmayr, E.P., Wegscheider, W. & Jaufmann, L. (1991) *J. Chromatogr.*, **550**(1/2), 27.
162. Shen, H.-L., Cui, H., Liang, Y.-Z. & Frank, L. (1998) *Huaxue Xuebao*, **56**(4), 378.
163. Lan, W.C., Chee, K.K., Wong, M.K., Lee, H.K. & Sin, Y. (1995) *Analyst*, **120**(2), 281.
164. De Braekeleer, K., De Juan, A. & Massart, D.L. (1999) *J. Chromatogr. A*, **832**(1/2), 67.
165. Outinen, K., Vuorela, H. & Hiltunen, R. (1996) *Eur. J. Pharm. Sci.*, **4**(4), 199.
166. Outinen, K., Vuorela, H. & Lehtonen, P. (1995) *Kem-Kemi*, **22**(8), 703–706.
167. Bowman, P.B., Hann, J.T., Marr, J.G.D., Salvat, D.J. & Thompson, B.E. (1993) *J. Pharm. Anal.*, **11**(11/12), 1295.
168. Bowman, P.B., Marr, J.G.D., Salvat, D.J. & Thompson, B.E. (1993) *J. Pharm. Biomed. Anal.*, **11**(11/12), 1303.
169. Colgan, S.T. & Pollard, E.B. (1991) *LC-GC*, **9**(11), 772, 774.
170. Gant, J.R., Vandemark, F.L. & Poile, A.F. (1990) *Am. Lab. (Fairfield, Conn.)*, **22**(8), 15.
171. Faulstich, R. & Catalano, T. (1991) *LC-GC*, **9**(11), 776.
172. Galushko, S.V., Kamenchuk, A.A. & Pit, G.L. (1995) *Am. Lab. (Shelton, Conn.)*, **27**(5), 33G.
173. Hamoir, T. & Massart, D.L. (1994) *Analytica Chimica Acta*, **298**, 319.
174. Anon (1999) *Pharmaceutical applications of LC-MS*. Anon. USA. *Chromatogr. Sci. Ser.*, 79 (Liquid Chromatography-Mass Spectrometry (2nd edn) 405.
175. Ermer, J. & Vogel, M. (2000) *Biomedical Chromatography*, **14**(6), 373.
176. Huang, W. & Chen, J. (1998) *Guangpuxue Yu Guangpu Fenxi*, **18**(2), 191.
177. Guiochon, G. (1997) *Book of Abstracts, The evolution of preparative – scale chromatography*. 213th ACS National Meeting, San Francisco, American Chemical Society, Washington, DC.
178. Potts, B.C.M., Albizati, K.F., Johnson, M. & James, J.P. (1999) *Magn. Reson. Chem.*, **37**(6), 393.
179. Crowe, E.A., Roberts, J.K. & Smith, R.J. (1995) *Pharm. Sci.*, **1**(2), 103.
180. Axelsson, B.-O., Jornten-Karlsson, M., Michelsen, P. & Abou-Shakra, F. (2001) *Rapid Comm. Mass Spect.*, **15**(6), 375.
181. Mcclure, G.L. (1999) *J. Pharm. Sci. Technol.*, **53**(3), 129.
182. Leal, W.S., Kuwahara, Y., Matsuyama, S., Suzuki, T. & Ozawa, T. (1992) *J. Braz. Chem. Soc.*, **3**(1/2), 2.
183. Maylin, G.A., Dewey, E.A. & Henion, J.D. (1987) *LC-GC*, **5**(10), 6–12, 14.
184. Basiuk, V.A. & Douda, J. (2001) *J. Anal. Applied Pyrolysis*, **60**(1), 27.
185. Meuzelaar, H.L.C. & Arnold, N.S. (2000) *Book of Abstracts*, 219th ACS National Meeting, San Francisco, CA, 26.
186. Ross, G.A. (2001) *LC-GC Europe*, **14**(1), 45.
187. Niessen, W.M.A. (1999) *Chimia*, **53**(10), 478.
188. Albert, K. (1995) *J. Chromatogr. A*, **703**(1/2), 123.
189. Schewitz, J., Pusecker, K., Gfrorer, P., *et al.* (1999) *Chromatographia*, **50**(5/6), 333.

190. Sepaniak, M.J., Nirode, W.F., Devault, G. & Lavrik, N.V. (2002) *Abstracts of Papers*, 223rd ACS National Meeting, Orlando, FL, United States.
191. He, L., Natan, M.J. & Keating, C.D. (2000) *Anal. Chem.*, **72**(21), 5348.
192. Wilson, I.D. & Morden, W. (1991) *J. Planar Chromatogr. – Mod. TLC*, **4**, 226.
193. Banno, K., Matsuoka, M. & Takahashi, R. (1991) *Chromatographia*, **32**(3/4), 179.
194. Koglin, E. (1989) *J. Planar Chromatogr. – Mod. TLC*, **2**(3), 194.
195. Poole, C.F., Poole, S.K. & Dean, T.A. (1988) *Recent Adv. Thin-Layer Chromatogr.*, 11.
196. Wong, S.H.Y. (1989) *Clin. Chem.*, **35**(7), 1293.
197. Bartle, K.D., Clifford, C.A. & Raynor, M.W. (1992) *J. Chromatogr. Libr.*, **53**, 103.
198. Roberts, J.K. & Hughes, M.J. (1998) *J. Chromatogr. A*, **828**, 297.
199. Guest, D.W. (1997) *J. Chromatogr. A*, **760**, 159.
200. Simcox, C. Unpublished work.

3 Chiral analysis of pharmaceuticals
W. John Lough

3.1 Significance of chirality in pharmaceutical R&D

That the resolution of the enantiomers of chiral drugs should be considered as a topic in its own right in a treatise on pharmaceutical analysis should come as no surprise. The importance of chirality in many fields of natural and applied science is well established [1]. In pharmaceutical analysis, this topic which commands its own nomenclature (Table 3.1) is especially important, as is apparent from the proportion of drugs on the market that are chiral (Table 3.2).

The chirality of these drugs is a feature which cannot be ignored. While the enantiomers of a chiral drug will have identical physical and chemical properties in an achiral environment, they can be distinguished in a chiral environment provided that there are suitable interactions with a chiral *selector* in a non-racemic form. In molecules such as proteins, enzymes and carbohydrates there is an abundance of such chiral selectors in the body. Drugs might interact with receptor proteins at their site of action. On the way to the site of action, they may be actively transported across membranes (e.g. intestinal membrane or cell wall membrane, leading into – or, in some cases, out of – the site of action) via an interaction with a protein in a so-called protein pump. Binding also takes place with proteins – predominantly albumin and α_1-glycoprotein – in plasma. In metabolism, the drug must first bind to an enzyme as its substrate before the enzyme can act upon it. Less obviously, chiral interactions may be involved in binding to tissue, or in secretion of drugs in fatty deposits, for example. Given this range of processes encountered by drugs entering the body which involve these selectors, it is seldom that the body is not able to distinguish between the enantiomeric forms of a chiral drug. Accordingly the drug enantiomers are very likely to have different pharmacological and toxicological properties so that to all intents and purposes the body *sees* the enantiomers as different drugs. The potential for the body to be able to distinguish between enantiomers has long been recognised [2] but it was only until the so-called Thalidomide Tragedy (Fig. 3.1) in the late 1950s early 1960s that the issue was brought into stark reality. While the sedative effect of thalidomide was attributed to the (R)-enantiomer and the adverse teratogenic effecy to the (S)-enantiomer, it is not always, in fact not often, a simple case of good enantiomer/bad enantiomer. Even in the, by reputation, classic case of thalidomide, things are

Table 3.1 Selected nomenclature used in the field of chirality

Chiral molecules	Molecules whose mirror images are not superimposable upon each other.
Achiral molecules	Molecules whose mirror images are superimposable upon each other.
Stereoisomers	Compounds, the molecules of which have the same atoms connected in order but differ from each other in the way the atoms are orientated in space.
Enantiomers	One of a pair of molecular species that are mirror images of each other and not superimposable.
Diastereomers	Stereoisomers with multiple chiral centres that are not enantiomers.
Epimers	Diastereomers differing in configuration at one of two or more chiral centres (less frequently encountered are instances when it is used to describe cases involving chiral axes or planes).
Meso-compound	Diastereomer with two or more chiral centres where the four groups on each of the chiral carbon atoms contains a plane of symmetry within the molecule.
Racemic mixture	1:1 mixture of enantiomers.
Racemates	Strictly speaking this describes a particular solid state which is distinguishable from *conglomerates* and *solid solutions*; now generally used as being synonymous with *racemic mixture* with *racemic compound* being reserved for the description of the distinct solid state.
d-	Dextrorotatory optical rotation at a specified wavelength, solvent and temperature, usually 589 nm (sodium D line emission); now supplanted by (+).
l-	Laevorotatory optical rotation at a specified wavelength, solvent and temperature, usually 589 nm (sodium D line emission); now supplanted by (−).
(+)	Dextrorotatory optical rotation at a specified wavelength, solvent and temperature, usually 589 nm (sodium D line emission).
(−)	Laevorotatory optical rotation at a specified wavelength, solvent and temperature, usually 589 nm (sodium D line emission).
D-	Configurational descriptor for carbohydrates and α-aminoacids based on correlation with (+)-glyceraldehyde; use for other kinds of chiral compounds is now obsolete (also, for amino acids the *R/S* nomenclature is now more commonly used).
L-	Configurational descriptor for carbohydrates and α-aminoacids based on correlation with (−)-glyceraldehyde; use for other kinds of chiral compounds is now obsolete (also, for amino acids the *R/S* nomenclature is now more commonly used).
Absolute configuration	The spatial arrangement of the stereogenic centres (atoms) in a chiral molecule.
Cahn-Ingold-Prelog (CIP) Rules	Accepted system (1982 version usually used) for the description of configuration of stereogenic centres in organic molecules; generally used to assign a stereogenic centre (atom) arising from e.g. a carbon atom with four different substituents as *R*- or *S*- but there are also rules for describing other types of chiral molecules.
R-	*Rectus*; stereochemical descriptor in the Cahn-Ingold-Prelog Rules.
S-	*Sinister*; stereochemical descriptor in the Cahn-Ingold-Prelog Rules.
Chiral	Not superimposable with its mirror image; may be applied to molecules, conformations, and macroscopic objects such as crystals; note also that its use has commonly been extended to describing compounds the molecules of which are chiral, even if the bulk sample of the compound is racemic.
Homochiral	Not to be confused with *enantiopure*; its use must be restricted to fundamental stereochemistry e.g. a sample of (+)-tartaric acid is made up of *homochiral* molecules (not a clear term and still controversially discussed and used).

Table 3.1 (continued)

Heterochiral	As counterpart to homochiral → unprecise term.
Enantiopure	Characterization of a compound that does not contain its enantiomer impurity according to available or applied analytical methods (limits of detection and qualification of the minor enantiomer needs to be specified).
Optical purity	Measure of the composition of a mixture of enantiomers determined by measurement of optical rotation (reference compounds and methods needed).
Enantiomeric excess (ee)	The percent excess of the enantiomer over the racemate in a mixture of a pure enantiomer and a racemate.

Compiled from E. Eliel and S.H. Wilen (1994) *Stereochemistry of Organic Compounds*, Wiley-Interscience, New York, USA; G. Helmchen (1996) *Glossary of Problematic Terms in Organic Stereochemistry*, Enantiomer 1; C.A. Challener (ed.) (2001) *Chiral Intermediates*, Ashgate Publishing Company, Burlington, USA.

Table 3.2 Proportion of single enantiomer chiral drugs in development and in licensed medicines

1996–2000 (of new chemical entities assessed by the UK Medicines Control Agency): 36% were achiral; 48% single enantiomers and 16% were racemates
ca. 1997 (of drug substances in all licensed medicines): 30% were natural or semi-synthetic (of these, 98% were single isomers, 1% were mixtures of isomers and 1% were non-chiral) 70% were synthetic compounds (of these, 4.5% were single isomers, 35.5% were mixtures of isomers and 60% were non-chiral) i.e. the % of chiral drug substances which were single isomers was 99% for natural/semi-synthetic drug substances and 11%* for synthetic drug substances
1998 (of all drugs under development): 80% were single enantiomer chiral drugs
1999 (of value of all dosage-form drug sales worldwide): 33% from single enantiomer chiral drugs
2000 (of value of all dosage-form drug sales worldwide): 40% from single enantiomer chiral drugs
2008 (projected value of dosage-form single enantiomer chiral drugs sales worldwide: projected from the 2000 figure of $133 billion (*cf* $360 billion for all dosage-form drug sales)): $200 billion

* This was an increase on, say, five years previous.
Sources: Caldwell (1999); Branch (2000); Challener (2001) [C11]; Stinson (2001).

not so simple as this [3–6]. Complications include *in vitro* racemisation, *in vivo* inversion, appropriateness of the original animal models used and additional toxicological issues other than phocomelia. There are a number of different scenarios which might ensue (Fig. 3.2) and looking at the case of verapamil (Fig. 3.3) it is apparent just how complicated the situation can get. Despite the possibility of encountering such complexity, there are still usually clear advantages in progressing a drug candidate as its single enantiomer. Even if one enantiomer is not significantly more active or significantly less toxic than the other, there may be advantages in one enantiomer being more selective

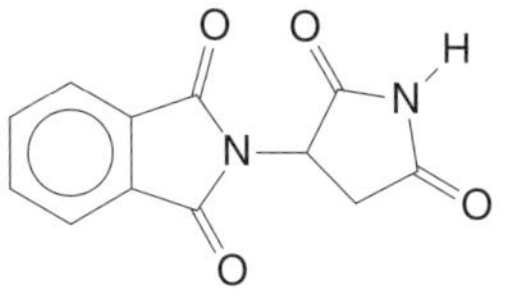

CHIRAL ANALYSIS OF PHARMACEUTICALS

Fig. 3.1 Thalidomide was developed as a non-addictive sedative as an alternative to barbiturates. Following its administration to pregnant women, it was found that the babies were born with truncated limbs (phocomelia). At the time this was perceived as a classic case of the pharmacological activity residing in one enantiomer (R-) and the undesirable toxicological effect residing in the other enantiomer. The situation is now known to be much more complex than this (see text and reference cited therein).

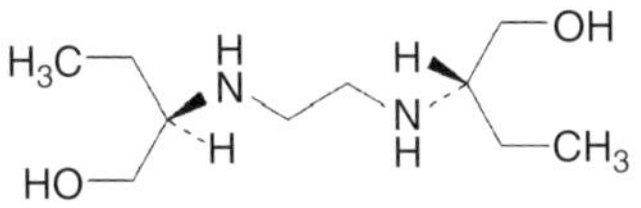

Have adverse toxicological effects e.g. ethambutolantituberculosis, while enantiomer may induce blindness (also, by reputation, thalidomide (R-sedative, S-teratogenic) but see qualification in text and references cited therein)

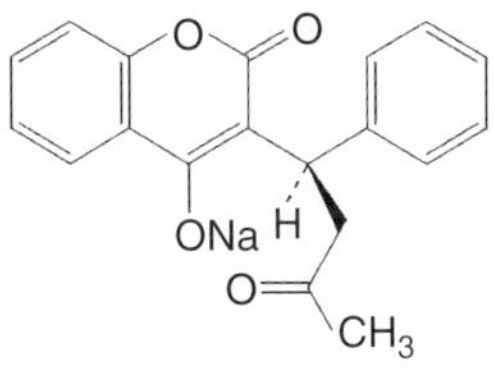

Have the same desired effect, but at a lower level e.g. warfarin (sodium) (anti-coagulant) (also propanolol (β-blocker)).

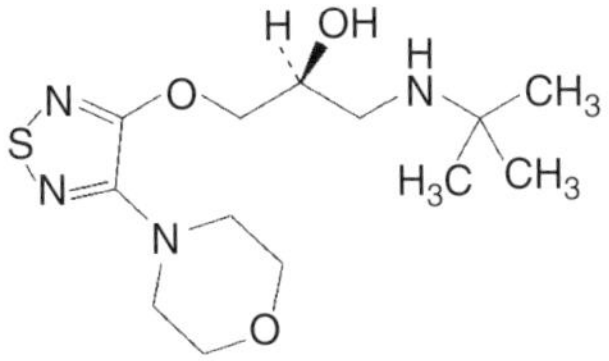

Have differing activity for more than one therapeutic effect e.g. timolol (R- is ¼ as potent at reducing intraocular pressure and 1/50 as active as a β-blocker so that in principle it might be a safer drug for the treatment of glaucoma than the S-).

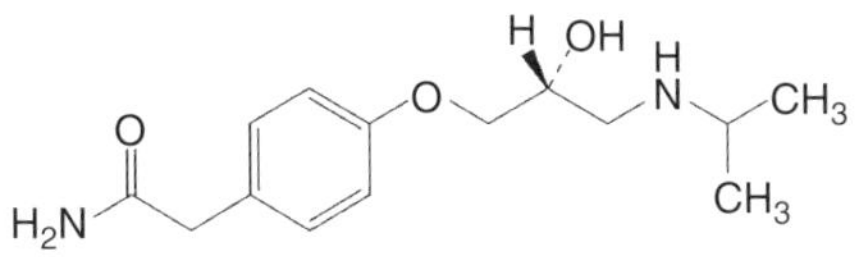

Be totally ineffective and harmless e.g. atenolol (β-blocker).

Fig. 3.2 Illustrations of some of the differences in activity of *unwanted* enantiomers that may arise.

S-enantiomer

- More potent Ca^{2+} channel blocker
- Both vasodilating and cardiac depressant
- More rapid clearance.

R-enantiomer

- Less potent Ca^{2+} channel blocker
- Principally vasodilating
- More bioavailable
- Na^+ channel and other cell pump actions.

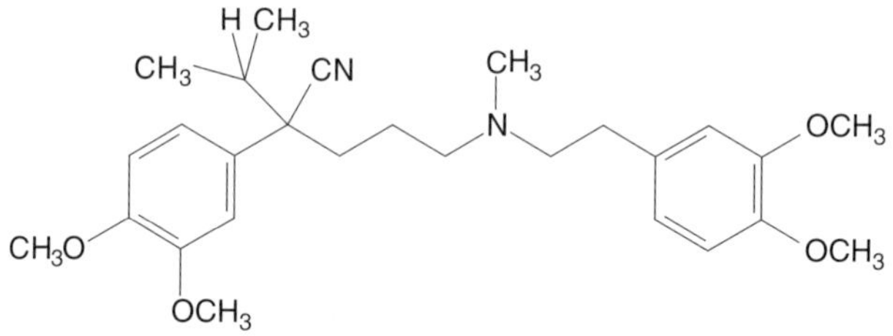

Fig. 3.3 Pharmacological action of verapamil enantiomers.

in its therapeutic action. Also, in the rare case of an enantiomer that, if it is possible at all, is considered completely inactive there is still a preference for developing a single enantiomer. The reason for this is that there is no justification for administering an equal amount of *isomeric ballast* to the active pharmaceutical ingredient when there is no need for it to be there. While there may be exceptional cases when the progression of a racemate or, for that matter, a mixture of enantiomers in a ratio other than 50:50 may be justified, it is now the case that, irrespective of the exact nature of the regulations, it is usually mandatory to develop a new chiral drug candidate as its single enantiomer.

Given this strong preference for single enantiomer drugs, it is not surprising that racemic chiral drugs already on the market have been revisited. The replacement of a marketed racemic chiral drug by one of its enantiomers as a superior drug is known as a chiral switch (or occasionally but less commonly as a racemic switch). In a recent review [6] on all aspects of the chiral switch approach a case was made that there are still opportunities in this area, e.g. single enantiomers of therapeutically active metabolites that are produced *in vivo* from a racemic drug. Nonetheless it is fair to say that the possibilities of gaining advantage from making a simple chiral switch are becoming exhausted with most of the obvious opportunities (Table 3.3) already having been exploited.

Table 3.3 Illustrative marketed chiral switches

Drug (trade name)	Company	Single enantiomer (trade name)	Company
Omeprazole (Losec)	AstraZeneca	Esomeprazole (Nexium)	AstraZeneca
Bupivacaine (Sensorcaine)	AstraZeneca	Levobupivacaine (Chirocaine)	Chiroscience
Ketoprofen (Orudis)	Wyeth	Dexketoprofen (Enantyum)	Menarini
Ofloxacin (Floxin)	Ortho-McNeil	Levofloxacin (Levaquine)	Ortho-McNeil
Cetirizine.HCl (Zyrtec)	Pfizer	Levocetirizine (Xyzal)	Sepracor

Source: Agranat et al. (2002).

Under the circumstances described above it is no surprise that the shift in proportion of chiral drugs is swinging back to single enantiomer drugs (Table 3.2). In the early days of the pharmaceutical industry most drugs came from natural sources and hence single enantiomer drugs predominated. As the proportion of synthetic drugs increased, the proportion of achiral and racemic drugs increased. Now, the proportion of single enantiomer drugs will continue to increase, irrespective of the exact nature of the regulations.

Therefore the main emphasis with respect to analysis is that stereoselective separative methods are needed primarily to resolve any unwanted trace enantiomeric impurity from the main enantiomer present in a drug substance or drug product sample. Chiral separative methods may also be required in drug discovery in order to isolate sufficient quantities of each enantiomer for pharmacological screening. Preparative scale chiral separations are even becoming a commercial proposition for the production of single enantiomer drugs. With respect to drug bioanalysis, stereoselective separative methods are still required, for example, to check whether or not *in vivo* chiral inversion takes place, as in the case of ibuprofen [7], or perhaps to check the enantiomeric distribution of metabolites.

With all these applications a chiral separation must be developed, but, after, under two decades of rapid progress in commercially available chiral separation technology, the situation has now been reached that developing a suitable chiral separation is often the easiest part of the overall method development.

3.2 Evolution of methodologies for chiral resolution

Just as modern liquid chromatography (LC) is the dominant separative technique in the analysis of pharmaceuticals, so, in the evolution of methodologies for the chiral analysis of pharmaceuticals, LC has emerged as the pre-eminent technique. As already intimated, discrimination between enantiomers requires the presence of a chiral selector. In the determination of enantiomers by LC this may be by a chiral derivatisation agent, a chiral stationary phase (CSP) or a chiral mobile phase additive (CMPA). A further prerequisite for discrimination is that the chiral selector must be able to form three point-to-point interactions with one or both enantiomers [8]. While this *three-point interaction rule* is a simplification in that the interactions will not involve *points* it was nonetheless very useful in aiding the design of early successful LC chiral selectors and rationalising others. CSP had clear advantages over the use of chiral derivatisation [9] and CMPA, which, for example, involves costly consumption of the chiral selector. Also the presence of the additive in the mobile phase might cause detection problems in that (a) for example, a UV-absorbing additive would give rise to high background noise for a UV detector, and (b) each enantiomer would pass the detector on average, at least in part, as its transient diastereomeric complex with the CMPA so that potentially the detector response arising from each of the enantiomers could be different. Accordingly, the vast majority of the commercial development was in CSP such that derivatisation and mobile phase additives are now only used for a few specific applications to which they are well suited.

The first wave of development of commercial CSP was in the 1980s at a time when analytical technology had advanced to the point when it was at last able to meet the need to address the study of the individual enantiomers of chiral drugs that had been sharply brought into focus by the thalidomide tragedy some two decades earlier. Progress made in the commercial development of CSP during the latter half of the 1980s has been well chronicled (ancillary reading list). The features of the eight classes of chiral selector that were brought into routine use during the 1980s are summarised in Fig. 3.4 a–h. Wainer [10] was able to reduce these to five different *types*. By the end of the 1980s, then, there were a number of well-defined classes of commercially available CSPs for LC and for each class of CSP there was a reasonable possibility of predicting what types of compounds, the enantiomers of which, could be resolved using that class.

(a) Ligand exchange

- Cheap, readily available pure, natural amino acid chiral selector, in this case proline;
- organic-aqueous buffer, pH 7.5–9.5 containing metal ions, often Cu^{2+};
- may be carried out on an achiral reversed-phase LC column, using a hydrophobic amino acid as a mobile phase additive.

Fig. 3.4 1980s chiral selectors.

Features

- High enantioselectivity for α-amino acids and α-hydroxy amino acids;
- several options for manipulating experimental variables to optimise resolution;
- occasionally poor chromatographic efficiency is encountered at ambient temperature because of slow mass transfer: in such cases LC at elevated temperature may be advantageously carried out.

(b) Synthetic multiple interaction (Pirkle-type)

- Modified natural amino acid chiral selector, in this case *N*-(3,5-dinitro-benzoyl) phenylglycine;
- *straight-phase* solvents used, typically propan-2-ol–hexane.

Features

- Initially claimed to be *broad spectrum* but essentially π-acidic selectors worked well for neutral compounds containing a π-basic group and essentially π-basic selectors worked well for neutral compounds containing a π-acidic group.
- Very good resolution may be obtained if use achiral derivatisation to add a π-acidic or π-basic group to the analyte as appropriate.

(c) Proteins

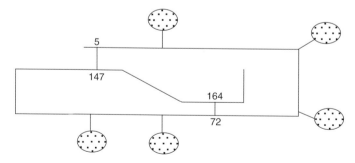

Schematic representation of α_1-acid glycoprotein showing the peptide backbone, two disulphide bridges and five carbohydrate units.

Fig. 3.4 (continued)

- Large molecular weight, natural chiral selectors; importantly, capable of chiral discrimination *in vivo*.
- α_1-acid glycoprotein and bovine serum albumin CSP first to be introduced and most commonly used.

Features

- Broad spectrum, especially when considering AGP and BSA together since AGP is particularly successful for basic analytes and BSA is best suited for acidic analytes.
- Physical and chemical stability issues, not the problem that many perceive but best not to stray too far from physiological pH of 7.4 and not to use too high an organic content in the mobile phase.
- Because of slower mass transfer arising at least in part from the size of the chiral selector, the chromatographic efficiency is reduced compared to that obtained when using for example, conventional achiral reversed-phase columns.

(d) Cyclodextrins

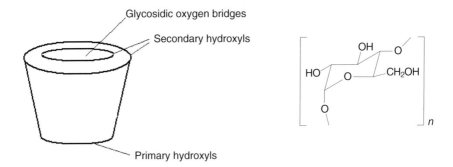

General cyclodextrin structure (e.g. for α-cyclodextrin, $n=7$); the glucose units form a bucket-like structure suitable for molecular inclusion complexation.

- Low/medium molecular weight chiral selector;
- obtained by enzymatic degradation of a larger, readily available natural chiral molecule.

Features

- Originally used only in the reversed-phase mode with reasonably good breadth of spectrum;

Fig. 3.4 (continued)

- good chromatographic efficiency;
- given that, in the reversed-phase mode, chiral recognition involves the inclusion of a hydrophobic group of appropriate size into the hydrophobic cavity of the cyclodextrin *bucket* (e.g. a naphthalene ring into a β-cyclodextrin), there may be a sharp difference in enantioselectivity arising from a slight difference in the structure of the analyte.

(e) Derivatised cellulose

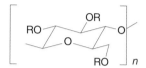

Cellulose triacetate
and benzoate

- Derivatised natural chiral selector;
- acetyl and benzoyl phases introduced initially.

Features

- Used with *straight phase* mobile phases;
- acetyl and benzoyl phases successful for aryl-neutral hydrophobic molecules;
- need to avoid chlorinated solvents.

(f) Synthetic polymers

Synthetic polymer, poly(triphenylmethyl methacrylate) forms a helical structure similar to that of cellulose; chirality arises from the twist of the helix with one enantiomer being formed because of the use of a chiral initiator in the polymerisation.

- Synthetic, large molecule chiral selectors;
- designed to mimic derivatised cellulose phases.

Fig. 3.4 (continued)

Features

- Very similar properties to derivatised cellulose phases;
- used with *straight phase* mobile phases;
- successful for aryl-neutral hydrophobic molecules.

(g) Crown ethers

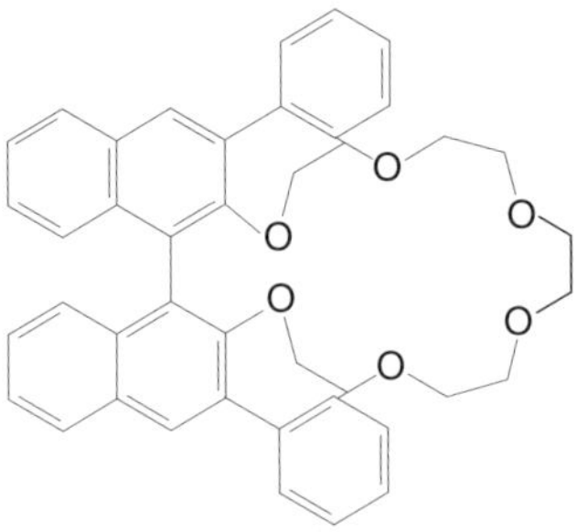

Crown ether; with this particular example the chirality arises from molecular overcrowding in the binaphthol moiety.

- Synthetic chiral selector;
- used with a very simple aqueous perchloric acid mobile phase.

Features

- Very limited breadth of spectrum; only works for primary amines as there is an ammonium ion;
- little scope for mobile phase optimisation;
- good chromatographic efficiency.

(h) Ion-pairing reagents

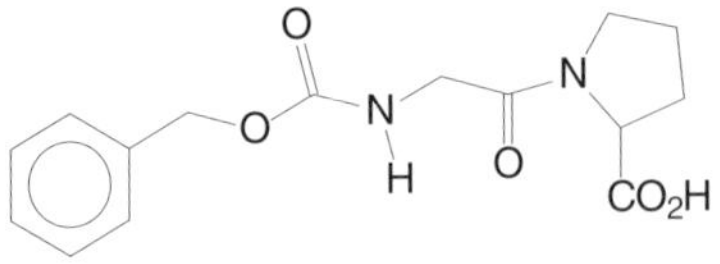

N-benzoxy-glycyl-proline (ZGP)

D-10-camphorsulphonic acid

- Modified natural chiral selectors;
- used as mobile phase additive under *straight phase* conditions.

Features

- Successful separations for β-amino alcohols;
- careful control of trace water content of mobile phase needed.

Fig. 3.4 (continued)

Following the rapid advances made during the 1980s, there followed a period of proliferation of CSP and the use of chiral selectors developed for LC in other analytical techniques. The proliferation, though, was not entirely without purpose. By designing additional complexity onto existing types of chiral selector it was hoped to bring about new chiral selectors that would still have good chromatographic efficiency but greater breadth of spectrum with respect to the range of compound classes for which they could bring about a chiral separation. A good illustration of this was the development of aryl carbamates (Fig. 3.5) of not only cellulose but also amylose to form coated stationary

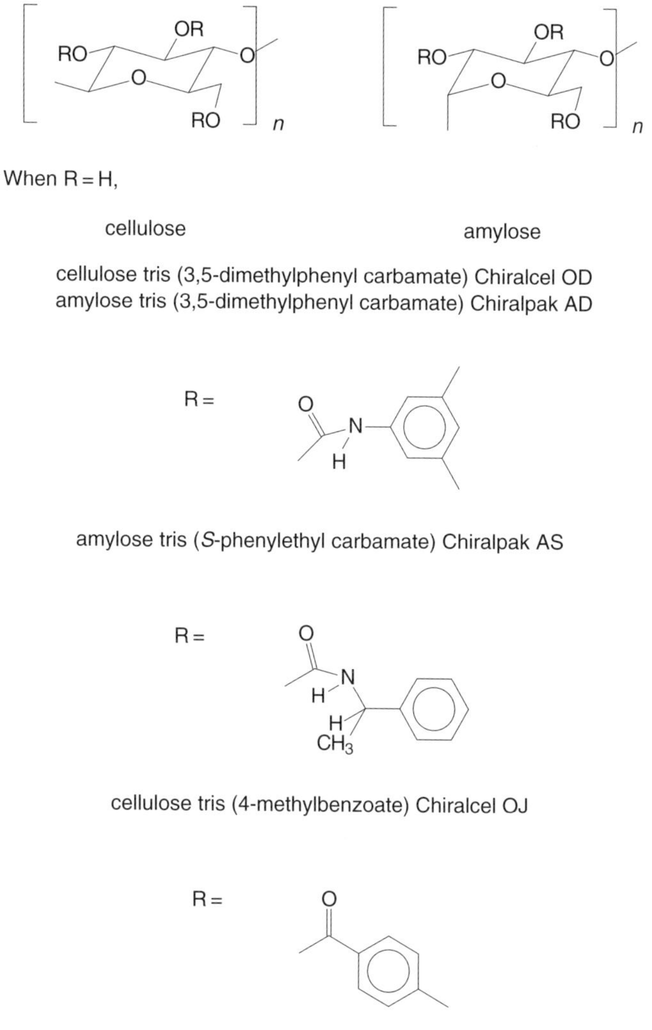

When R = H,

cellulose   amylose

cellulose tris (3,5-dimethylphenyl carbamate) Chiralcel OD
amylose tris (3,5-dimethylphenyl carbamate) Chiralpak AD

R =

amylose tris (S-phenylethyl carbamate) Chiralpak AS

R =

cellulose tris (4-methylbenzoate) Chiralcel OJ

R =

Fig. 3.5 Popular Daicel derivatised polysaccharides.

phases that were highly successful when using *straight-phase* mobile phases such as *n*-hexane-propan-2-ol. These CSP gave the desired good chromatographic efficiency and chiral resolution for a much broader spectrum of compound classes than the earlier cellulose triacetate and cellulose tribenzoate-coated CSP. A similar strategy was adopted in attempting to enhance the performance of cyclodextrin CSP by derivatising the secondary hydroxyl groups on the *rim* of the cyclodextrin *bucket* (see Fig. 3.4(d) for the general shape of cyclodextrins) to produce naphthylcarbamate groups. This CSP gave excellent enantioselectivity for specific analytes such as dinitrobenzyl-*R,S*-1 (1-naphthyl)ethylamine for which the dinitrobenzoyl group gives a π–π interaction with the naphthalene rings on the CSP, and the naphthalene rings on the analyte fit into the hydrophobic cyclodextrin *bucket* in the CSP. Different enantioselectivity was therefore obtained from the equivalent native cyclodextrin CSP but there were no general improvement in breadth of spectrum. Similarly, the earlier generations of Pirkle-type or synthetic multiple interaction CSPs were modified by introducing additional charge-transfer facilitating groups and structural rigidity. In general, these CSPs were an improvement on the earlier such CSPs but they did not have the same breadth of spectrum as the derivatised polysaccharide CSP. These developments and the introduction of macrocyclic antibiotic CSP in which the additional structural complexity is natural but which are smaller than proteins and so give better chromatographic efficiency have been well reviewed (see Ancillary reading list or [9] for an overview).

The CSPs that have been introduced are in the main, more effective, and taken together offer the analyst much greater diversity of options in deciding which CSP to use to resolve the enantiomers of a particular compound. To a limited extent these benefits are offset by the fact that with these newer CSPs it is more difficult to predict what chiral compounds may be separated into their enantiomers unless sophisticated approaches such as molecular modelling are adopted. As discussed, simple classifications of CSPs by mechanism of chiral recognition and types of chiral analyte that may be resolved, as could be made in the 1980s, can no longer be made. Therefore the current trend in the practice of chiral separations is towards a screening approach whereby a number of different CSPs are evaluated in an automated system. These approaches are so successful that, in fact, in terms of simply being able to develop a chiral separation, it could be argued that technology has progressed as far as it needs to progress. Certainly the field of chiral separations is now relatively mature. Despite this, the practices of chiral separation continue to evolve in light of the changing needs for methods for the determination of enantiomers, particularly in the pharmaceutical industry, and the ongoing proliferation of new commercially available chiral selectors. This will continue to be the case given the strategic importance of chirality in drug development and the sheer size of the chiral drugs market. That technology, that has progressed as far as it needs to, could not be said for specific applications that arise in pharmaceutical R&D or similar environments. The determination of a trace enantiomeric impurity in a sample of a single enantiomer drug substance in the presence of a range of other structurally related impurities and a large excess of the major enantiomer remains challenging. Similarly, in chiral drug bioanalysis being carried out now more in discovery rather than in development and in preparative chiral LC being considered for production, there will always be a case where there will be a need for new developments to try to improve throughput and drive down costs. These issues are discussed in more detail later.

3.3 Recent developments in commercial CSP for LC

3.3.1 Polysaccharide-based CSP

The polysaccharide-based CSP manufactured by Daicel Industries [C1] have been amongst the most successful and useful ever since the early 1990s. Within the wide range of these phases there are a few that can be identified as being the most successful. This has been acknowledged by Daicel by coining the phrase *gold medal* to describe their OD, AD, OJ and AS CSP (Fig. 3.5). However, almost all of this success was achieved using organic *straight-phase* mobile phases. A reversed-phase version of the OD CSP was introduced but it was only until recently that a range of reversed-phase compatible derivatised polysaccharides was developed. It is now possible to purchase CHIRALPAK® OD, OJ, AD, AS CSP in -RH versions which are not only suitable for use with polar-organic and *reversed-phase* organic-aqueous mobile phases but are supported on 5 µm particles, thus warranting the *high resolution* tag to distinguish them from earlier Daicel materials based on 10 µm and 20 µm (primarily for semi-preparative work) particles. -H versions, based on 5 µm particles, but only suitable for *straight-phase* use, are also available. It remains to be seen just how successful the CHIRALPAK-RH CSP will prove to be since they have not been on the market long enough to have been comprehensively evaluated. However, although the driving force for their development seems to have been a simple *me-too* attempt to obtain a slice of the lucrative *reversed-phase* chiral market, there is already enough evidence to suggest that they represent a useful addition to the range of commercial chiral LC options. Increasing retention is usually accompanied by increasing resolution. Accordingly, chiral resolution may be increased by decreasing the proportion of organic solvent in organic-aqueous mobile phases. Enantioresolution may be increased for basic chiral drugs by increasing mobile phase pH. Enantioresolution may be increased for acidic chiral drugs by decreasing mobile phase pH. Similarly, as with the earlier introduced CHIRALPAK OD-R CSP, retention and resolution may be increased by increasing ionic strength. Interestingly, the nature of the salt used to regulate ionic strength, e.g. sodium perchlorate vs sodium phosphorus hexafluoride, may also influence retention, suggests an element of ion-pair interactions contributing to the retention mechanism. Reduction in temperature too may be used to increase retention and resolution but, as in most other cases, the use of sub-ambient temperatures would not be the first option to go for in developing a method for routine use.

This all seems like very good news and would be especially good if the newer RH-versions could be used for all applications. However, it seems that these phases exhibit a memory effect such that it is recommended that one column should be used for one mode and another for the other mode (i.e. reversed-phase or straight-phase). This memory effect has been noted by users and is definitely not just a ploy to boost sales.

3.3.2 Macrocyclic antibiotic CSP

Just as Daicel Industries have been responsible for the commercialisation and success of polysaccharide CSP borne out of the work of Okamato, Advanced Separation Technologies Inc. [C2] has been the commercial conduit for CSP and chiral selectors developed by the American chiral separations pioneer, Armstrong. This began with the exploitation of cyclodextrin products, but the later introduced macrocyclic antibiotics from Astec are now more popular. These have been well established for some time now and, like the cyclodextrins, may be used in both reversed-phase and polar-organic modes. As such they regularly feature in the chiral method development screening strategies employed by pharmaceutical R&D analysts, particularly when enantioselective separations for polar compounds are being sought.

One significant commercial development which has been taking place is that it has been demonstrated that macrocyclic antibiotic CSP may be used successfully with normal phase solvents, mixtures of n-hexane and ethanol being the most frequently used. Most of the illustrative examples cited in the commercial literature feature non-polar analytes not all of which are of pharmaceutical interest (Fig. 3.6) but nonetheless it is a good selling point that these CSP may be used with the complete range of mobile phase polarities.

The most recent addition to the suite of commercialised macrocyclic antibiotic chiral selectors is teicoplanin aglycone, which, as its name suggests, is teicoplanin with the three carbohydrates removed (Fig. 3.7). This selector does genuinely offer something different particularly in comparison to teicoplanin itself. As might be expected, there are differences such as some acidic molecules showing increased enantioselectivity and a number of neutral molecules showing enhanced resolution, unusually, in single solvent mobile phases such as methanol, ethanol and acetonitrile. However, the most characteristic advantages are the significantly enhanced resolution displayed for many of the amino acids and

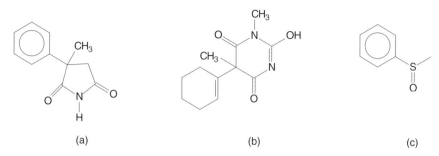

Fig. 3.6 Illustrative analytes resolved on macrocyclic antibiotic CSP in the straight phase mode (a) phenyl-α-methyl-α-succinimide, hexane–ethanol (30:70, v/v) on teicoplanin CSP (Chirobiotic T) (b) hexobarbital, hexane–ethanol (40:60, v/v) on vancomycin CSP (Chirobiotic V) coupled to teicoplanin CSP (Chirobiotic T) as part of a coupled short column method development screen (c) methyl-phenyl sulphoxide, hexane–ethanol (40:60, v/v) on teicoplanin agylcone CSP (Chirobiotic TAG).

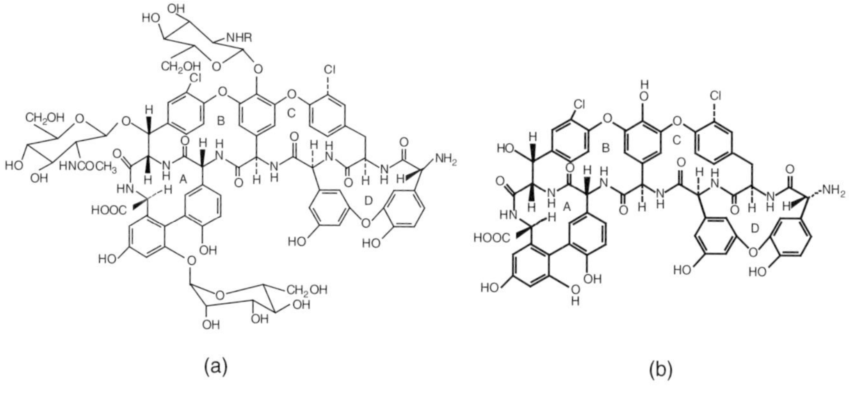

Fig. 3.7 Structure of chiral selectors (a) teicoplanin, and (b) teicoplanin aglycone.

for sulphur-containing compounds including the amino acids, methionine, histidine and cysteine.

Another notable development in the use of macrocyclic antibiotic CSP is that it is being advocated [C2, C3] that in screening for method development using these phases three short columns (e.g. vancomycin, teicoplanin, ristocetin) in series are used instead of three conventional length columns in parallel with automated column switching between runs. The short columns in series approach might be faster and be carried out at lower cost but there is a possibility, surely not a remote one, that a chiral separation on one column might be missed because of the resolution being cancelled out by a separation on another column with similar enantioselectivity and opposite retention order. Since method development screens are often carried out overnight on automated systems, speed is not often an issue so long as the exercise is completed by the following morning. Given this, the minor risk of a false negative in the short columns in series approach might often be a risk not worth taking.

3.3.3   Synthetic multiple-interaction CSP

The *flagship* of the synthetic multiple-interaction or Pirkle-type class of CSP is the Whelk-O 1 phase. This has a highly suitable shape and electronic distribution to resolve naproxen and accordingly naproxen enantiomers may be separated on it using reversed-phase conditions [C4, C5]. However, in almost all the other applications that have been reported, straight phase conditions, typically using *n*-hexane in combination with propan-2-ol, are used. A notable exception is the chiral separation of warfarin using a methanol–water (75:25, v/v) mobile phase to give an enantioselectivity, α, of 1.49 with the first eluting peak having k 2.53. As indicated earlier, these new generations of CSPs have evolved by designing in additional points of interaction and structural complexity onto the earlier generations. Several of the higher generation Pirkle-type CSP that have emanated from Prof. Pirkle's research group have been commercialised. As well as the Whelk-O 1 phase, there are others such as the DACH–DNB, α-Burke 2, Pirkle 1J and β-GEM 1 (Fig. 3.8). The Whelk-O 1 and DACH–DNB CSP have both π-acceptor and π-donor properties while the α-Burke 2, Pirkle 1J and β-GEM 1 CSP are π-acceptors. The Whelk-O 1 CSP has the broadest spectrum of applicability but each of the others is claimed to be particularly suitable for certain applications. DACH–DNB is recommended for sulphoxides, phosphonates and phosphine oxides. α-Burke 2 is recommended for benzodiazepines. Pirkle 1J is recommended for non-steroidal anti-inflammatory drugs (NSAIDs). β-GEM 1 is similarly recommended for NSAIDs. It is also recommended for a wide variety of carboxylic acids and for anilide derivatives.

The number 1 in the title of a Regis CSP designates linkage of the chiral selector to the stationary phase via a single Si–O linkage while 2 designates a *trifunctional* version with each silicon atom on the selector terminus being linked to three, in theory, oxygen atoms on the silica surface. This latter linkage is said to improve resistance to hydrolysis while using strong organic modifiers such as trifluoroacetic acid (TFA). The Whelk-O, α-Burke and β-GEM in particular seem to be available in both 1 and 2 forms.

The synthetic multiple-interaction or Pirkle-type class of CSP is not covered by patents in the same way that some other classes of CSP are. Accordingly researchers other than Pirkle's group have made Pirkle-type CSP worthy of commercialisation. Lindner has made tartramide coated phases, as well as derivatives [11], phases derived from cinchona alkaloids [12] utilising

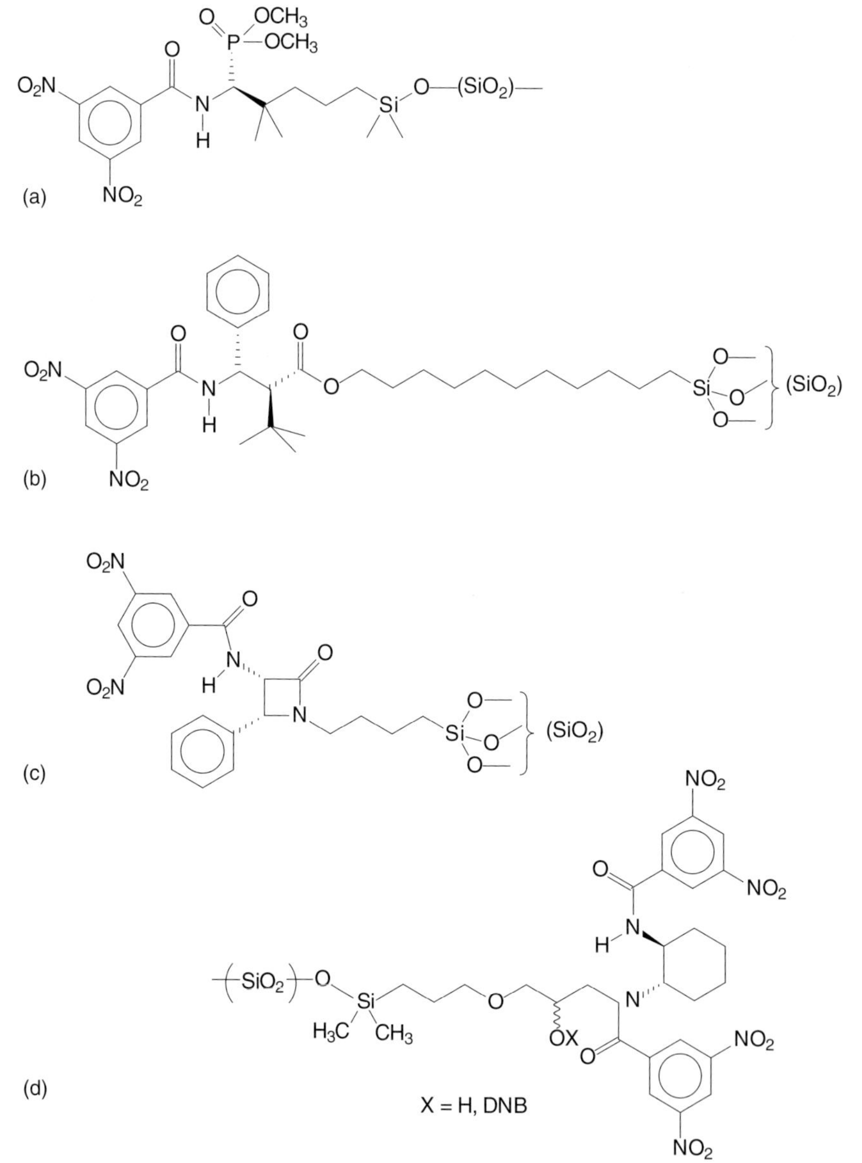

Fig. 3.8 Popular CSP manufactured by Regis (a) α-Burke 2, (b) β-Gem 1, (c) Pirkle 1J, and (d) DACH–DNB.

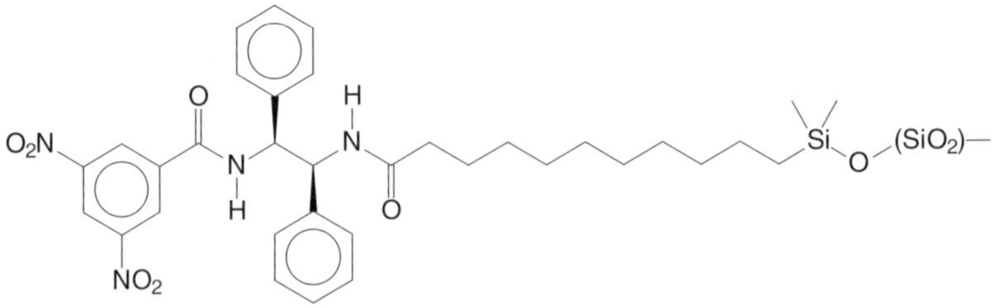

Fig. 3.9 The ULMO CSP manufactured by Regis.

ion–ion attraction as one of the multiple interactions, and was involved with co-workers' development of a CSP derived from (R,R) or (S,S)-1,2-diphenyl-1,2-diaminoethane [13, 14] that was to be named ULMO (Fig. 3.9) by Regis. The ULMO CSP works well for many of the compound classes for which the Whelk-O CSP is used and works especially well for resolving the enantiomers of aryl carbinols. It is ranked in order of preference by Regis just behind the Whelk-O and ahead of the DACH–DNB, α-Burke, β-GEM, Pirkle 1J, leucine, phenylglycine (i.e. Pirkle-type 1A) and naphthylleucine in that order.

Just as Pirkle does not have a monopoly on synthetic multiple-interaction CSP, similarly Regis do not have a monopoly on their commercialisation. Kromasil [C6] market a range of CSP based on the work of Allenmark's research group [15] in which a derivatised tartramide chiral network polymer is covalently bonded to silica. As discussed later, these products are geared towards the preparative chiral LC market. In this market the advantage of all synthetic multiple-interaction CSP that the retention order of enantiomeric analytes may be switched by switching from the CSP based on the R-chiral selector to that based on the S-chiral selector or vice versa is particularly pertinent. This feature will ensure that there remains a future for synthetic multiple-interaction CSP even in the face of successful developments of CSP with a broader spectrum of enantioselectivity.

3.4 Role of *historical* CSP

The current success and dominance of the commercial CSP market by derivatised polysaccharide, macrocyclic antibiotic and, to a lesser extent, synthetic multiple-interaction Pirkle-type materials is such that it could be considered questionable whether or not there is a need to maintain an awareness of the capabilities of some of the earlier developed CSP. Indeed in a rather shrewd marketing move some Diacel products have been dubbed *historical* CSP. This suggests, at the same time, that the newer CSP represent a major advance in technology but that the older ones are well worth preserving. The latter point is especially true. No matter how successful multi-column screening approaches

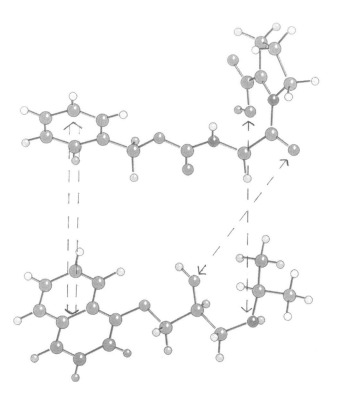

Fig. 3.10 Possible three points of interaction between *N*-benzoxy-glycyl-l-proline, ZGP, and propranolol. There will be a strong attraction between the acidic functional group of the chiral selector and the amino-group of the analyte and a potential H-bonding interaction between a carbonyl group of the ZGP and the hydroxyl group of the propranolol. A third interaction is possibly a π-stacking interaction, or steric repulsion, between the naphthalene ring of propranolol and the benzene ring of the ZGP.

for chiral separations – method development have been, there is no one who claims a success rate greater than 80%. This suggests that for the time being there will remain a need for these older CSP as alternatives. More importantly, given that the mechanism for older CSP classes has been very well characterised and it is known which types of compound may be well resolved on them, it ought often to be possible to proceed to using one of them right away without needing to resort to conducting a screen.

The enantiomers of β-amino alcohols such as the β-blocker drugs are well separated by chiral ion-pair chromatography using either benzoxyglycyl-proline (ZGP) [16] or D-10-camphorsulphonic acid [17]. There is a strong attraction between the acidic functional group of the chiral selector and the amino group of the analyte and a potential H-bonding interaction between a carbonyl group of the chiral selector and the hydroxyl group of the analyte (Fig. 3.10).

diaminopimellic acid (and potential enzyme inhibitor analogues)

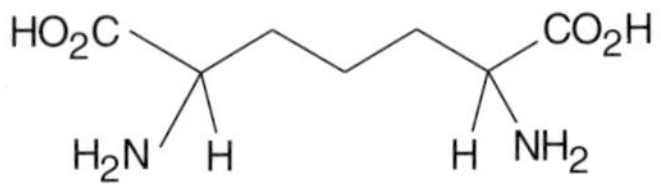

$$HO_2C-CH(NH_2)-(CH_2)_3-CH(NH_2)-CO_2H$$

CrownPak CR(+), aq. perchloric acid, pH 1.3
$\alpha_{ll/dd}$ 5.12 (4.82 for diethyl ester)

(while only partial resolution obtained in one set of conditions in a multi-column screen)

Fig. 3.11 Example of high enantioselectivity using crown ether CSP.

Column: TSK Enantio L1
m. phase: 1 mM $CuSO_4$ (aq.)
flow: 1.0 ml min^{-1}
detector: UV, 215 nm

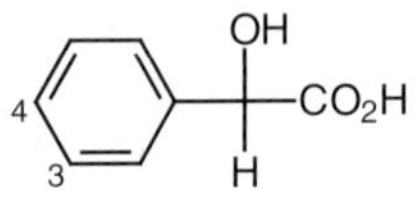

Substituents		k_1	k_2	α
3-H,	4-H	0.79	1.25	1.58
3-H,	4-OH	0.68	0.82	1.21
3-H,	4-Cl	1.40	2.32	1.66
3-CH$_3$O,	4-OH	0.72	0.94	1.31
3-OH,	4-CH$_3$O	1.28	1.79	1.40

Fig. 3.12 Chiral resolution of mandelic acids by ligand-exchange LC.

Accordingly there is little needed by the way of a third interaction to bring about the required chiral separation. Similarly the CROWNPAK CSP from Daicel are very well suited for the separation of the enantiomers of any chiral compound that contains a primary amino group. The mobile phase required, aqueous perchloric acid, is very simple, efficiency is good, and enantioselectivity is generally high. For example, these conditions give good selectivity between d,d- and l,l-diaminopimellic acid (Fig. 3.11). There is no shortage of CSP that will resolve the enantiomers of amino-acids but the ligand-exchange CSP which were amongst the first to be commercialised are very reliable. They are also very reliable for the enantioresolution of α-hydroxy-carboxylic acids. As can be seen from the data for mandelic acid and related compounds (Fig. 3.12) the selectivity obtained is fairly consistent given that in each case the transient diastereomeric complex is formed via the hydroxyl group and the carboxylate anion acting as a bidentate ligand.

3.5 Chiral drug bioanalysis

As suggested earlier, there is still a need for chiral drug bioanalysis even although chiral drugs are now being developed almost exclusively in single enantiomer forms. There is a need to support some preliminary pharmacokinetics in drug discovery at the point where the decision is made as to which enantiomer is to be progressed. Also, during development it is necessary to ascertain whether or not there has been any *in vivo* racemisation or inversion of stereochemistry. Such phenomena are more common than imagined, arising for example in such well known drugs as ibuprofen [7] and thalidomide [5].

Since the mid-1980s when the whole question of chiral drug bioanalysis was first seriously addressed by pharmaceutical companies, a perfectly satisfactory range of approaches to such assays has been developed. Chiral derivatisation strategies have frequently been used and there are an abundant number of cases reported in the literature in which chiral derivatising agents incorporating a strong UV chromophore or a fluorophore have been used in order that, as well as there being a separation of diastereomers, detection at low levels of analyte is facilitated [18]. It must be noted that some of the critical disadvantages of using diastereomer formation [9], such as enantiomeric purity of the chiral derivatising agent, do not have the same potentially detrimental impact on assessing the ratio of diastereomers in a biological fluid as they would in the case of accurately quantifying a trace enantiomeric impurity in a single enantiomer drug substance. An approach that was prevalent when large numbers of samples from pharmacokinetic and other studies on racemic development and marketed drugs were still being processed, was to carry out an achiral separation followed by an chiral separation on-line using column switching technology. In this way the expensive chiral column is exposed only to the drug or its metabolites' enantiomers and not to the many polar endogenous compounds that might pass down the achiral column [19]. Later, when chiral drug bioanalysis was being carried out in studies involving fewer samples, it could be argued that it was important to have chiral drug bioanalysis methods that could be easily developed. The use of achiral derivatisation, for example, with the use of derivatisation by achiral fluorogenic agents and subsequent LC on a Pirkle 1A column [20], fulfils this criterion. Conventional sample clean-up may be used (the resolution of enantiomers when exploiting $\pi-\pi$ interactions between dinitrobenzene and naphthalene or higher polyaromatics very rarely fails) and very low limits of detection may be obtained with both UV and fluorescence detection. In the context of speed, Krull [21] demonstrated that pre-column achiral derivatisation could be carried out on-line. The simplest solution of all is of course to carry out a sample clean-up that is more extensive than would normally be needed so that a fairly *clean* sample of the drug enantiomers may be injected onto a chiral column. The degree of clean-up might even extend to eliminating drug metabolites since, with hydrophobicity not always being a major factor in retention on CSP, a drug and its metabolites might elute closely together in LC using a chiral column.

While the methods mentioned above may be fine in their own right, they do not address the fact that today a very high proportion of drug bioanalysis is carried out by LC-MS often with minimal sample preparation and using short columns. Accordingly, much of the recent development in chiral drug bioanalysis is orientated around the need to have LC-MS compatible methodology. For example, in the use of the polar-organic mode for cyclodextrin and macrocyclic antibiotic CSP, it is not only being acknowledged that there is more than one (formerly anointed *magic*) mobile phase that will give polar-organic or *new* polar-organic separations, but also the use of alternative acidic and basic additives and varying ratios of acidic additive to basic additive ratios is being advocated as a means of enhancing sensitivity in LC-MS. With these modifications it is possible to carry out rapid short column LC separations on cyclodextrin and macrocyclic antibiotic CSP with MS detection. However, it does not follow that it is possible to carry our rapid chiral drug bioanalysis on such CSP. It remains to be seen. A more pragmatic approach to chiral drug bioanalysis using LC-MS has been adopted by Imrie and Noctor (poster presentation at 14th International Bioanalytical Forum, Guildford, Surrey, UK (2001)).

3.6 Preparative chiral separations

There may not have been any recent revolutionary new developments in preparative chiral LC but practice has evolved to the point that it can be viewed as an option for the production scale isolation of enantiomers of drug substances or intermediates as well as for the smaller quantities required in discovery and development. The use of 10 μm particles and techniques such as *recycle and shave* are still being advocated [C2]. Similarly it is still worth considering using unusual solvent compositions in order to optimise enantioselectivity and, often more important, analyte solubility. In illustrations of preparative LC on the relatively new range of macrocyclic antibiotic CSP it has been pointed out that since chiral separations on these CSPs in the reversed-phase mode may be carried out with mobile phase compositions often containing under 15% organic solvent, fractions eluting from the chiral column may be concentrated up by passing through a C18 column (and subsequently eluted in a small volume of methanol which is easy to evaporate off). This methodology is similar to *on-column sample focusing* which may be used in drug bioanalysis [22] and even more reminiscent of the preparative work using β-cyclodextrin as a CMPA carried out by Cooper and Jefferies [23] in the early 1990s. Like cyclodextrin CSP, macrocyclic CSP may be used in the polar organic or new polar organic mode. This mode is especially commended for preparative scale use. Methanol is suitably volatile and the use of trifluoroacetic acid as the acidic additive and ammonium hydroxide as the basic additive at compositions ranging from 0.5 to as low as 0.005 parts per 100 parts of methanol is advised.

An important factor that must always be taken into account in preparative chiral LC is that, because of peak tailing, the second eluting enantiomer is always more likely to be contaminated with the second eluting enantiomer than vice versa. Because of this, despite the popularity of the Daicel derivatised polysaccharide CSP and to a lesser extent the macrocyclic antibiotic and cyclodextrin CSP for preparative work, there will always be a place for Pirkle-type CSP. Because these columns are available in their different enantiomeric forms, one enantiomer of the analyte may be purified as the first-eluting component on one column and the second enantiomer may be freed from its enantiomeric contamination by passing it through the column of opposite chirality (on which it will be the first eluting enantiomer). Preparative LC on the Regis range of Pirkle-type or synthetic multiple-interaction CSP is well established but of more recent prevalence the use of aroyl tartaric acid based network polymeric phases on Kromasil silica has been demonstrated. The use of the chiral network polymer is purported to give rise to a long lifetime through high chemical stability, high capacity through a large number of chiral sites, high enantioselectivity and universality through the three-dimensional structure and high efficiency and enantioselectivity through a reduction of non-chiral interactions. While the principle is sound, and certainly a high loading capacity would be expected there is insufficient evidence as yet to suggest that these materials may be used successfully for as wide a range of compound classes as Daicel and Regis CSPs for straight phase conditions. Indeed it is difficult to make a comparison at all as the preferred mobile phases used, typically t-butylmethyl ether–ethyl acetate, are quite different from the more familiar n-hexane–propan-2-ol.

As in analytical chiral LC, Daicel derivatised polysaccharide CSPs are the most frequently used materials in preparative scale chiral separations. Recently CSPs have been prepared in which derivatised polysaccharides have been covalently bonded to the solid support rather than coated on as in the Diacel materials. The rationale for this is that it is advisable to reduce the chance of the chiral selector leeching off the column in trace amounts to contaminate samples of chiral drugs isolated by production scale LC. However, the extent to which the Daicel coated CSPs are now used in production scale chiral LC would tend to suggest that such a problem, if it exists, is not a very significant risk.

The most discernible recent trends in preparative chiral resolution have been the increasing use of supercritical fluid chromatography (SFC) and simulated moving bed (SMB) chromatography, and the fact that Daicel through its US and European Chiral Technologies subsidiaries have offered these as a custom service has played a role in this.

Generally in SFC supercritical carbon dioxide is used as the mobile phase along with a small fraction of modifier solvent to enhance analyte solubility. Analyte solubility in such mobile phases is good, the low viscosity of supercritical carbon dioxide gives rise to good efficiency and allows for higher flow rates, occasionally enantioselectivity might be improved by dipole interactions with the carbon dioxide and, probably most importantly, the bulk of the mobile phase is very easily removed instantly by evapouration at atmospheric pressure. SFC is not uniquely carried out on Daicel derivatised polysaccharide CSP. It may also be carried out on Pirkle-type synthetic multiple-interaction CSP and macrocyclic antibiotic CSP as promoted by PDR-Chiral, using a Berger SFC and a laser polarimetric detector as well as UV to aid method development [C7] and by Berger Instruments themselves [C8].

To quote the exact words used to promote the Separation Services of Chiral Technologies in their own brochure [C9], 'SMB chromatography is a continuous, counter-current separation method and, as a binary separation technique, is much more efficient in the use of stationary and mobile phases than a batch chromatography system'. Thus, SMB separations produce more concentrated product streams while consuming less solvent. Typically, a chiral separation with SMB is achieved with five to eight columns of short length (~10 cm). A contributory factor to the success of this technique is the very fact that Chiral Technologies, and to a lesser extent others, have built up a high level of expertise in this technique and offer it as a service rather than there being a case of every potential new user having to build up their own expertise.

The logical conclusion of these recent trends in preparative chiral chromatography ought to be that the next step would be to carry out SFC in an SMB system. Unfortunately it seems that the technological difficulties would almost certainly outweigh any potential benefits (personal communication [C10]).

3.7 Present and future perspectives

3.7.1 Alternatives to chiral LC

LC is undoubtedly the most important technique in the chiral analysis of pharmaceuticals. However, that is not to say that chiral LC of pharmaceuticals is synonomous with chiral analysis of pharmaceuticals. The chiral selectors utilised in LC may be usefully deployed in a range of other analytical techniques. Despite this, chiral LC is more dominant now than it ever was. The early commercial developments in chiral analysis were in LC. The application of similar selector systems was demonstrated in other techniques but now with the notable exception of chiral capillary electrophoresis (CE) there is a strong reliance again on LC.

Chiral GC was under development and was being used successfully, for example for chiral drug bioanalysis, at the same time as commercialised CSP for LC were breaking through. The technique received a boost when cyclodextrin GC phases came onto the market. However, with most drug bioanalysis being carried out by LC, and LC-MS in particular, chiral GC is more the preserve of application areas such as the analysis of odourous compounds [24] and the field of insect pheromones [25].

Successful methodology for chiral LC can be adapted for use in TLC more readily than for other analytical techniques. All CSPs and CMPAs used in LC can in principle be used in TLC systems and many have been used in practice. TLC plates for chiral ligand exchange using a CSP are the longest established of the commercial products and cyclodextrins have frequently been used as CMPA for TLC. Despite the advantages of TLC over LC in certain situations (cheap, easy to use, all sample components may be detected therefore making it suitable for investigation of mass balance problems, many samples may be analysed simultaneously) it is now being used much less frequently in analytical support for pharmaceutical development. To exacerbate matters, this less frequent use has led to a loss of expertise in the technique. In chiral TLC there are even fewer reasons to prefer it over chiral LC, especially since, spots are visualised against a chiral background and so may not give an equal response, unless the fortuitous situation arises in which the chiral selector can be eluted off the plate leaving the mixture components behind. Things are moving towards the situation where chiral TLC will only be done in the situation where the TLC enthusiast wishes to demonstrate that it can be done.

Chiral NMR is a different matter. While chiral selectors that are used in LC, particularly cyclodextrins, may be used quite effectively in NMR, the chiral selectors used are usually ones that are especially suited to NMR. Large chemical shifts were produced by *lanthanide shift reagents* but these are not used so frequently nowadays because of the signal broadening that accompanies the induced shifts. With modern high-field instrumentation the differential shift between enantiomers brought about by chiral solvating reagents such as 2,2,2-trifluoro-1-anthrylethanol (Fig. 3.13) is sufficient to give chiral resolution. Provided the instrumentation is available in the first place, it can be a very simple means of obtaining quantitative information. However, it is more likely to be used for monitoring a chiral resolution or analysing a drug intermediate in which the levels of enantiomeric contamination are greater than 2% than for determining trace levels of enantiomeric impurity in a single enantiomer chiral drug substance.

In chiral CE, screening approaches can be so facile that there is often little need for a rational approach. Just how successful CE screening can be may be gauged by studying the work of Nussbaum [26] who used a screen involving five cyclodextrins, each at a high or low concentration, and

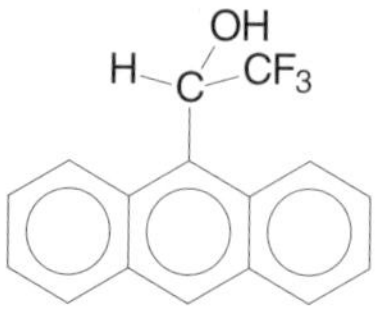

Fig. 3.13 2,2,2-Trifluoro-1-(9-anthryl) ethanol, a chiral solvating agent commonly used in ^{1}H NMR. Because of the strong aromatic ring current, a shift is induced in compound to which it is added in the NMR tube. The single enantiomer solvating agent gives different shifts for enantiomers for a wide range of compounds even when no obvious *3-point interaction* may be observed.

was able to obtain a 100% success rate for his own class of basic development compounds. In terms of simply being able to develop a separation, it could be argued that chiral CE methodology has progressed as far as it needs to progress. Without recourse to the scientific literature, it is usually possible, similar to Nussbaum, to set up a quick screen using a few cyclodextrins that are available in a laboratory to achieve one or more chiral separations which may easily be optimised. Within a few simple experiments one can obtain one or more chiral separations, which may easily be optimised. One possible limitation is that with some pharmaceutical analytes having a weak UV chromophore, special measures (e.g. *bubble* cells, extended path length cells, high resolution with high loading perhaps via stacking techniques) might need to be taken to be able to detect down to 0.05 or 0.1% of a trace enantiomeric impurity. However, this is not always an issue, and often low enough limits of detection may be obtained with a conventional 50 μm capillary and normal loading.

Chiral CE is used primarily in the early stages of the development phase of pharmaceutical R&D with the degree of uptake often depending on the expertise and preferences of the individual analysts involved. However, despite the obvious advantages of chiral CE including its speed of method development, low cost (one capillary and small quantities of a range of chiral selectors, usually cyclodextrins, does it all) and versatility, its use seems to have reached a plateau with little penetration of chiral CE into methods included in regulatory submissions and even less in methods used in production sites. The reason for this is more to do with the overall uptake of CE in industry for the analysis of small molecules than to do with chiral CE itself. Generally LC is doing a perfectly good job and for most applications any advantages that might be had by using CE are insufficient to warrant the wholesale replacement of LC equipment with CE equipment.

This limitation applies even more so to chiral capillary electrochromatography (CEC). This technique [27] is essentially LC in a fused silica capillary using an electric field rather than a pressure drop to drive the flow. As such the separations may be achieved by chiral derivatisation, by use of a CMPA or a CSP just as in chiral LC. Greater efficiency may be had in chiral CEC compared to chiral LC and very low amounts of CSPs or CMPAs are needed. However, the technique does not share the convenience of ease of use that is enjoyed by chiral CE and has not been developed to the point where it could be considered suitable for routine use.

Both CE and CEC would stand a better chance if the tendency to reduce dimensions in LC had progressed to the point where the standard LC column was at least of 1 mm i.d. and could be used in a system that was also compatible with running CE and CEC. Such a hybrid system has existed as a prototype but was probably ahead of its time.

3.7.2 Fit for intended purpose?

One thing that is certain with the developments in chiral selectors and their applications that have been described is that they will continue, whether it be by gradual evolutionary fine-tuning in an already mature area or by some yet unforeseen major advances. What is of more importance is how these developments currently shape up with respect to the applications presented by the chiral analysis of pharmaceuticals.

The most common application of chiral LC is the determination of a trace enantiomeric impurity in a sample of a single enantiomer chiral drug. There are certainly enough options when it comes to developing the required separation in the first place. The success rate of screening approaches to chiral LC method development using a selection of the most effective CSP is usually reckoned to be in the order of 80%. It will generally be possible to take care of the other 20% by using the CSPs that would not be included in those chosen for the screen. Obtaining sufficient resolution to carry out the determination is therefore not normally a problem. Nor is it normally a problem achieving a limit of detection low enough to quantify 0.05–0.1% w/w of the enantiomeric impurity. (The guidelines set by the International Conference on Harmonisation (ICH) (www.ich.org) levels down to which related substances in samples of active pharmaceutical ingredient should be qualified, identified, or reported are about to change, and new guidelines should be in place by 2004.) If the resolution is a problem, then there are plenty options for optimising the separation including using a CSP of opposite chirality in order to reverse the retention order. In fact, a more common problem is that, due to the successes in developing CSP with high chiral recognition properties, excessive chiral resolution is obtained. For example, the first enantiomer might have a very low k value while the second enantiomer might elute with a very high k. The former occurrence might lead to specificity problems with respect to the first eluting peak and peaks arising from other fast eluting sample exponents. The latter occurrence leads to unnecessarily long analysis times. What is needed is to *de-tune* the separation. A solution for this for synthetic multiple interaction, ligand-exchange and crown ether CSP is to use the chiral column in conjunction with the racemic version of the phase. While this strategy is sound in principle, it is rarely used. For work on derivatised polysaccharide and macrocyclic antibiotic CSP detuning may be carried out by altering the ratio of acidic to basic mobile phase additives.

Perhaps there is scope for further developments in CSP in order to more frequently achieve high enough enantioselectivity for facile preparative isolation to be carried out. However, most of the current developments in terms of preparative chiral separations seems to be in developing technolgies for SFC and SMB and, to a lesser extent, CSP in which there is not even an ultra-trace leakage of the chiral selector.

Chiral drug bioanalysis is an area that could be better served by existing technology. There is still work to do, on achieving good mass spectrometric detector response when using some of the mobile phases that are needed to obtain good separations of enantiomers on certain CSPs. Further, rapid analyses will not be possible without attention being given to the sample preparation step. In general, especially considering column prices, having a relatively *clean* sample solution to inject is more of an issue when working with a CSP than when working with an achiral phase such as a C18 bonded phase.

3.7.3 The future

Future developments should follow on automatically from current needs. Hopefully this will be the case in chiral drug bioanalysis, where rapid, sensitive methods still cannot be easily developed. This might come, for example, through a marriage of chiral LC technologies with biofluid-compatible LC technologies or this might be an area where CE, with analyte focusing methodologies and/or sensitive detection systems, can play a greater role.

What is more likely to happen is that despite the maturity of the field, there will be further developments in new CSPs. Even now there could still be a market for new types of broad spectrum CSPs. What would be more useful would be the design of easily constructed cheap chiral selectors tailored to specific analytes in order to give such high enantioselectivity that preparative isolation of the individual enantiomers was a simple one-step batch process. Again looking a bit further into the future, it is worth noting that the determination of the enantiomeric impurity in a single enantiomer chiral drug is almost always a separate assay from the determination of other structally related substance impurities. With a continuing drive to achieve *more for less* in pharmaceutical R&D it would be useful to have methodologies or new stationary phase technologies that would allow these determinations of the enantiomeric impurity and other impurities to be carried out from the same chromatographic runs. While the chiral analysis of pharmaceuticals is a mature area which is no longer the major challenge that it once was, there is still more to come.

Ancillary reading

Wainer, I.W. (1987) Classification of chiral stationary phases, *Trends in Anal. Chem.*, **6**, 125–134.
Welch, C.J. (1994) Evolution of chiral stationary phase design in the pirkle laboratories, *J. Chromatogr. A*, **665**, 3–26.
Armstrong, D.W. (1997) The evolution of chiral stationary phases for liquid chromatography, LC*GC, Current Issues in HPLC Technologies (Suppl.), S20–S28.
Lough, W.J. (1998) Chiral resolution for pharmaceutical R&D – Beyond the Final Frontiers?, *Eur. Pharm. Rev.*, **3**, 48–55.
Lee, J.T., *et al.* (1998) Enantiomeric impurities in chiral catalysts, auxilaries, synthons, and resolving agents, *Tet. Asymmetry*, **9**, 2043–2064.
Caldwell, J. (1999) *Modern Drug Discov.*, **2**, 51–60.
Lee, J.T., *et al.* (1999) Enantiomeric impurities in chiral catalysts, auxilaries, synthons, and resolving agents, *Tet. Asymmetry*, **10**, 37–60.
Soo, E.C., Salmon, A.B. & Lough, W.J. (1999) Separate ways for chiral molecules, *Chemistry & Industry*, **6**, 220–224.
Soo, E.C., Lough, W.J. & de Biasi, V. (1999) Current frontiers in chiral separations by capillary electrophoresis, *Pharm. Sci. Tech. Today*, **2**, 422–426.
Branch, S.K. (2000) in *Chiral Separation Techniques. A Practical Approach,* 2nd edn (ed. G. Subramanian), Wiley-VCH, Weinheim, 317–341.
Lough, W.J. (2000) Pharmaceuticals/chiral separations: liquid chromatography, *Encyclopaedia of Separation Science*, **III**, Academic Press, London, 3714–3719.
Maier, N.M., Franco, P. & Lindner, W. (2001) Separation of enantiomers: needs, challenges, perspectives, *J. Chromatogr. A*, **906**, 3–33.
Stinson, S.C. (2001) *Chem. Eng. News*, **79**, 79–97.
Lough, W.J. (2002) Separation of chiral compounds – from crystallisation to chromatography in *Chirality in the Natural & Applied Sciences*, (eds W.J. Lough & I.W. Wainer), Blackwell Publishing Ltd, Oxford, pp. 179–202.

References

1. Lough, W.J. & Wainer, I.W. (2002) (eds) *Chirality in Natural and Applied Science*, Blackwell Science Ltd, Oxford.
2. Fischer, E. (1894) *Ber. Dtsch Chem. Ges.*, **27**, 2985–2993.
3. de Camp, W.H. (1989) Importance of enantiomer separations, in *Chiral Liquid Chromatography* (ed. W.J. Lough), Blackie & Son Ltd, Glasgow, 14–22.
4. Knoche, B. & Blaschke, G. (1994) *Chirality*, **6**, 221–224.
5. Meyring, M., Chankvetadze, B. & Blaschke, G. (1999) *Electrophoresis*, **20**, 2425–2431.
6. Agranat, I., Caner, H. & Caldwell, J. (2002) *Nature Reviews*, **1**, 753–768.
7. Hutt, A.J. & Caldwell, J. (1983) *J. Pharm. Pharmacol.*, **36**, 693–704.
8. Dalgliesh, C.E. (1952) *J. Chem. Soc.*, 3490–3492.
9. Lough, W.J. (2002) Separation of chiral compounds – from crystallisation to chromatography, in *Chirality in Natural and Applied Science* (eds W.J. Lough & I.W. Wainer), Blackwell Science Ltd, Oxford, 179–202.
10. Wainer, I.W. (1987) *Trends in Anal. Chem.*, **6**, 125–134.
11. Heldin, E., Lindner, K.J., Pettersson, C. & Lindner, W. (1991) *Chromatographia*, **32**, 407–416.
12. Lämmerhofer, M. & Lindner, W. (1996) *J. Chromatogr. A*, **741**, 33–48.
13. Uray, G. & Lindner, W. (1990) *Chromatographia*, **30**, 323–327.
14. Uray, G. & Maier, N.M. (1996) *Enantiomer*, **1**, 211–217.

15. Allenmark, S.G., Shalini, S., Moller, P. & Sanchez, D. (1995) *Chirality*, **7**, 248–256.
16. Pettersson, C., Karlsson, A. & Gioeli, C. (1987) *J. Chromatogr.*, **407**, 217–229.
17. Pettersson, C. & Schill, G. (1981) *J. Chromatogr.*, **204**, 179–183.
18. Gao, C.-X. & Krull, I.S. (1989) *J. Pharm. Biomed. Anal.*, **7**, 1183–1198.
19. Lough, W.J. & Noctor, T.A.G. (1994) Multi-column approaches to chiral bioanalysis by liquid chromatography, in *Pharmaceutical and Biomedical Applications of Liquid Chromatography* (eds C.M. Riley, W.J. Lough & I.W. Wainer), Pergamon, Oxford, 241–258.
20. Lough, W.J., Groves, S.-J., Law, B., Maltas, J., Mills, M.J. & Saeed, M. (1996) *Methodological Surveys in Bioanalysis of Drugs*, **24**, 142–146.
21. Bourque, A.J. & Krull, I.S. (1991) *J. Chromatogr.*, **537**, 123–152.
22. Mills, M.J., Maltas, J. & Lough, W.J. (1997) *Chromatographia*, **45**, 275–283
23. Cooper, A.D. & Jefferies, T.M. (1991) *J. Pharm. Biomed. Anal.*, **8**, 847–851.
24. König, W.A. (2002) Chirality in the natural world – odours and tastes, in *Chirality in Natural and Applied Science* (eds W.J. Lough & I.W. Wainer), Blackwell Science Ltd, Oxford, 261–284.
25. Mori, K. (2002) *Chirality in the natural world: chemical communications*, in *Chirality in Natural and Applied Science* (eds W.J. Lough & I.W. Wainer), Blackwell Science Ltd, Oxford, 241–260.
26. Liu, L. & Nussbaum, M. (1999) *J. Pharm. Biomed. Anal.*, **19**, 679–694.
27. Lämmerhofer, M., Svec, F., Fréchet, J.M.J. & Lindner, W. (2000) *Trends in Anal. Chem.*, **19**, 676–698.

Commercial literature

C1 Daicel Chemical Industries, Ltd, Tokyo, Japan; Chiral Technologies – EuropeSARL, Ilkirch, France.
C2 *Astec* Advanced Separation Technologies Inc. (www.astecusa.com): Advanced Separation Technologies Inc., 37 Lesley Court, P.O. Box 297, Whippany, N.J. 07981, USA; Advanced Separation Technologies, 1 Blake Street, Congelton, Cheshire CW12 4DS, UK.
C3 *Chirobiotic™ Handbook (4th Edition); Chromatography Product Guide, 2000.*
C4 *Chiral Application Guide III; Chromatography Catalog.*
C5 Regis Technologies Inc. (www.registech.com): Regis Technologies Inc., 8210 Austin Avenue, Morton Grove, IL 60053-0519, USA.
C6 *Kromasil® Chiral*
C7 *Fully Automated Chiral Method Development Stations for HPLC and SFC* (includes a Berger SFC and Chiral Advanced Laser Polarimeter), Gary W. Yanik, PDR-Chiral.
C8 Berger Instruments Inc. (www.bergersfc.com): Berger Instruments Inc., 130 Executive Drive, Ste. 2A, Newark, DE 19702, USA.
 Your Answer to Chiral Separation and Purification, 2001; *MultiGram™*; *Analytical SFC; Why SFC.*
C9 *Separation Services; Industrial SMB Applications; Chiral Columns.*
 Eka Chemicals/AKZO Nobel (www.kromasil.com): Eka Chemicals AB, Separation Products, S-445 80 Bohus, Sweden; Eka Chemicals Inc., 1775 West Oak Commons Court, Marietta, GA 30062, USA.
C10 Chiral Technologies Inc. (www.chiraltech.com): Chiral Technologies Inc., Exton, PA, USA; *Technical Support, Products and Services for Chiral Analysis and Separation* (2002 Edition).
C11 Challener, C.A. (ed.), *Chiral Intermediates*, Ashgate, Aldershot, 2001.

4 Nuclear magnetic resonance spectroscopy in pharmaceutical analysis

Richard J. Smith and Andrew J. Edwards

4.1 Introduction

There are many excellent texts on the theory, practice and general application of nuclear magnetic resonance (NMR) to structure elucidation [e.g. 1–3]. The object of this chapter is to highlight the use of NMR in pharmaceutical development and to examine some of the techniques within the NMR armoury that are particularly appropriate for aiding the progression of drug candidates. Biofluid NMR, while an increasingly important aspect of the use of NMR in drug development, is beyond the scope of this chapter. However, two reviews have been written recently which provide a very good overview of this area [4, 5]. Likewise, the binding of drugs to proteins is not covered here; a review of this area can be found in reference [6].

NMR is still the premier method for structure determination of small molecules. This is the primary use of NMR spectroscopy in pharmaceutical development. The structures of both the drug candidate and its impurities will be subject to great scrutiny as the drug moves through development. The section on structure elucidation, which includes a discussion of the uses of multinuclear NMR, will deal with this area specifically.

The increasing need for identification of low-level (0.1%) impurities has lead to the development of specific methods for microgram amounts of analyte and for mixture analysis, and now key among these methods is the hyphenation of NMR to separation techniques such as high performance liquid chromatography (HPLC). LC/NMR is now such a widespread technique that a separate section is devoted to this topic.

There are also an increasing numbers of analyses that make use of NMR quantitative, either for limit tests, quantitation of impurities or for assays of components in drug substance or drug product. Section 4.4 discusses the requirements for quantitative NMR and the practical aspects, including validation of NMR methods.

Finally, Section 4.5 deals with the principles of solid state NMR and an indication of the future prospects in this technique, which is gaining widespread acceptance as an important method in the characterising of solid state form and the understanding of structure in the solid state.

4.2 Structure elucidation

4.2.1 Background and historical perspective

NMR is such a powerful and comparatively fast technique that it is unrivalled in its utility for structure elucidation. Single crystal X-ray has great utility for structure determination too, especially for determining absolute stereochemistry – but it doesn't have the speed, and insufficient small molecules crystallise to allow it to be competitive with NMR. Of course there will always be cases, albeit a minority, when single crystal X-ray is the *only* way to determine a structure analytically.

Mass spectrometry (MS) is faster than NMR and has much better sensitivity; the result (e.g. a molecular weight) can also be easier to interpret automatically – so MS may be the method of choice for structure *confirmation*, particularly in surveying large numbers of samples such as those that result from combinatorial or array chemistry. However, MS rarely gives sufficient information to *elucidate* fully the structures of unknowns and NMR is required, especially to settle questions of regiochemistry.

NMR structure elucidation tools can obviously be applied to new drug substances – often referred to as the active pharmaceutical ingredient (API) – as well as to impurities, metabolites and degradants. The guideline on impurities in new drug substances (CPMP/ICH/142/95 Part IIC) requires structural characterisation of organic impurities at or above an apparent level of 0.1%, although this limit may be lower if toxicity data indicate the need for greater control. Reference [7] reviews the use of NMR as applied to case studies in impurity identification and determination. The identification of low-level compounds in general – whether from synthesis, metabolism or degradation – represents one of the greatest challenges in NMR spectroscopy in drug development, and it has been the driver for the adoption of many technical advances such as high field magnets, LC/NMR, high-Q probes and cryogenically cooled probes, as well as the development and exploitation of new pulse sequences designed to give better quality data from low sample amounts.

Before considering the modern NMR experiments in routine use today in the pharmaceutical industry, it is worth gaining a little historical perspective on NMR to appreciate how these modern experiments have revolutionised analytical chemistry.

In the early 1960s, structure elucidation at the cutting edge would have used 60 MHz ^{1}H NMR together with the vibrational and electronic spectroscopies and electron impact or chemical ionisation MS to draw conclusions about the molecular structure. However, the information content of the combined spectroscopies was nowhere near sufficient to determine the structural complexities of natural products and in consequence there would also have been a heavy reliance on chemical methods to identify functional groups and produce easily identifiable degradation products. The assembly of the fragments thus identified was a difficult process and would have taken months or years to complete for complex cases.

By the early 1980s, NMR technology and methods had progressed to such an extent that ^{1}H–^{1}H and ^{1}H–^{13}C connectivities could be established relatively quickly, and this combined with advances in ionisation methods in MS, meant that *wet chemical* methods declined in importance. Molecular and electronic spectroscopies also became of less importance for structure elucidation, although they obviously could still be used to confirm the presence of a carbonyl or nitrile group and they were later to find other applications of importance in reaction monitoring and in investigations of the solid state. Structure elucidation would have made use of 200–500 MHz ^{1}H NMR spectroscopy, with 2D correlation experiments like COSY, NOESY and HETCOR more routinely available (see Table 4.1 for a glossary of experiment names), and even complex structures would have been soluble in weeks or months, with greater degrees of certainty than had been possible 20 years before.

An example of the effects of this progress in technology is that of the structure elucidation of the antibiotic pulvomycin (also called labilomycin, because of its great lability). In the 1960s the structure 1 was ascribed to this antibiotic on the basis of degradative, NMR, UV and IR spectral data [8, 9]. Specific tests for functional groups, degradative and synthetic chemistry and NMR were all required to deduce the substitution pattern and stereochemistry of the labilose sugar ring and UV, IR and NMR data were used to reconstruct the structure 1 from fragments identified after further degradative work. In the 1980s, this structure was revised to give structure 2, considerably more complex than had been expected from the early work and deduced using mostly 400 MHz NMR, but crucially also with molecular weight data from a then relatively new mass spectrometric technique, fast atom bombardment mass spectrometry (FAB-MS) [10]. The relative stereochemistry data shown in 2 were obtained from nOe [11] and coupling data [3].

4.2.2 The move to higher fields

By today's standards, the structure elucidations of the 1980s now appear ponderous and inefficient. Experiment times have decreased markedly and

1

Table 4.1 A glossary of NMR experiment names[a]

Acronym/term	Expansion	Application/explanation/definition
COSY	Correlated spectroscopy	Homonuclear experiment to detect coupling. Used mostly for ^{1}H but for example can be used for ^{19}F too.
CPMAS	Cross polarisation magic angle spinning	Cross polarisation is the transfer of polarisation from, for example, proton to carbon. This is used in ^{13}C solid state NMR to enhance the magnetisation of the ^{13}C spins (and hence increase S/N).
DEPT	Distortionless enhancement via polarisation transfer	Commonly used to detect protonated ^{13}C and edit the ^{13}C spectrum. Sensitivity greatly improved relative to ^{13}C direct observe.
DOSY	Diffusion-ordered spectroscopy	Pulsed field gradients are used to map the physical location of molecules and molecular diffusion is characterised along the direction of the applied field gradient (usually the z-axis). Used to distinguish signals from different molecules.
DPFGSE	Double pulsed field gradient spin echo	A preparation sequence that produces pure phase magnetisation of a selected spin, whilst non-selected spins are dephased. It is essentially the equivalent of an ultra clean selective pulse.
HETCOR	Heteronuclear correlation	An X-detected experiment for e.g. ^{1}H–^{13}C 1-bond correlation, which suffered from lack of sensitivity (in the case of ^{13}C). Now supplanted by HMQC or HSQC.
HMBC	Heteronuclear multiple bond correlation	Inverse detected heteronuclear experiment. Cross peaks show correlations between carbons and protons usually 2–4 bonds away. A critical experiment in structure elucidation.
HMQC	Heteronuclear multiple quantum coherence	Inverse detected heteronuclear experiment. Cross peaks show correlations between carbons and their attached protons (1-bond).
HSQC	Heteronuclear single quantum coherence	Inverse detected heteronuclear experiment. Cross peaks show correlations between carbons and their attached protons (1-bond). Cross peaks are sharper (and therefore easier to detect) than in HMQC because ^{1}H–^{1}H couplings are removed. Usually performed as sensitivity enhanced and with editing.
INADEQUATE	Incredible natural abundance double quantum transfer experiment	Establishes correlations between like spins of low natural abundance (e.g. ^{13}C–^{13}C). In the case of ^{13}C, the experiment only observes 1 in every 10 000 molecules, so large sample amounts are needed.
MAS	Magic angle spinning	Spinning at 54.7° (the *magic angle*) from static field axis to achieve averaging of the chemical shift anisotropy to zero. This sharpens the signals of solid state spectra to solution-state-like widths. Suppression of ^{13}C dipolar couplings is also achieved. Also used in gel-phase NMR (e.g. beads) and in NMR of tissue samples.

NOESY	Nuclear overhauser effect spectroscopy	2D method of obtaining (transient) nOe data for a compound.
ROESY	Rotating-frame overhauser effect spectroscopy	Method of measuring rotating-frame nOes (or rOes). Of importance for mid-sized molecules, for which nOes can be zero whereas the rOes will still be measurable. Gives approximately the same information (though more care is required in interpretation).
TOCSY	Total correlation spectroscopy. Also called HOHAHA	All protons in a spin system are correlated, allowing spin networks to be deduced. Also used in HSQC–TOCSY, in which all protons in a spin system where one proton is attached to a particular ^{13}C are correlated.

[a] The information in the table was abstracted in part from references [1], [267], [268].

2

new experiments, impossible in the 1980s, are commonplace. These advances have been made possible by a variety of computing, hardware and software improvements over the years, including better and more reliable electronics and more sensitive probes. Some of the developments in probes will be mentioned briefly below, where they are relevant to the theme of structure elucidation.

One of the key advances in NMR over the years has been the availability of higher and higher field strengths. The importance of high fields can be seen by exemplifying the impact they have had on sensitivity and resolution. NMR is an inherently insensitive technique – it is not, for example, the technique you would immediately choose for trace analysis – so any steps you can take to improve the signal-to-noise (S/N) of the technique are important. One way is to use signal averaging – collecting and averaging n spectra. However, S/N only

Table 4.2 The effect of field strength on S/N (2 mmol sucrose)

Field strength/MHz	500	600	700	800
S/N	84	110	145	175
S/N, relative to that at 500 MHz	1	1.3 (1.31)*	1.7 (1.65)	2.1 (2.02)

Notes:
Conditions: 2 mmol sucrose in D_2O in a selective inverse 4 mm flow probe, with Z-gradient. This is a single pulse single scan experiment, processed with an lb = 1. S/N measured on anomeric proton and noise over 200 Hz.
* Figures in parentheses are the theoretical ratios of S/N, calculated from the relationship S/N $\propto B^{3/2}$.

rises as $\sqrt{n}$, so this can become a slow process. The S/N for a given nucleus is also proportional to $B^{3/2}$, where B is the static field of the magnet. This relationship is demonstrated by the data in Table 4.2, which show how the S/N for a 2 mmol sucrose solution increases with field strength, going from 500 to 800 MHz [12].

The impact of this S/N improvement may not be apparent but doubling the S/N quarters the time it takes to collect the same quality data, and in really low-level samples, even the move from 500 MHz to 700 MHz can make the difference between obtaining usable data and just having noise.

As well as S/N improvements, higher fields give better resolution as each ppm (part per million, the conventional measure of chemical shift) corresponds to a greater span in frequency (measured in Hertz) as the field strength increases. Figure 4.1 shows this effect for the high field part of the ^1H spectrum of

Gramicidin, 3

PHARMACEUTICAL NMR 111

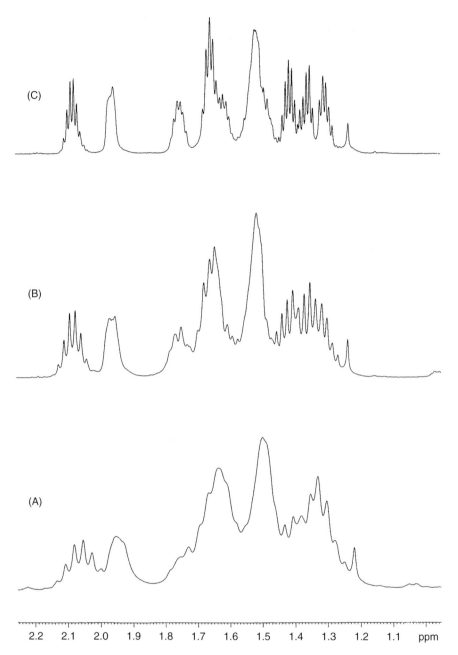

Fig. 4.1 Spectra (A)–(C) show the high field expansion of proton spectra of a 50 mM solution of Gramicidin-S (3), in DMSO-d_6 recorded at field strengths of 250, 500 and 700 MHz, respectively. The increase in dispersion, hence information content, is self-evident as field strength increases.

gramicidin-S, 3, at 250, 400 and 700 MHz. It is clear that the spectrum at 700 MHz contains far more useful information that that at 250 MHz.

However, higher fields are not necessarily a simple answer to the problems of NMR. For example, exchange processes in the molecule being studied can make structure elucidation much more difficult and higher fields may not simplify this and may indeed exacerbate it, depending on the rate of the exchange between the species. *Exchange* here could for example, be tautomeric exchange or interconversion of rotameric forms. The latter is particularly prevalent in pharmaceutical NMR as many molecules of pharmaceutical interest contain tertiary amines or similar systems which are prone to restricted rotation [13]. Tumbling rates of molecules also have a big impact on the information that can be extracted using NMR; there is a field-dependent range of molecular tumbling rates which will result in very weak or even zero nOes – typically molecules of molecular weights of around 1000 daltons, particularly in viscous solutions, will fall into this problematic range. The use of higher fields actually makes it more likely that molecules of this size (or even smaller) will fall into this range where measurement of nOes may not be possible [1]. Measurement of nOes in the rotating frame (*rOe*) using a ROESY or experiment may be one way of getting around this problem – rOes remain positive for all molecular tumbling rates [14].

4.2.3 Modern ^1H NMR experiments

Successive technical improvements in NMR hardware over the last ten years have resulted in vastly increased data quality and sensitivity, but perhaps the most important single innovation has been the introduction of pulsed field gradients (PFGs) with actively shielded probeheads [15, 16]. PFGs can be used to select the required coherence pathways, reducing the need for lengthy phase-cycling and thus cutting experiment times [17, 18]. They can also be used very effectively in combination with soft radiofequency pulses for very selective excitation, and for suppression of unwanted resonances e.g. solvent suppression [19]. A detailed description of PFGs and their mechanism of action is beyond the scope of this chapter (see [20]) but we shall consider their impact on modern NMR methods.

COSY, NOESY and ROESY experiments are well-established means of obtaining connectivity information (COSY) and through space correlations (NOESY, ROESY). Of these, COSY in particular finds extensive use in structure elucidation. There are many different versions of this experiment, aimed variously at simplifying the cross-peaks or extracting more information from them (e.g. coupling constants), suppression of singlets, enhancing the detection of small couplings, etc. [1, 21]. The commonly implemented versions now use PFGs to reduce acquisition times to just a few minutes, while also reducing T_1 noise in the spectra.

NOESY – or the gradient-selected version *gs-NOESY*, which dramatically reduces the time taken for the experiment – may be used for the detection of nOes in drug molecules, particularly in automated NMR [22]. Increasingly, however, the 1D gradient nOe experiment is chosen [23–25] because of its quality. This experiment uses double pulsed field gradient spin echoes (DPFGSE) to provide very clean, selective excitation in the nOe experiment. It typically gives high quality results essentially free from subtraction artefacts, and thus in which even small nOes can be detected reliably. The older *steady state* method of obtaining 1D nOes (1D nOe difference method) is not so common these days, mainly because of the length of time required to obtain good quality *on* and *off* resonance spectra and the inevitable presence of subtraction artefacts. Figure 4.2 shows a comparison of steady state and DPFGSE nOes using the strychnos alkaloid brucine 4 as an example. A similar approach in the use of gradients yields the DPFGSE-ROE experiment, which is useful for what by typical drug molecule standards are larger molecules [26, 27].

Brucine, 4

Most small molecule NMR will make use of highly deuterated solvents. However, there are times when protonated solvents have to be used, for example when examining intact biofluids, or in LC/NMR, and in these cases efficient suppression of the protonated solvent signals is imperative if the solutes are to be sensitively detected. This is a key use of PFGs and a series of pulse sequences have been devised, such as WATERGATE [28], WET [29], and excitation sculpting [30] to achieve solvent suppression. More will be said about these methods in relation to LC/NMR in Section 4.3.1.3.

Molecular diffusion can also be studied using PFG-based experiments, although it is more usual to make use of the differing diffusion properties of molecules to differentiate them in a mixture, than it is to derive diffusion constants *per se*. A field gradient has the effect of spatially encoding (defocusing) the NMR signal from a nucleus; provided that the nucleus does not diffuse before a decoding (refocusing) field gradient is applied, the NMR signal can be recovered in full. This is usually a deficiency in standard gradient-assisted spectroscopy but diffusion ordered spectroscopy (DOSY) [31, 32] makes use of the loss of

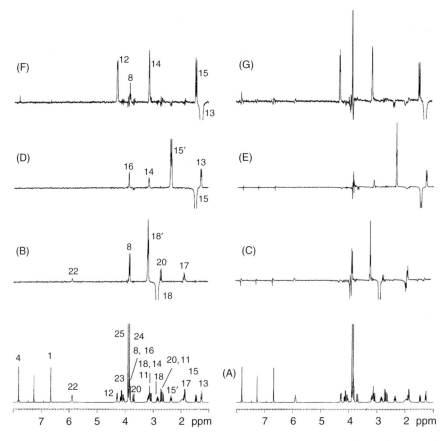

Fig. 4.2 Comparison of nuclear Overhauser spectra of Brucine (4) recorded under steady state and DPFGSE conditions at 400 MHz. (A) Control proton spectra of brucine (ca. 15 mg mL^{-1} in CDCl3) showing 1H assignments. (B), (D) and (F) showing the spectra obtained using DPFGSE excitation of protons 18, 15 and 13 respectively. Each spectrum was recorded in 20 min using a 35 ms Gaussian pulse and a mixing time of 1 s. (C), (E) and (G) show the difference spectra obtained for protons 18, 15 and 13 respectively under steady state conditions. Each spectrum is the result of subtraction of a *control* from an *on resonance* spectrum, obtained by irradiation of the target resonance for 3 s, using a field of 3 Hz. Each on resonance and the single control spectrum were recorded for 1 h. Comparison of the two sets of spectra show that the DPFGSE spectra are of superior quality. For instance, the nOe from proton 15 to 16 is very clear in spectrum D (DPFGSE), but obscured by a subtraction artefact in the corresponding steady state spectrum E.

signal that occurs if molecule *does* diffuse in the period between the first defocusing and the second refocusing gradient pulse. By changing the power of the gradient pulses the differential decay of each resonance can be measured and fitted to a diffusion constant. It is important that the NMR resonances of the different species are not overlapped, as current processing algorithms (e.g. inverse

Laplace transform) cannot solve multiexponential decays. The resonances are then plotted on a pseudo-2D representation with chemical shift along one axis (normally the abscissa) and the measure of the diffusion constant along the other (ordinate). All proton resonances from a single molecule will have the same diffusion constant and appear at the same y co-ordinate in the pseudo 2D diffusion map, allowing resonances from each molecule of a mixture to be differentiated. Figure 4.3 shows an example of using DOSY editing to differentiate individual components in solution. By combining DOSY editing with common 2D NMR experiments it is possible to overcome to some extent the limitations that overlapping resonances impose. These techniques yield pseudo 3D experiments such as COSY-DOSY, DOSY-NOESY and DOSY-HMQC [33, 34, 35] in which a series of 2D correlation planes are extracted for a range of diffusion constants. The experimental considerations and many of the biofluid applications of DOSY have been recently reviewed [36], and there has been a short review of its use in affinity NMR [37].

4.2.4 Nuclei other than the proton

^{13}C NMR data are usually a *must have* in structure elucidations where little is known about the molecule. In analysis of degradants and metabolites, ^{1}H data may be sufficient to relate the unknown to the parent drug molecule but in other cases, ^{13}C data will usually prove essential to obtain sufficient connectivity data. Apart from ^{1}H and ^{13}C, there are several other elements commonly found in drugs which can be detected and quantitated directly by NMR. Table 4.3 lists some of these basic properties for nuclei of significant utility in pharmaceutical NMR. The relative sensitivity for these nuclei gives some indication of the potential problems that are encountered in heteronuclear NMR. For

Table 4.3 Properties of nuclides

Nuclide	Spin	Natural abundance (%)	Magnetogyric ratio $(\gamma) \times 10^7$ rad T^{-1} s^{-1}	NMR frequency at 9.4T (MHz)	Relative sensitivity compared to ^{1}H
^{1}H	1/2	99.98	26.7522	400	1
^{2}H	1	0.015	4.1064	61.4	1.45×10^{-6}
^{3}H	1/2	0	28.5335	426.7	1.2*
^{13}C	1/2	1.11	6.7283	100.6	1.76×10^{-4}
^{14}N	1	99.63	1.9324	28.9	0.001
^{15}N	1/2	0.37	−2.7126	40.5	3.85×10^{-6}
^{17}O	5/2	0.037	−3.6266	54.2	1.08×10^{-5}
^{19}F	1/2	100	25.1815	376.3	0.83
^{31}P	1/2	100	10.8394	161.9	6.63×10^{-2}

Notes:
* Assuming 100% ^{3}H labelling.
Data taken from refs [1] and [38].

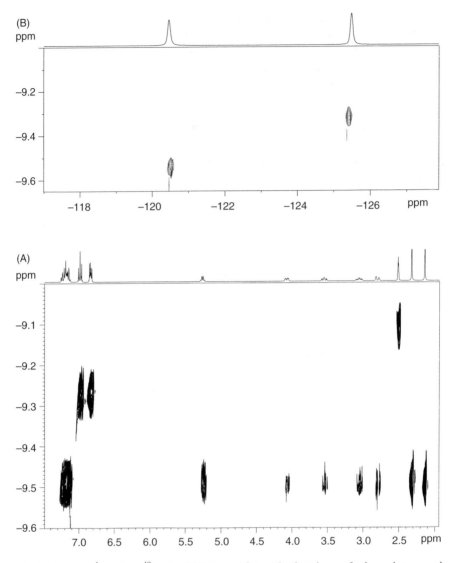

Fig. 4.3 400 MHz ^{1}H (A) and ^{19}F (B) DOSY spectra of an equimolar mixture of a drug substance and its starting material (both contain a mono-fluoro aromatic ring) dissolved in DMSO-d_6. The resonances of the drug substance and the starting material are clearly resolved in the diffusion dimension of both proton and fluorine DOSY spectra. Spectrum (A) also shows the different diffusion characteristics of the residual solvent resonance at 2.5 ppm (DMSO-d_5).

example, ^{19}F and ^{31}P have for many years been exploited to provide information complementary to that of ^{1}H, but ^{15}N has only recently found widespread use with the development of new *inverse* methods. Chemical shift, coupling, and much other data have been comprehensively gathered together for ^{19}F, ^{31}P, ^{17}O and ^{15}N in a recent book [38].

4.2.4.1 ^{13}C

As noted above, detection of carbon is usually critical to structure elucidation. In particular, the long-range couplings (over 2–4 bonds) between ^{1}H and ^{13}C nuclei, expressed as ^{nJ}HC, are often key to elucidating the framework of an organic molecule. The one bond coupling (^{1J}HC) may also contribute to assignment of the ^{13}C spectrum.

The NMR methods of the 1980s would routinely have relied on *direct* detection of heteronuclei such as ^{13}C and these ^{nJ}HC, with the consequent requirement for relatively large sample amounts and hence, for natural products or impurities, extensive isolation would have been needed to produce these amounts. Direct ^{13}C-acquire experiments still have an important place in pharmaceutical NMR as they provide good resolution of closely spaced signals, which is not usually attainable in 2D correlation experiments, although selective HSQC and HMBC experiments do offer a way of approaching this resolution [39]. Polarisation transfer experiments such as DEPT [40] are also still very useful for their greater sensitivity and editing capabilities. However, DEPT is still a ^{13}C-detected experiment and greater sensitivity still is attainable with modern ^{1}H–^{13}C correlation experiments operating with routine *inverse* detection of ^{13}C and with *gradient enhancement* or selection by pulsed-field gradients.

What is inverse detection and why does it matter? The intrinsic sensitivity of a nucleus depends on the magnetogyric ratio, γ, which is constant for a given nuclide (Table 4.3). The magnetogyric ratio determines many aspects of the magnetic properties of the nuclide, including its resonance frequency for a given applied magnetic field, but in this context it is important that the S/N of an NMR experiment is proportional to $\gamma^{5/2}$ [1]. In general, when more than one nuclide is involved in a sequence,

$$\frac{S}{N} \propto \frac{NA}{T} B_0^{3/2} \gamma_{exc}^{3/2} \gamma_{obs}^{3/2} T_2^* \sqrt{ns} \tag{4.1}$$

Where
N no of molecules in observed volume;
A is a term representing the abundance of the NMR-active spins involved in the experiment;
T is the temperature;
B_0 is the static magnetic field;

γ_{exc} γ of initially excited nucleus;
γ_{obs} γ of observed nucleus;
T_2^* effective transverse relaxation time.

Modern inverse-detection experiments achieve increased S/N relative to ^{13}C direct detection by both exciting and detecting the high γ proton nucleus. These methods require the suppression of the 98.9% of ^{13}C–^{1}H signal in favour of 1.1% ^{13}C–^{1}H signal, (satellite excitation) which is possible only because of the high degree of coherence selection provided by PFGs. Several versions of the gradient selected one-bond ^{13}C–^{1}H inverse correlation experiment are available such as HMQC and variations of the newer HSQC, including editing and sensitivity enhancement [41–43].

The high S/N available allows an increase in the quality of a spectrum for the same acquisition time or a substantially reduced experiment time for the same quality. In practice both these options are used – typically, fast ^{1}H–^{13}C correlation experiments (ten minutes) are used for *open access* applications, e.g. HSQC experiments designed for walk-up use by chemists, so that ^{13}C data can be obtained without compromising sample throughput unduly [44]. Spectroscopists, however, may accept experiment times which approach those of more conventional ^{13}C spectroscopy to get high quality data to solve problems.

Modern NMR probes are designed to take advantage of the great prevalence of inverse detection methods, by placing the proton coil as close to the sample as possible (greatest filling factor), with the X nucleus (decoupler) coil further removed. This makes them poorer for direct ^{13}C-observation although older-style direct X detection probes (with the addition of PFG coils) are still available for this purpose. Direct-observe ^{13}C sensitivity in fact has improved little over the last ten years – in contrast to that of ^{1}H – and today's *broad-band* probes, with the X-coil next to the sample, have very similar S/N specifications to their forebears. One way of addressing increasing sensitivity without going to higher magnetic field is through the use of recently available cryoprobe technology. By cooling the observation coil with liquid helium, thermal noise in the receiver is decreased, yielding a greatly increased S/N figure. These probes are available in a variety of configurations, both inverse and direct observation with PFG coils. A dual ^{1}H/^{13}C cryoprobe can give direct observe ^{13}C data on 1 mg of sample (of typical molecular weight) in less than an hour with no special sample preparation or presentation, whereas inverse cryoprobes can obviously be used to get exceptional ^{1}H sensitivity [45]. Alternatively, microcoil probes have been designed to get the most out of mass-limited samples [46, 47].

^{1}H–^{13}C HMBC [48, 1] is one of the key experiments available to an NMR spectroscopist. It can give key information about carbonyl assignment, ring junctions, and other areas of a molecule involving a quaternary carbon, which otherwise might be difficult to probe. It is *tuned* to the long-range couplings between ^{1}H and ^{13}C that arise over two or three (or even four) bonds and this is both a strength and a weakness – the experiment gives many correlations between a proton and its neighbouring carbons and hence a large amount of structural detail can be available from an HMBC experiment, but care needs to be taken in piecing the evidence together. An example of the use of one and multibond ^{1}H–^{13}C correlation experiments recorded on brucine (4) is shown in Fig. 4.4. Recent modifications of the HMBC experiment – ACCORD–HMBC [49], IMPEACH–MBC [50], and CIGAR–HMBC [51] – have been reported, although the basic (gradient selected) experiment still remains popular.

4.2.4.2 ^{19}F

The relative sensitivity of ^{19}F, relative to ^{1}H (0.83) makes ^{19}F an attractive nucleus for NMR studies. Further, since the homonuclear couplings between ^{19}F nuclei separated by four or five bonds are still sizeable (100–200 Hz; see, for example, [52] and references therein), it can be a useful probe of structure and conformation [53]. Much ^{19}F chemical shift and coupling constant information has been gathered together in a relatively recent review [54] and is available in extensive tables [55].

^{19}F NMR spectroscopy has the great advantage over ^{1}H and ^{13}C spectroscopy of producing relatively simple spectra with only one or at most a few discrete resonances for typical drug molecules in a ^{1}H-decoupled ^{19}F spectrum. It therefore finds great utility in detecting and quantitating fluorine-containing impurities in drugs as the ^{19}F resonances of the impurities will often be distinct from the parent and not obscured by other resonances. These quantitative applications will be covered in Section 4.4.

For structure elucidation, ^{19}F–^{19}F or ^{19}F–^{1}H correlation methods are useful adjuncts to homonuclear ^{1}H methods, see Fig. 4.5 for an example of the use of ^{1}H–^{19}F heteronuclear nOe in spectral assignment.

Long range ^{19}F–^{15}N shift correlation has also been demonstrated for the first time on a mixture of 2- and 3-fluoropyridines [56], using a modified IMPEACH–MBC pulse sequence. This may have application in the structure elucidation of heterocyclic systems lacking appropriately-placed protons.

4.2.4.3 ^{31}P

^{31}P NMR spectroscopy is perhaps not quite so popular as ^{19}F in pharmaceutical development because of the lower number of drug substances containing phosphorus. However, it has importance in biochemistry and the same potential for impurity quantitation as ^{19}F, and there are many structural problems where it can help, for example in Wittig reactions, ylides, phosphate esters, nucleotides and phospholipids [57]. Two-dimensional methods for phosphorus have been reviewed [58]. The latter reference also reviews ^{1}H–X, X–X and X–Y correlation experiments for other heteronuclei. Given the relative dearth of tertiary literature on phosphorus NMR in chemistry, the book by Berger *et al.* [38] is particularly useful.

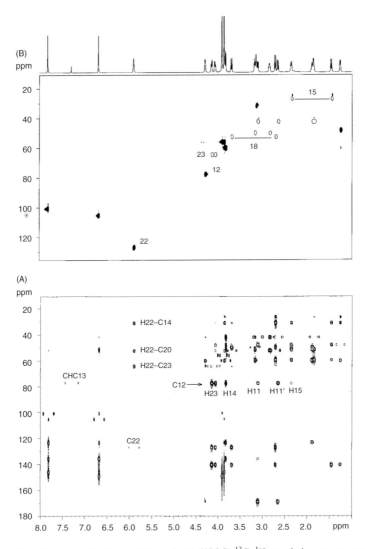

Fig. 4.4 (A) multi-bond (HMBC) and (B) one-bond (HSQC) $^{13}C-^{1}H$ correlation spectra recorded at 700 MHz on a 15 mg ml^{-1} solution of brucine (4) in CDCl$_3$. In (B), the filled circles represent correlations form carbon atoms of odd multiplicity (CH and CH$_3$), whereas those shown with open circles show methylene carbons (quaternary carbons do not have directly attached protons so do not appear in the HSQC spectrum). Assignments of selected resonances are indicated, for instance, the two protons of the methylene carbon 15, can be seen to resonate nearly 1 ppm apart. This editing of HSQC spectra obviates the need to record separate DEPT spectra in most cases. The HMBC spectrum (using a long-range correlation time of 60 ms) in (A), shows correlations between protons and carbon atoms up to four bonds apart. For instance, the olefinic proton, H22 at 5.9 ppm shows correlations to carbons 14, 20 and 23 and represent 3, 3 and 2 bond correlations respectively. The one bond correlation seen to C22 at ca. 128 ppm appears as a doublet, due to the fact that no ^{13}C decoupling is used in the HMBC experiment. Correlations from carbon 12 are shown parallel to the proton axis and are observed to protons: 23, 14, 11, 11' and 15. There is a wealth of information in the HMBC experiment. Indeed the HMBC correlations observed from and to quaternary carbon atoms are often key to their assignment.

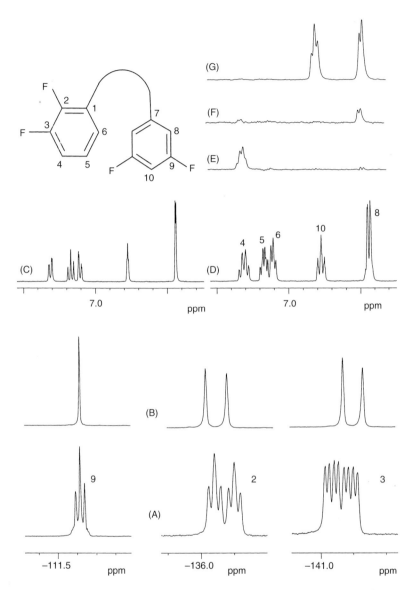

Fig. 4.5 Spectra (A) and (B) represent proton coupled and proton decoupled regions of the ^{19}F spectrum of the compound shown. Fluorine atoms 2 and 3 couple together, so that in the proton decoupled spectrum (B), each appears as a doublet. (C) and (D) represent the proton spectrum with and without fluorine decoupling respectively. Spectra (E)–(G) show the proton difference spectra resulting from steady state heteronuclear nOe experiments, with ^{19}F irradiation at d^{19}F −140.9, −135.9 and −111.4 ppm. In (E), nOe is seen from F3 to proton 4, and (G) shows nOes from F9 to H8 and H10 – all examples of intra-ring nOes. Spectrum (F) however, shows an inter-ring nOe from H2 to H8, revealing that the molecule can adopt a conformation in which the two aromatic rings are close in space.

4.2.4.4 ^{17}O

^{17}O has found limited application in structural studies, mainly because of the experimental difficulties of observing it with typical (unenriched) compounds. However, it can find use in electronic and conformational studies, even at natural abundance, if there is sufficient compound available [59, 60, 61]. For example, in compounds of general structure 5, the ^{17}O chemical shift for the carbonyl oxygen could be used as an indicator of bond order and hence torsional angle and electronic distribution [62].

5

For these natural abundance experiments, typical experimental conditions require suffcient of a 0.5 M solution of the compound to fill the sample volume of a 10 mm tube and 30–200K scans, although the short acquisition time and relaxation delay do help to make these experiments feasible.

Applications of ^{17}O NMR in natural product chemistry have been reviewed [63], as have applications to other structural problems, hydrogen bonding, tautomerism and mechanistic studies [64].

4.2.4.5 ^{15}N

^{15}N is increasingly of importance in structure elucidation. Nitrogen has long been regarded as *spectroscopically silent* because of the practical difficulties of working with the quadropolar nucleus ^{14}N and the highly insensitive ^{15}N. Nitrogen atoms in molecules thus became gaps to bridge with experiments like ^{1}H–^{13}C HMBC and arguments based around chemical shifts. However, with the advent of inverse methods, ^{15}N chemical shift and ^{1}H–^{15}N coupling data are now accessible on realistic amounts of material (10 mgs) in realistic experiment times (overnight) at realistic field strengths (500 MHz for ^{1}H). Cryogenic NMR probes are also contributing to the utility of ^{15}N, producing very significant time savings in ^{1}H–^{15}N correlation experiments at natural abundance [65].

The first application of gradient selected long-range ^{1}H–^{15}N heteronuclear shift correlation (HMBC) was to ajmaline [66] at 500 MHz. Long-range couplings were observable in reasonable experiment times (4.5–9 h) using approximately 19 mg of ajmaline in 650 µl of 99.96% d_6-DMSO. Other applications of this technique have included the *Strychnos* alkaloids, including brucine [67] (4), cryptolepine [68], the bis-indole anticancer drug Navelbine® [69], isoquinaline alkaloids [70], in the differentiation of regioisomers [71] and in characterising degradants [72].

One of the problems of utilising long-range ^{1}H–^{15}N coupling constants as a structural probe is the variation in size of those long-range couplings. This has an impact because the magnitude of the coupling must be estimated in order to calculate the long-range delay in the pulse sequence. The corollary is that the experiment will not detect couplings which vary widely from the estimate, yet the long-range couplings to the two nitrogens in strychnine, for example, vary from <2 to 16 Hz [67]. One method of addressing this is by use of *accordion* excitation [73, 49], which has been implemented for long-range ^{1}H–^{15}N couplings in ACCORD–HMBC [74] and IMPEACH–MBC [50]. These methods seek to sample a range of potential long-range heteronuclear couplings and have been compared in a recent paper [75]. Figure 4.6 shows a comparison of three ^{15}N–^{1}H long-range correlation methods recorded on a sample of brucine (4).

Particularly useful applications of ^{15}N NMR include the elucidation of structure of N-oxides and N-sulphates, which may result from metabolism studies. N-oxidation, for example, may not cause changes in neighbouring ^{1}H or ^{13}C shifts that are readily or reliably interpretable, and even in MS the oxidation may be difficult to detect. However, using ^{15}N as the reporter, N-oxidation can be detected and located to a particular nitrogen atom. For example, the effect of N-oxidation on the ^{15}N chemical shift of N5 in cryptolepinone 6 was studied by Martin *et al.* [76]. These workers found that oxidation of N5 (by DMSO) resulted in a downfield shift of +84.7 ppm for this nitrogen.

In another example, Farlet *et al.* found that long-range ^{1}H–^{15}N heteronuclear coupling could be used to establish the ^{15}N chemical shift of an oxidised nitrogen and hence unequivocally locate the site of N-oxidation, where *conventional* spectroscopic methods had failed [77].

Another of the problems of ^{15}N spectroscopy is the relative lack of chemical shift data; it may well be the case that model data have to be generated to answer specific questions. References [78, 79, 38], give some very useful shift data and references to other sources. Reference [80] gives an excellent introduction to the area, with a discussion of ^{15}N referencing, ^{15}N chemical shift ranges and practical experimental detail, in addition to a survey of long-range correlation experiments.

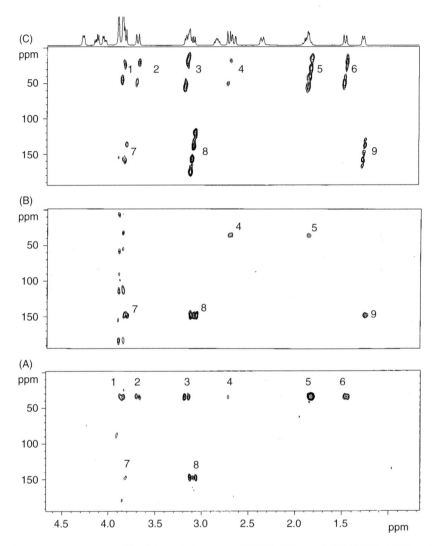

Fig. 4.6 (A)–(C) show the ^{15}N–^{1}H correlation spectra HMBC (100 ms), IMPEACH (3–16 Hz) and ACCORD (3–16 Hz) respectively recorded on a 15 mg mL^{-1} solution of brucine (4) in CDCl$_3$ at 500 MHz. The correlations are labelled 1 to 9 and represent: N19 (δ 36 ppm) to: (1) H16; (2) H20; (3) H18; (4) H20′; (5) H17; (6) H15 and N9 (δ 148 ppm) to (7) H8; (8) H11; (9) H13. The ACCORD experiment shows all correlations expected, and can be easily differentiated from the noise by the characteristic *skew* in the f1 (^{15}N) dimension, casued by ^{1}H–^{1}H coupling. Both the ACCORD and the IMPEACH spectra sample a range of long-range ^{15}N–1H coupling constants (3–16 Hz), whereas the HMBC is optimised for a value of 5 Hz. Even so, only the correlation N9 to H13 is absent from the HMBC spectrum.

4.2.5 Computer-assisted structure determination

The rate-determining step in structure elucidation of an unknown is usually not the data acquisition but the data interpretation. It is therefore logical that much work has been put into developing software that will speed up this process. This type of software would find particular use in natural product structure elucidation, where the structure of the sample may be truly unknown. In pharmaceutical development, however, it is usually the case that significant parts of the *unknown* – for example an impurity or degradant in a drug substance batch – may be known or guessed. It is also the case throughout chemical programs aimed at both discovery and development of new drugs, that syntheses are targetted at specific outcomes and thus when NMR is used to confirm the outcome of a synthetic scheme, the aim is structure *confirmation* rather than structure *elucidation*. Thus there are perhaps two discrete areas of interest in computer-assisted interpretation, namely automated interpretation of spectra in terms of a proposed structure, and automated derivation of a structure from the data presented (i.e. true computer-assisted structure elucidation, CASE).

4.2.5.1 Computer-assisted interpretation

The goal of computer-assisted interpretation is the automated interpretation (and presentation) of spectral data in terms of a structure. Tables of substituent chemical shift values [81] and collections of spectra [82] have been in use in chemistry laboratories for many years. *Computer-assisted* interpretation has built on these additivity rules and substituent chemical shifts for classes of compounds [83–85] or has used large databases of empirical shifts [86–91] to predict chemical shifts for given environments. Users have built their own databases [92] and software is also now available which allows automated comparison of predicted and experimental data (e.g. [89]). Comparisons of prediction software are available for a limited number of compounds [93, 94]. In practice, ^{13}C predictions are usually found to be more accurate than those for ^1H because ^1H shifts are more solvent-dependent than ^{13}C shifts and more dependent on 3D structure. Thus ^{13}C shift predictions have found utility for some years, whereas ^1H predictions have found more limited application. This is particularly unfortunate when one considers that probably around 99% of 1D spectra-run in a pharmaceutical company will be ^1H spectra (for example, in high-throughput open access environments), for which computer-assisted interpretation is more difficult. Customisation of databases, using more closely related compounds from which to make predictions, or by using solvent-specific databases, may help in this, as will more advanced algorithms for the predictions but there is the remaining problem of stereoisomers and diastereotopic protons which current connectivity-based systems and databases cannot solve.

Perhaps one way of addressing this is to use semi-empirical calculations of ^1H shifts [95, 96]. These methods make explicit use of 3D structure and corrections for solvent can be made; they also do not rely on similar structures being found in the database as long as the effects of all appropriate functional groups on the chemical shifts of neighbouring protons have been previously determined and parameterised. The semi-empirical scheme CHARGE calculates ^{1}H chemical shifts by modelling the effects of the atomic charge, steric effects, electrostatic effects of nearby bonds, anisotropic contributions from alkyne bonds, for example, and ring current and π-electron density of nearby aromatic rings and carbon atoms [97, 98]. The shift δ of a particular proton is given by the general equation:

$$\delta_{total} = \delta_{charge} + \delta_{steric} + \delta_{anisotropy} + \delta_{electrostatic} + \delta_{ring\_current} + \delta_\pi \qquad (4.2)$$

The major interactions which determine proton chemical shifts in organic molecules have been quantified, and good agreement with observed shifts can be achieved for many proton environments. Neural networks have also been applied to prediction of ^{1}H chemical shifts, with a recent strategy including geometric descriptors of the 3D environment of the protons, in addition to physicochemical and topological descriptors [99]. This system was found to give approximately the same magnitude of error as commercial packages, with the advantage of distinguishing diastereotopic protons. However, this scheme, like the others based on 3D structures, does have the weakness that it does not take into account any molecular flexibility, which is often a feature of pharmaceutically active compounds.

Where sets of spectra for related structures or substructures are available, neural networks or pattern recognition methods can be used to confirm molecular structure [100], although these methods have been more commonly used in the realm of biofluid analysis (e.g. [101–106]) where ^{1}H data can be used as indicators of metabolic or toxicological processes. The lack of reliable proton-based automated structure verification methods and the low intrinsic sensitivity of ^{13}C have been recognised in an alternative verification system for combinatorial libraries based on 2D HSQC [107]. This has the advantages of the use of ^{13}C shifts for verification, the dispersion of two dimensions and the sensitivity of an inverse experiment. It does require, however, that the spectra of the component parts of the compounds are available, and its applicability to the verification of new structures is therefore limited. However, the general approach of using HSQC to provide ^{13}C data, which can then be used in a ^{13}C prediction/comparison system is perhaps more generally useful, though HSQC data do have the significant drawback that quaternary carbons are absent and ^{13}C shift information will be less precise.

4.2.5.2 Computer-assisted structure elucidation
The goal of computer-assisted structure elucidation (CASE) is the derivation of all possible structures that are consistent with a set of spectroscopic data, with the minimum of intervention by the spectroscopist. If computer-assisted

interpretation is difficult, it can be readily understood that automated elucidation is more challenging still. Approaches usually involve the integration of components which can analyse 1D and 2D NMR spectra with a structure generator and a component that can rank the possible structures in order of likelihood. They may also include integration of data from MS and infra-red spectroscopy. There are recent reviews of this area which give a good overview of the approaches [108–110]. Structure elucidation methods (or components) have been based on 1D ^{1}H data [111], or on ^{13}C and/or 1D DEPT data [112, 113] without the use of molecular weight or molecular formula data, but most make use of 2D NMR data *and* molecular formula information [114–116]. Recently, schemes have been introduced which make use of ^{15}N – ^{1}H HMBC correlations to improve the information set available and reduce the number of possible structures compatible with a given data set [117, 118]. Some of these schemes make explicit use of experimental and predicted 1D ^{13}C spectra to judge the likelihood of the possible structures, but use of such data – especially when combined with the need for a molecular formula – makes their application in *real-world* structure elucidations on isolated low-level impurities problematical. One common theme from all these publications is that the productivity of a spectroscopic laboratory is not measured in terms of the amount of data generated but in terms of the number of problems solved and that, in these terms, the expert spectroscopist is more productive than current automated approaches.

4.3 On-line separations

4.3.1 LC/NMR

LC/NMR has become one of the routine tools of pharmaceutical NMR. Early LC/NMR [119–121] was applied to components which were at fairly high concentrations, in non-protic solvents, and at a maximum field strength of 100 MHz. Protic solvents were first approached in 1981 [122] and soon after, the use of higher fields both improved the sensitivity of the experiment and illustrated the difficulties of continuous flow NMR using a gradient chromatographic method [123]. It has taken a combination of improvements in probe design, in console electronics and particularly in computing technologies for LC/NMR to become a practical technique for routine use.

Solvent suppression is a particular problem in LC/NMR and has been a theme throughout its development. Early methods for suppression of the protonated solvent signals which otherwise dominate the NMR spectrum made use of *binomial pulse sequences* [124–126]. Methods in use today either use fully deuterated solvents, or make use of solvent suppression schemes such as the NOESY presaturation technique [127], WATERGATE [28, 128], WET [29, 129], or excitation sculpting [30, 130, 131]. These methods have for some time made it possible to study relatively low-level (several %) impurities [132, 133]. The need for identification of even lower level components such as metabolites or impurities in drug substance at the 0.1% level, has driven practitioners to ever higher field strengths [134, 135], and users, probe designers and instrument vendors to new ways of achieving concentration of fractions for LC/NMR [136], of improving probe sensitivities [137] and of utilising techniques such as microbore [137] and capillary [138] HPLC to introduce samples into the NMR flow cell at higher concentrations. Good reviews of LC/NMR are available [139–141], but here we will emphasise applications in pharmaceutical development.

4.3.1.1 Chromatographic considerations

The principal assets of LC/NMR are speed of analysis, particularly for complex mixtures, and the freedom from the impurity peaks generally associated with the chromatographic purification of low-level components. Its principal drawbacks are the restrictions on chromatographic loadings and the presence of protic solvents, which may both lead to restrictions in the amount and type of data which can be obtained. The corollary of these drawbacks is that some chromatographic expertise is required to get the most out of the technique.

Before LC/NMR is considered as the approach for a particular problem, its use should be compared with semi-preparative chromatography followed by tube NMR or by LC/NMR if only partial purification is possible. Generally, if there is more than one component of interest and speed of analysis is critical, LC/NMR will be the preferred route, but there are further decisions required, which are typically based on the column loadings achievable (see Scheme 1).

The basis of the decision tree in Scheme 1 is that (at 500 MHz) a good quality 1D ^{1}H spectrum can be obtained from 1 µg of sample in the flow cell, good quality 2D homonuclear experiments like TOCSY or NOESY can be obtained from 10 µg in the flow cell, and if inverse correlation data like HSQC or HMBC are required, at least 100 µg of the component will be required in the NMR flow cell. These amounts may seem large but it must be remembered that they are not detection limits – they represent the amount required to get good quality data from which reliable conclusions can be drawn. Scheme 1 emphasises that some pre-knowledge of the type of problem to be solved can aid in choosing the correct route to resolution. For example, if in a metabolite study hydroxylation is suspected at a site on an aromatic ring, often the coupling pattern of the aromatic protons in the 1D ^{1}H spectrum is sufficient to distinguish the substitution position. Similar precepts hold for degradation studies or for related impurities.

For impurity analysis, it should be borne in mind that if an impurity is present at only 0.1% of the sample, a column loading of 1 mg is required to obtain 1 µg of the impurity in the NMR flow cell; this may or may not be possible, depending on the quality of the separation and the conditions employed. It is usually the case that choosing the optimal chromatographic conditions for LC/NMR requires a trade-off between supra-optimal column loading and the concomitant peak broadening this brings. If a peak broadens quickly with increased loading, the

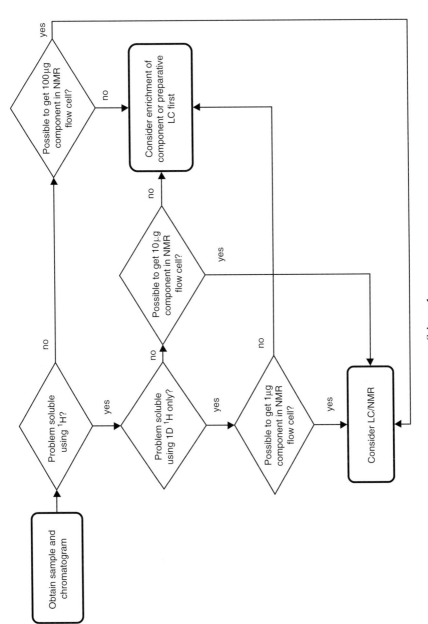

Scheme 1

peak concentration actually reduces and no further improvement in NMR S/N will result. The optimal peak width (for stopped-flow work) is achieved when the chromatographic peak width at half height is approximately equal to the NMR flow cell volume [142], which is typically 60–120 µL. This usually implies that gradient HPLC methods will be used for components which are retained for more than a few minutes to ensure that later-eluters will not be excessively broad. It may also lead users to narrow bore HPLC columns (e.g. 2.1 mm id) which, using lower flow rates, produce smaller peak widths. Narrow bore columns are often optimal for LC coupling to NMR but their lower volume also leads to lower maximal loadings and once again a compromise has to be reached. Chromatographers with experience of preparative and semi-preparative HPLC will be useful colleagues when analysing low-level impurities by LC/NMR.

Eluent choice is also critical for a successful LC/NMR analysis. Throughout the above, it has been assumed that D_2O will be used in place of H_2O for the chromatographic separation – this simplifies the solvent suppression task greatly for a relatively small investment. The organic solvent and any buffer or acid/base needed will usually be determined by the chromatographic problem to be solved but the NMR properties of these components must be borne in mind when planning the analysis. For example, if critical ^1H resonances are expected around 2 ppm, then acetonitrile is unlikely to be a good choice of organic solvent, despite its excellent chromatographic properties. Of course, deuterated organic solvents may be used in cases such as this; despite their expense, they provide increased detectability close to their solvent resonances. d_1-Methanol (i.e. CH_3OD) is a particularly good replacement for undeuterated methanol as it is relatively inexpensive but again simplifies the solvent suppression needed and increases the spectral window available. Where buffers and acids or bases are required, unprotonated versions are always preferred. d_1-trifluoroacetic acid is very useful, and the ready availability of d_2-formic and d_4-acetic acids provides good alternatives, should the chromatography demand them. The properties of commonly used solvents, buffers and additives has been summarised in reference [143].

4.3.1.2 NMR mode

The decision to use LC/NMR involves a decision also as to the mode of NMR detection to be used in the analysis. The alternatives are stopped flow, loop capture, or continuous flow. The latter is probably the most obvious use of LC/NMR to a chromatographer – the NMR becomes a detector in the way that a photodiode array detector or MS might be used. Structural information will obviously be available from this use but typically the constraints of continuous flow preclude the structural analysis of low-level components. For example, in a typical continuous flow experiment at $1\,mL\,min^{-1}$, the need is to acquire as many NMR spectra per unit time as is possible to give sufficient S/N in the NMR spectrum while retaining adequate time resolution in the chromatographic dimension. Thus 32 scans may be acquired per time increment, resulting in a time increment of around 30 s. To achieve this number of scans in such a relatively short time, compromises must be made in the quality of the NMR data obtained and for example, the number of datapoints is usually limited to 8k (instead of 32 or 64k), resulting in degraded spectral resolution, and the pulse repetition rate may lead to incomplete relaxation of spins and therefore some NMR signals in the spectrum may be artefactually diminished. Further, for many low-level components, 32 scans may simply be insufficient to obtain meaningful NMR data. However, continuous flow is sometimes indispensible when analysing very labile components and it can also be very useful in scouting a mixture for components of interest. Quantitative on-flow LC/NMR has been considered by Haw [144] and later by Godejohann [145]; many of the requirements for good quantitation can be extrapolated from the discussion in Section 4.4 below, but these authors also examined the effect of flow and the use of reference materials for quantitation.

One way of improving the S/N in continuous flow LC/NMR is to reduce the flow rate. Godejohann *et al.* [146] reduced the flow to 17 µL min^{-1} to allow more time for spectra accumulation, leading to much improved S/N in each spectrum and the identification of ground water pollutants at the microgram per litre range. This technique can be applied to impurity profiling in drugs too, and using ^{19}F LC/NMR. In the example given in Fig. 4.7, the flow rate was reduced by a factor of 10 (to 0.1 mL min^{-1}), while maintaining the chromatographic gradient at the original % change/mL, to create the opportunity to collect around 300 scans for the same effective time resolution. Detection of fluorine-containing impurities can be achieved at levels well below the levels of the ^{13}C satellites of the main component.

Another way of improving S/N is to stop the chromatographic flow for NMR data collection. The simplest method for this is to *time-slice* the chromatogram and stop the flow in a time-dependent manner – in a sense the ultimate in slow flow. This allows more scans to be acquired before moving to the next time slice and the use of longer acquisition times, thus giving higher spectral resolution. There is no requirement to calibrate the time taken for a peak to move from UV detector to the NMR flow cell and indeed there is no need for peaks to be UV-active at all McCrossen 1998. Time-slicing is usually used in scouting experiments and in investigations of chromatographic peak purity [147]. An example of time-slicing is shown in Fig. 4.8 for a model system of 0.1% (w/w) phenyl acetate in *m*-toluenesulphonate methyl ester, under conditions chosen so that the components approximately co-eluted. The data show that with sufficient scans at each time-slice, such low-level impurities are detectable although one should be careful to recognise that sufficient chemical shift differences between major and minor components are required to detect the minor one at these levels.

The other methods of improving S/N and spectral resolution are to employ *peak-directed* stopped flow LC/NMR or loop-capture. The former is a

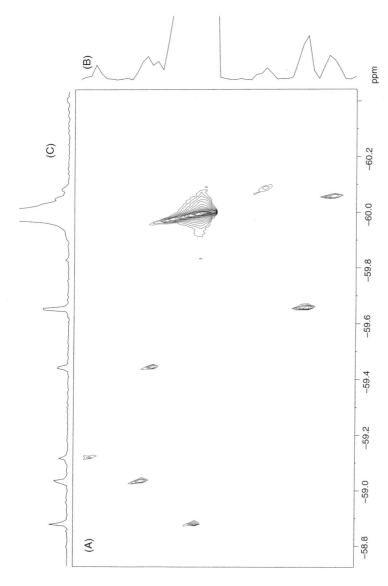

Fig. 4.7 ^{19}F LC/NMR impurity profile of a batch of drug substance, shown as a pseudo-2D plot (chemical shift vs time) A, a projection of the total NMR data on the time axis, B, and a projection of the total NMR data on the frequency axis, C. Note the change in chemical shift of each peak with the change in eluent composition as they elute. ^{19}F NMR data were acquired with ^{1}H-decoupling on a Bruker DRX500 spectrometer interfaced to a Bruker BPSU-36 LC/NMR apparatus. Chromatography was performed using a Waters Symmetry C18 column, eluted with a gradient of protio-acetonitrile and ammonium formate-d$_5$/formic acid-d$_2$/deuterium oxide at a flow rate of 0.1 mL.min^{-1}.

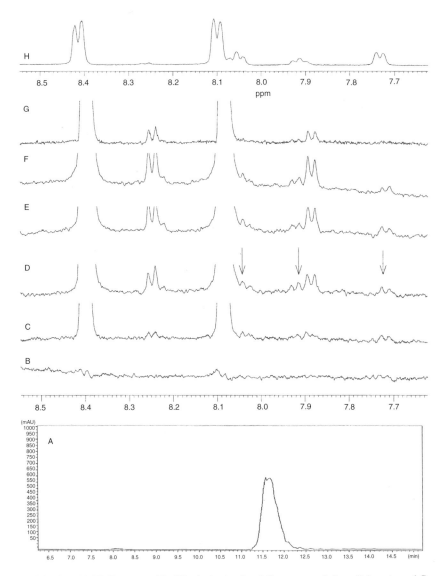

Fig. 4.8 500 MHz NMR spectra (B)–(G) obtained using fully automated time-slicing stopped-flow LC/NMR. Spectra correspond to timeslices taken every 15 s across the peak in the chromatogram shown in (A), for a mixture of 0.1% (w/w) phenyl acetate in methyl *p*-toluenesulphonate, starting at 11.5 min for (B), through to 12.75 min for (G); (H) shows an example spectrum for the same system except with 10% (w/w) phenyl acetate. Spectrum (D) shows the appearance of peaks (arrowed) corresponding to the phenyl acetate. Other impurity peaks are obvious in this and succeeding spectra. Chromatography conditions: Bruker LC22 pump controlled by Hystar software, Waters Symmetry C18 5μ column, 150×3.9 mm i.d., isocratic elution at 60:40 acetonitrile:D_2O v/v, 40°C, 0.2 mL min^{-1}, UV detection at 254 nm, 100 mL injection. NMR conditions: Bruker DRX500, 4 mm SEI z-gradient LC probe (120 mL active volume), 2k scans per slice, solvent suppression using WET.

simple procedure that requires only that the time taken for the peak maximum to move from the UV detector of the LC to the middle of the active volume of the NMR flow cell is known. This can be simply measured with several identical injections of a given material and varying the delay between the time of detection of the UV maximum and the flow cut-off. The optimum timing gives the maximal NMR S/N and allows a margin of error for slight misplacement of the peak. With LC/NMR interfaces not specifically designed for the purpose, stopping the pump does not stop the flow immediately and due allowance has to be made for this before NMR measurements start. With specifically designed interfaces this process is considerably simplified.

The flow may be stopped for as long as necessary to collect adequate S/N and without the need for reduced spectral resolution. Flow can be restarted and stopped for multiple peaks as required. However, it should be recognised that while the flow is stopped, diffusion of analytes remaining in the LC system will occur along the length of the LC column and associated tubing, resulting in broader peaks and lower peak concentrations and therefore in lower S/N in the NMR spectrum [143]. In practice, this is not a critical issue, particularly if a gradient LC method is being employed.

Stopping the flow in a gradient method can bring its own problems, however, as some LC contollers will not allow the user to restart the gradient where it was halted. One method of avoiding this is to use capillary loops to capture the components as they elute from the UV detector. A loop capture device can be built quite easily from two 6-port valves and appropriate (equal) lengths of capillary LC tubing. The volume of each loop is matched to be the same as the active volume of the NMR flow cell being used plus a margin to allow more easy placement of the eluent slug from the capillary centrally in the active volume of the NMR flow cell. In this case the timing of the transfer of the UV peak maximum to the loops and the timing of the transfer from the loops to the NMR flow cell must be measured. The former can be measured by injecting identical solutions into the LC system, varying the time delay from UV peak maximum to the time of loop switching and observing the minimal peak height from a UV detector placed after the loops; the latter is measured by filling the loops with identical solutions (e.g. by syringe filling) and proceeding as for stopped flow timings.

Loop capture makes uninterrupted gradient elution possible and is particularly suitable for the trapping of multiple components or for when long NMR acquisitions are required for each chromatographic peak. It also works better than simple stopped flow when peaks are not well separated chromatographically. Back-mixing in the NMR flow cell usually means that a sample takes longer to leave the flow cell than it did to enter it. This has the effect that poorly separated peaks can become mixed in the NMR flow cell in simple stopped-flow work. Using loop capture provides the opportunity to rinse the flow cell clean between each sample, removing this problem. Loop capture does, however, have the disadvantage that the transfer process inevitably dilutes the components slightly, and lower NMR S/N is typically observed, relative to an equal amount of a component analysed by stopped flow.

Automation of the above LC/NMR procedures has now advanced to a stage where, with the appropriate interface and software, the analyst can inject a sample and walk away, leaving the LC interface to detect and store (or not) peaks appropriately, and the NMR spectrometer to collect data with S/N-dependent collection of 2D data as desired.

4.3.1.3 Solvent suppression in LC/NMR

The simplest and most popular method of solvent suppression is presaturation of the solvent signal; a continuous weak irradiation at the solvent frequency before each acquisition. Indeed, this method may well be used in combination with other methods, for example, to suppress the small residual HOD peak of D_2O while another technique is used to suppress the large signal due to the organic modifier. Simple presaturation of a large signal, however, is typically neither efficient nor clean enough for difficult applications in LC/NMR and a more effective procedure is required. A popular choice is the NOESY-presat sequence, which uses a non-selective NOESY sequence to reduce the effect of regions outside the homogenous field of the active region of the coils (and flow cell). Other techniques have been introduced and used widely, such as the WATERGATE [28, 128] and WET [129] methods, or excitation sculpting [30, 130, 131]. The latter method appears to be particularly an effective method as shown in Fig. 4.9.

One practical outcome of these improved solvent suppression methods is that it has now become important to suppress the ^{13}C satellites of the solvent signal as these may be the largest signals left in the spectrum after the suppression of the ^{12}C solvent signal [131].

Another aspect of solvent suppression in on-flow LC/NMR is the need to match the suppression frequency to the exact frequency(ies) of the solvent(s). In gradient LC/NMR the frequency of the solvent signals vary quite widely with increasing organic content of the eluent [123]. The usual means of accounting for this in on-flow LC/NMR is to perform a quick scouting scan without suppression, peak pick the largest signals and suppress these in the immediately following acquisition. This can now all be achieved under automation, making even sophisticated on-flow solvent suppression possible. This automation also has advantages in automated stopped-flow LC/NMR.

4.3.1.4 Applications of LC/NMR

The testament to the success of LC/NMR is the burgeoning number of publications in the area. Impurity profiling, as has been pointed out in a recent review of the area [148], has been relatively lightly reported, with only some relatively early industrial examples [132, 133], which used home-built interfaces and a field strength of 400 MHz, through a comparison of LC/NMR with other methods

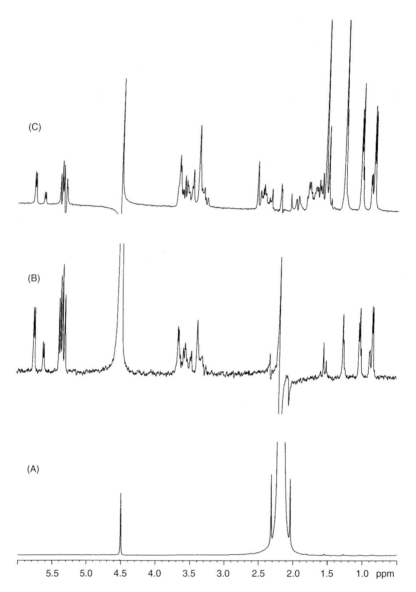

Fig. 4.9 (A)–(C) ^{1}H spectra recorded at 500 MHz on a 4 mM solution of a compound in a 60:40 (v/v) mix of D_2O and CH_3CN. (A) Simple pulse and collect 1D ^{1}H spectrum showing the ^{13}C satellites of acetonitrile; (B) NOESY presat spectrum with presaturation of the MeCN signal applied during the relaxation delay (2 s) and during the mixing time (200 ms) using a field of 90 Hz; (C) Excitation sculpting sequence using a 1 s presaturation of water and a selective proton π pulse of 4.25 ms. Compare the spectral region around the residual acetonitrile resonance in (B) and (C) – excitation sculpting results in the obliteration of far less of the spectrum near the suppressed solvent resonance, which in this case contains many solute resonances.

for detecting non-UV detecting impurities at 500 MHz [149] and a structural study of a bulk drug impurity in a glycinamide ribonucleotide transformylase inhibitor [150], also at 500 MHz, to impurity profiling [147] and the identification of degradation products [151] at 600 MHz. The relative dearth of examples of impurity identification by LC/NMR should not, however, be taken as an indication of a lack of applicability in this area; it is more that LC/NMR has become a routine application and that publication in this area is often constrained by intellectual property considerations.

LC/NMR has also been applied to combinatorial chemistry [152] and to natural product struture determination [153–155], in which area, the relative merits of LC/NMR in the analysis of complete unknowns in crude mixtures have been discussed [156]. Metabolite identification by LC/NMR has been an area of much research [134, 135, 157–161]. ^{1}H LC/NMR has been used in conjunction with ^{19}F LC/NMR to provide complementary data on synthetic materials [162, 133] and, more recently on metabolites [163–165], and ^{2}H detection has been used as well [166].

4.3.1.5 LC/NMR/MS

From the foregoing it is obvious that LC/MS is a necessary partner to LC/NMR; NMR can only give the relative number of each type of proton – we need a molecular weight to be able to make use of this information. Many users have therefore taken the conceptually obvious, but practically more difficult, step of hyphenating LC/NMR to a mass spectrometer and producing LC/NMR/MS [167]. The practical aspects of this have been discussed by Taylor et al. [168]. The advantages of this extra hyphenation are typically speed of analysis – e.g. the fraction examined by LC/NMR does not have to be recovered and re-analysed by LC/MS – and increased certainty that the correct component is being analysed by both techniques, a certainty that neither retention time nor UV spectrum can necessarily give in all cases. The mass spectrometer can also be used as a very sensitive detector, which does not rely on a UV chromophore in the compound, to give mass-directed LC/NMR. Representative examples of this technique are given by references [165, 169, 141].

4.3.1.6 Capillary LC/NMR

Improvement in the mass-sensitivity of LC/NMR can be accomplished by reducing the eluting volume with smaller diameter columns. With mass-limited samples, capillary HPLC offers higher concentrations in the eluting peaks. The coupling of capillary HPLC to NMR also offers an advantage over standard bore LC/NMR in reduced solvent consumption, making the use of fully deuterated solvents feasible throughout. Wu et al. [137] reported the use of solenoidal detection coils wrapped around a 355 μm o.d. capillary as the detector, attached by a short length of Teflon tubing to the outlet of a 1.0×150 mm microbore LC column. The rapid transit of the sample through the flow cell allowed rapid pulsing and this, together with the solenoidal design of the coil, produced good

S/N for a flow cell of only 50 nL volume. The main drawback was rather wide NMR linewidths. Behnke *et al.* demonstrated capillary LC/NMR in 1996 [138], using flow cells of 50 nL and 900 nL and a standard saddle coil, and achieved sensitivities approximately 10 times greater than conventional LC/NMR. The approach, however, did not incorporate a UV detection cell so that stopped-flow NMR measurements had to be made by reference to on-flow NMR measurements.

Capillary HPLC has also been applied to the separation and identification of terpenoids [170], using a commercial capillary-based HPLC system and a solenoidal coil detection cell of 1.1 µL volume. This arrangement produced a detection limit of 37 ng for α-pinene in an on-flow example. Commercialisation of systems of this type will increase the applicability and ease of use of this technique.

4.3.2 Hyphenation to other separation techniques

Conventional LC/NMR has been by far the most popular coupling to date. Supercritical fluid chromatography (SFC)/NMR was first reported in 1994 [171] and has been applied to monomeric acrylates [172] and vitamin A acetate isomers [173]. However, its use has not been widespread, perhaps more due to the lack of popularity of SFC than the undoubted advantage SFC/NMR offers in the lack of solvent signals [174].

The coupling of capillary zone electrophoresis (CZE) and capillary electrochromatography (CEC) to NMR, at first sight, seems to be very difficult as the typical flow cell volume used in LC/NMR at around 100 µL is typically three orders of magnitude greater than that used in CZE or CEC. However, much progress has been made in recent years, in parallel with the progress in hyphenation of capillary LC. Nanolitre volume NMR microcells have been developed and used to obtain detection limits of less than 50 ng for amino acids [175–179] and CZE/NMR and CEC/NMR have been applied to metabolite analysis [180]. The influence of gradient chromatography conditions [181] and of pressure [182] in CEC/NMR have been investigated. Sample focusing using capillary isotachophoresis has also been developed to focus sample volumes of several microlitres down to volumes of around 100 nL, providing an improved method of getting such small volumes into the NMR detection cell [183].

For some applications there is a need simply for final clean-up of samples before they are analysed by NMR. In these cases, sophisticated separation methods are not warranted and solid phase extraction may suffice. This can, of course, be achieved off-line but on-line coupling of SPE to NMR [184] may be useful as an automated means of removing polar additives from previous chromatographic steps in a manner that will not introduce further impurities from sample handling. From this it is but a short distance to having no separation technique at all but merely to injecting samples using an HPLC autosampler directly into an NMR flow probe. This technique is widely used in applications where sample throughput is critical, such as combinatorial chemistry [185] and biofluid analysis [186, 187].

4.4 Quantitation

4.4.1 The basics of quantitation

The area or integral of an NMR signal is directly proportional to the molar amount of the species giving rise to the signal. Thus,

$$\frac{A_1}{A_2} = \frac{N_1}{N_2} \qquad (4.3)$$

Where A_1 and A_2 are the areas of the two signals and N_1 and N_2 are the respective numbers of protons generating the signals. For cases where protons in the same molecule are being quantitated, N_1 and N_2 represent the number of protons in the functional groups, n_1 and n_2. In general for two species we can say that

$$\frac{A_1}{A_2} = \frac{n_1 m_1}{n_2 m_2} = \frac{n_1 W_1}{n_2 W_2} \cdot \frac{M_2}{M_1} \qquad (4.4)$$

Where m_1, m_2 are the numbers of moles for the two species, W_1 and W_2 are the masses and M_1 and M_2 the molecular weights of the two species.

There are no response factors such as those in UV-detected HPLC, and the basis for detection is different so that non-chromophoric components may be detected or quantitated, depending only on the presence of a specific NMR signal from an NMR-active nucleus in the component. Further, since the area of the response is linearly related to the number of atoms in the probe, calibration of response is straightforward. Thus it may be assumed that NMR may be used quantitatively in a general and simple manner. Indeed, NMR is used routinely in a quantitative manner for structure elucidation when we calculate from the integral the relative numbers of protons responsible for particular signals. However, for this purpose, the quantitation is rarely required to be accurate to better than ±10% and this is easily achievable without special care. On the other hand, if we want to use NMR quantitatively with better accuracy – for example for main component assay or impurity quantitation – we must employ a set of experimental parameters, which are not necessarily those we would use for *routine* NMR spectroscopy, to minimise the errors inherent in the measurement and in sample preparation. Further, depending on the use to which the quantitative measurement will be put, we may need to validate the method and demonstrate that it is specific, accurate, precise, reproducible and robust. Thus quantitative NMR, where accuracy better than a few percent is required, is neither simple nor general – these methods are specific and require evidence that appropriate care has been taken in the development of the method.

4.4.2 Optimising the experimental parameters

There are many literature references to the use of NMR for quantitating species, particularly in formulated products as the specificity of the NMR measurement is of great benefit here. Most use ^{1}H but ^{31}P and ^{19}F methods are also common [188, 189]. For example, methocarbamol (3-(2-methoxyphenoxy)-1,2-propandiol 1-carbamate) and paracetamol have been quantitated at 400 MHz in paracetamol–methocarbamol mixture dosage forms, using maleic acid as the internal standard [190]. Recovery was better than 99%, and the authors reported a better standard deviation ($n=5$) than the USP XXIII HPLC method. Azathioprine (6-(1-methyl-4-nitroimidazole-5-ylthio)-purine) may be quantitated in tablets using fumaric acid as the internal standard at 400 MHz [191]. Again, good recovery and errors were reported. Ketoprofen has been quantitated in tablets [192], using mandelic acid as standard and ranitidine has been quantitated in solid dosage forms, using chloroacetophenone as standard [193]. A recent review [194] gives many other examples of the quantitative application of NMR, illustrating the range of applications possible.

USP 761 describes the NMR general method, noting that 'if appropriate instrument settings for quantitative analysis have been made, the areas (or intensities) of two signals are proportional to the total number of protons generating the signals'. However, even in the relatively recent references above there is little information about the NMR experimental parameters and no discussion of what effect they have on the measurement. So what are *appropriate instrument settings* and how do we achieve accurate integration? The key factors are discussed below, with particular concentration on quantitation using ^{1}H spectroscopy as this is the most frequently used.

Of prime consideration in setting appropriate parameters for data acquisition is the longitudinal (or spin-lattice) relaxation time constant T_1 [2]. Each nucleus in the sample will have a characteristic T_1, which is dependent on molecular motion (in turn dependent on molecular size and structure, temperature, solution viscosity, etc.). The longitudinal relaxation rate determines how long it takes for a particular nucleus to reach thermal equilibrium after an excitation pulse, and each nuclear environment may well have a different rate constant. It is essential for Equation (4.4) to hold that all nuclei have relaxed equally before each acquisition pulse – but since this is not attainable in practice for nuclei in differing environments – full relaxation to thermal equilibrium of all spins is required to minimise errors. As T_1s may be of the order of multiple seconds and relaxation is an exponential process, attaining thermal equilibrium for all spins is a non-trivial requirement and if 90° pulses are used, pulse intervals of tens of seconds will be required to ensure full relaxation, especially for small molecules. 60 s is a commonly chosen, if arbitrarily long, pulse interval. 90° pulses have the advantage of maximising S/N [195]. If smaller tip angles (e.g. 45°) are chosen, the pulse interval can be shortened and a delay of tens is commonly used and may be appropriate. However, when developing an NMR method for accurate quantitation, demonstration that the pulse interval is sufficient to allow essentially complete relaxation before the next acquisition pulse is an important step. A tip angle of 45° will reduce the S/N of the measurement (relative to a 90° pulse with the same pulse interval) but so long as this can be made sufficient (through increased number of scans or sample concentration) a 45° tip angle is usually preferred [196].

Tip angles of less than 90° also reduce the off-resonance effects resulting from incomplete excitation of resonances away from the carrier frequency, effects which increase with increasing field [196]. Other procedures for reducing the effect of the choice of carrier frequency include placing the carrier midway between the two signals of interest (if there are only two), or summing spectra with differing carrier frequencies [197].

Typical peaks in NMR spectra are of the order of 0.5–1 Hz wide and may only be characterised by a few data points in *routine* data collection conditions. Too few data points across a peak will lead to integration errors and hence errors in quantitation. However, it appears that a digital resolution of around 0.2 Hz/point is sufficient to reduce errors in ^{1}H spectroscopy, which is easily achieved [2, 196]; in heteronuclear NMR, achieving a similar number of points per peak may not be so straightforward but line broadening (with an appropriate window function) may help here. Line broadening in ^{1}H spectroscopy may also be beneficial as it has the effect of increasing S/N in the spectra, while the broadening of the peaks makes it easier to get adequate digitisation. However, quantitation of resolution-*enhanced* spectra should be performed with extreme care as the relative areas of peaks of differing linewidth can be changed substantially.

Signal-to-noise is another critical consideration for quantitative NMR; for lowest errors, the target should be for a S/N of better than 250:1. This is supported by the finding in one study that a ^{1}H S/N of >200:1 was a suitable criterion to judge whether the analysis would result in a relative standard deviation of <2% [197]. This corresponded to main component concentrations of 20 mM or more (^{1}H NMR, using 64 scans). Signal averaging over many scans is one way of improving S/N, but since S/N is only proportional to the square root of the number of scans, collection of sufficient scans, particularly with 10–60 s pulse intervals, can be a slow process. Obviously, for easiest achievement of a suitable S/N, the components of interest should be at a reasonable concentration – for typical drug molecules around 10–20 mg mL^{-1} is often used for ^{1}H spectroscopy.

The nature of the NMR lineshape means that to integrate a peak accurately the integral width either side of peak must be substantial – approximately 300 times the peak width in both directions to obtain 99.9% of the true area. This kind of integral width is usually impractical and the only way that quantitative NMR produces acceptable results is that the errors on each peak cancel each other out. However, integral widths large enough to provide meaningful results will always involve accounting for spinning sidebands around the peak and will also require the deliberate inclusion or exclusion of ^{13}C satellites. Spinning sidebands may be removable on a modern probe by the simple expedient of not spinning the sample – modern probes will often provide sufficient resolution non-spinning. Otherwise, spinning sidebands at least need to be identified – perhaps by spinning at two different spinning rates [198] to exclude the possibility that they are actually due to impurities. The inclusion or exclusion of ^{13}C satellites will probably depend on how congested the spectrum is but the choice should be made consistently for all peaks being compared.

Baseline and phase correction are also important factors in determining the accuracy of the NMR result – it is usually the case that these have to be done manually to obtain best results. It is also quite likely that different operators will perform these functions differently and this may be a cause of reproducibility problems (see Section 4.4.5 below). A recent paper [199], though not aimed at quantitation directly, has a useful discussion of the difficulties of obtaining precise integrals, particularly with respect to the phasing of signals.

For an assay using an internal standard, attention should be paid to the purity of the standard used. This is perhaps self-evident but if high purity samples are to be assayed, the standard should also be of high – and known – purity. The standard also needs to have a signal in a relatively clear area of the spectrum and be chemically stable. Maleic and fumaric acids are popular choices as standards. ^{1}H NMR has also been demonstrated as a method to quantitate crude samples in 96-well plates after parallel synthesis, a task which is otherwise difficult because of the low sample amounts and lack of adequate standards for the more traditional LC/UV approach [200]. In this case, 2,5-dimethylfuran was used as an internal standard because it is easily removable (bpt 92°C) and has two conveniently separated singlets (2.2 and 5.8 ppm), which not only increases the chances of one of these signals being well separated from analyte signals but also provides an internal check on the quality of the quantitation, serving to indicate sufficient relaxation of the analyte.

Lastly, the assignment of the spectrum should indicate some protons which should not be used for quantitation – those labile NH and OH protons which are exchange broadened and therefore more difficult to integrate accurately, but which may also partially deuterate in solvents such as D_2O or CD_3OD. Keto-enol tautomerism may also cause exchange for aliphatic protons beta to a carbonyl group, for example, and some aromatic protons may also be surprisingly labile, as shown in Fig. 4.10.

All the considerations above apply especially to *assay* by NMR, that is, strength determinations, usually using a well-characterised internal standard. Minimisation of errors leads to strength precision of ±1% [196]. Other literature too demonstrates that good precision can be obtained. For example, the fluoroquinolones pefloxacin, norfloxacin, ofloxacin have been assayed by both ^{1}H and ^{19}F NMR in commercial preparations [201] with good recoveries for

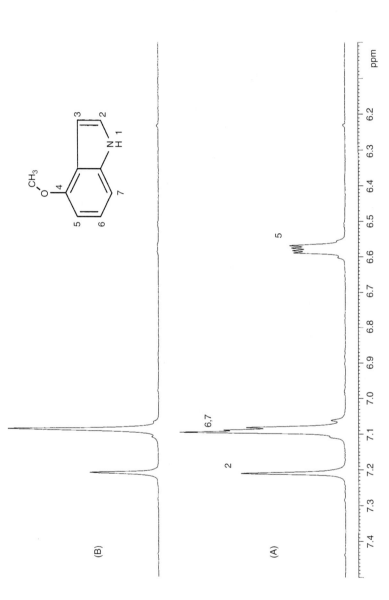

Fig. 4.10 (A) and (B) show ^1H spectra recorded at 500 MHz of a solution of 4-methoxy indole in 50:50 (v/v) D_2O (containing 0.1% trifluoroacetic acid-d_1) and methanol. Spectrum (A) shows the spectrum very soon after dissolution – the indole NH and 3 protons have already exchanged with solvent. Spectrum (B), obtained after a weekend in solution, shows the resonance of proton 5 has been completely exchanged. The 5 proton signal in (A) forms the X part of a tightly coupled ABX spin system.

each nucleus, good correlation with declared amounts and with relative standard deviations of <1%. The instrumental parameters are also well specified in this paper and the authors kept to the USP method in that they measured the area under the peaks by integrating not fewer than 5 times, using the average areas in their calculations.

4.4.3 Quantitative impurity determinations

For the estimation of impurities by NMR, the experimental stringencies associated with assays may be relaxed somewhat as larger errors in the values obtained may be acceptable. For example, whereas a relative standard deviation of 3% in an assay result may not allow an adequate conclusion about strength to be drawn, a similar error in a low-level impurity (compound-related or solvent) may not have the same consequence; 3% (or even 10%) of a 0.5% w/w result may not be significant, for example. Obviously, for impurity profiling, detection limits (DLs) and quantitation limits (QLs) are correspondingly more important than they are for assays. An indication of the compromises to be made in achieving a good DL is given by Lankhorst *et al.* [202]. A DL is often specified as that level of analyte which is detectable at a S/N of 3:1; the corresponding QL is the level corresponding to a S/N of 10:1. Given the discussion above regarding S/N requirements for good precision, it will be appreciated that assay-type precision will not be achieved near the QL. A more practical approach to defining the QL for a given analysis may therefore be to determine the level of analyte which under the experimental conditions used, gives an *acceptable* relative standard deviation.

Quantitative impurity profiling is an important aspect of the use of NMR in development – it is often the case that early batches of a compound entering development will be profiled by NMR for solvents content and/or related impurities, particularly if the corresponding GC or LC methods which might be used are as yet undeveloped [198, 194]. As long as the drug substance purity is high enough, it is usual to quantitate the impurities directly against a resonance of the main component. Alternatives are to add a standard for direct quantitation [202, 203] or to make use of the ^{13}C satellites that every proton attached to carbon will display. ^{13}C is present in organic compounds at 1.1% (Table 4.2) and the coupling between the ^{13}C and the attached proton(s) gives rise to characteristic sidebands or satellites symmetrically about the main peak (the latter representing protons attached to ^{12}C). Each satellite will be half of the 1.1%, i.e. 0.55%, of the central peak. These satellites provide a particularly simple method for estimating or quantitating low-level impurities [204].

As noted in Section 4.2.4.2, ^{19}F NMR may be particularly useful for impurity profiling because of the simplicity of the fluorine spectrum. Mistry *et al.* showed that fluorine-containing impurities in bulk drug could be detected and quantitated down to approximately 0.1 mole% (by comparison with HPLC-UV traces) [205].

These workers also measured the diffusion constants for each ^{19}F resonance to discriminate between monomeric and dimeric impurities.

^{19}F NMR is also very useful for detecting and quantitating non-drug-related impurities or counterions, such as trifluoroacetate or PF_6^-. Ion chromatographic methods may also be used for this purpose but ^{19}F NMR provides a simple method, provided that sufficient relaxation time is allowed between acquisitions as these species have T_1s of 2–4 s (d_6-DMSO, 300 K).

^{19}F NMR has also been used to quantitate resin-bound species in solid-phase synthesis, by use of of a polymeric support bearing ^{19}F as an internal standard [206], and has been reported as a rapid screening tool to quantify the biotransformations of fluorine-containing model drugs [207].

Estimation of chiral purities may also be performed by NMR, as an alternative to chiral HPLC, CE or GC. This work requires the addition of some chiral solvating or complexation agent to achieve the discrimination; *(S)*-(+)-2,2,2-trifluoro-1-(9-anthryl) ethanol (TFAE) [208, 209], cyclodextrins [210] and chiral lanthanide shift reagents [211] are common reagents. In general, it may be expected that the achievable chiral resolution in NMR will not be complete and the peaks of interest may be slightly overlapped, making accurate quantitation difficult, especially for low amounts of one or other of the enantiomers. However, as chromatographic resolutions of enantiomers may not be perfect either and the methods can be more time consuming to develop, NMR may well be competitive with chromatographic methods in some cases.

4.4.4 Summary of experimental considerations

In summary, when developing a quantitative NMR method, the following principles should be observed: use approximately a 45° tip angle and check that the interpulse delay is sufficient for full relaxation of all measured signals; ensure sufficient digital resolution (typically 0.2 Hz/point or better for ^{1}H spectroscopy); ensure sufficient S/N to ensure adequate precision for the purpose (e.g. 250:1 for an assay); choose appropriate, sharp, well-resolved signals for integration; check that automatic baseline correction and phasing are sufficient for the purpose or perform these manually; ensure an adequate region is chosen for the integral range such that as large a proportion of the peak is integrated as possible; make a conscious decision to include or exclude ^{13}C satellites and check for spinning sidebands; for best accuracy (e.g. for an assay), run at least five replicates and use the average integral figures.

4.4.5 Method validation

Having developed a quantitative NMR method, how should one assess whether it works well enough for routine use in quality critical applications? As implied above, the extent of validation required will depend on what the method is to be used for – a method where accuracy and precision are vital (e.g. a strength assay) will require more validation than a method essentially used as a limit test for a known impurity. However, the principles of validation are common to all methods – what varies is the acceptable limit for a given parameter and hence the extent to which the various aspects of validation are pursued.

Once again there is little in the way of literature to help decide how to validate an NMR method. There is a useful overview of the NMR in relation to the regulatory field [212] and some recent publications which indicate the scope of validation required [213, 197], but most literature reports simply conclude with a relative standard deviation to indicate how repeatable the assay is, giving no indication as to how the result may vary with instrument, operator, timelapse between analyses, or with laboratory. Further, few reports discuss the need to check the robustness of the method, that is the behaviour of the method as parameters deviate from the ideal. Bauer et al. [213] found a significant laboratory effect in a collaborative multilaboratory study and the quantitative values obtained were in some cases significantly different from the true values. These workers also suggested that the use of true replicates is better than simple reintegration of a single sample and that this method should be used in preference to obtain an average integral figure. While investigating the robustness of a quantitative chiral assay (using peak heights rather than areas), Lacroix et al. [209] studied the effect of small concentration changes in analyte and chiral solvating agent and also noted the need to check for interferences and compound stability. This study also found that the result could vary depending on the batch of chiral solvating agent used, emphasising the need for vigilance in an assay.

Maniara et al. [197] made a systematic exploration of experimental precision, accuracy, specificity, linearity, DL, QL, and robustness, for NMR assays of agricultural chemicals. They concluded that 'carefully implemented QNMR can compete effectively with chromatographic methods. The level of the major chemical ingredient can be determined with accuracy and precision significantly better than 1%. Impurities at the 0.1% or lower may also be quantified'. For the robustness study, these workers chose the number of scans and the relaxation delay, then operator to operator and instrument to instrument variabilities were tested. Unlike Bauer et al., they found that purity values were 'Not significantly dependent on analyst or instrument', but perhaps this was because the work was performed in one laboratory, rather than several disparate laboratories, each of which may have their own way of working. The work of Maniara et al. therefore tested *intermediate precision*, whereas that of Bauer et al. tested *reproducibility*.

Method validation is not the same as method development; in the former one tends to be seeking a *suitable* set of experimental parameters for the analysis, whereas in the latter one is demonstrating that these selected parameters are *sufficient* for the purpose. In general, where an NMR method is used in the specification for a drug substance, it should be validated – as would a chromatographic method – according to the ICH guidelines on method validation (CPMP/ICH/281/95 Part II, all sections). For NMR this typically means that – depending on the stage of development – validation of the following aspects of the method would be considered:

Specificity – does the signal being used for the assay represent only the species in question and is it sufficiently well separated from other signals? This is not necessarily a straightforward assessment if the signals of interest are in a crowded region of the spectrum. It may require high field investigations or changes in solvent to check this. Further, can you be sure that degradation of the sample in the NMR solvent doesn't produce a peak under the one of interest?

Detection and quantitation limits (DL and QL) – are these sufficiently low to allow adequate detection or quantitation of the analyte? For an analysis with a defined impurity limit, the QL should be, at maximum, half the limit value.

Linearity – although the area of the response observed in NMR is directly related to the amount of the analyte in the probe, there are method-related reasons why linearity must be checked – for example, one of the components involved may be near its concentration limit in the NMR solvent chosen, especially if the method has been developed to maximise S/N.

Precision – integration, particularly at low sample amounts, may be difficult and an assessment of the precision of the experiment is needed. As alluded to above there are three levels of precision testing; repeatability, intermediate precision and reproducibility. Repeatability expresses the precision under the same operating conditions over a short interval of time. This is frequently the only precision information provided in literature reports. Intermediate precision expresses within-laboratory variations across different days, different analysts, different equipment, etc. and it is a key indicator of how an assay will perform under real conditions. Reproducibility expresses the precision between laboratories and typically only becomes important if a method is transferred between laboratories – for example, from an R&D site to a manufacturing facility.

Accuracy – the closeness of the measured value to the true value needs to be checked. The *true* value may have been established by other techniques such as HPLC, or may be known from the gravimetric procedure used to prepare the sample. Alternatively, the method of standard addition may be used [202]. If an extraction procedure has been used during sample preparation, how well does the analyte quantity in the extract reflect the analyte quantity in the original sample, i.e. does the method have good recovery? If there are significant and/or variable losses, the method will not be reliable.

Stability of analytes (and standard) over the analysis period – self-evidently the system under test must not change during the test if the results from the test are going to be meaningfully related to the original sample.

Robustness – depending on the environment in which an NMR method is to be used, an appropriate level of robustness should be demonstrated with respect to the variables that might change in that environment. For example the sensitivity of the measurement towards variation in in-probe temperature, field homogeneity (shimming quality), spinning speed, relaxation delay, and pulse width accuracy may need to be determined. Another factor may be the quality of solvent – e.g. $CDCl_3$ may be more or less acidic, depending on the supplier and grade; DMSO may contain small but variable amounts of D_2O again depending on the source. Control measures may need to be put in place for all these parameters.

System suitability – to show that the control measures required have been complied with for a particular analysis on a particular day, a system suitability check is required. Such a check on the performance of the spectrometer and method may be used, for example, to ensure that the expected specificity and sensitivity can be achieved. One of the advantages of the use of NMR as a quantitation method is that the sample itself may provide such a system suitability test by, for example, making use of line-width or S/N data in the sample spectrum.

Obviously the above generalisations have to be considered in the light of what the analysis is going to be used for – in some cases less work will be required to produce an acceptable test and each case must be judged on its demands – which brings us back to the need to view quantitative NMR tests as specific and not necessarily simple.

4.5 Solid state NMR

4.5.1 Introduction

Preparation, control and understanding of the solid state form of a drug is one of the key goals of the drug development process. What is the most stable polymorph? How many polymorphs are there? Is the chosen form stable or does it change on formulation or storage? What about solvates? These are some of the key questions to be answered by the analysis of the solid state form of a drug candidate because different polymorphs will generally have different physico-chemical properties such as solubility, density, rate of dissolution, bio-availability, etc. This information is a key part of a drug-development company's intellectual property. The importance of the control of polymorphic form, even when the dosage form is not solid, is illustrated by the case of Ritonvir [214].

The front-line analytical techniques of X-ray powder diffraction (XRPD), vibrational (infra-red and Raman) spectroscopy and thermal analysis, and their application to solid state issues are discussed elsewhere in this book. Solid state NMR is a very sensitive reporter of molecular conformation, mutual interaction, dynamics and form. In this section, we will discuss the basics of solid state NMR and in particular the methods that can be used in the study of this state of matter.

4.5.2 Basic theory of solid state NMR

There are several excellent texts and articles covering both introductory [215, 216] and advanced solid state NMR [217, 218], but a brief explanation of some of the more important theoretical and technical issues are given here.

In solution, molecules are generally free to undergo translational and rotational movement such that their orientation with respect to each other and the static magnetic field is effectively random. This results in the averaging of any interaction that is dependent on the orientation of the magnetic field. However, solids usually have a well organised structure in which the relative positions of atoms have a defined mutual relationship. The spins of NMR active nuclei in solids are thus held in distinct environments and can interact with each other in a way that is not motionally averaged. These interactions are manifest in the very broad, often featureless NMR spectra of solids.

The direct magnetic interaction between two spins of magnetogyric ratio γ_1 and γ_2 (the dipole–dipole interaction – D) depends on their distance apart (r_{12}) and the angle (θ) between their inter-nuclear vector and the static B_0 field in the following way:

$$D \propto K\gamma_1\gamma_2 \frac{3\cos^2\theta - 1}{r_{12}^3} \qquad (4.5)$$

The geometric term in Equation 4.5 vanishes when $\theta = 54°44'$ the *magic angle*. The dipolar interaction (also referred to as dipolar coupling) should average to zero if the sample is spun at this angle with respect to the static field. However, to completely eliminate the dipolar interactions the speed of rotation must be several times the static linewidth, typically several 100 kHz which is not currently possible. Most commercial MAS systems are capable of rotating 4 mm sample rotors at a speed of 10–15 kHz or so.

For ^{13}C nuclei (with which we are mostly concerned in pharmaceutical solids) the main cause of dipolar broadening is the heteronuclear interaction with protons. In order to eliminate the residual heteronuclear dipolar coupling that remains during MAS conditions, high power proton decoupling is used (also called dipolar decoupling). The powers used need to be very high (by solution state standards) and make great demands on the hardware. It is necessary, for instance, to carefully control the acquisition time of the experiment, so that the high power decoupling is active for the shortest possible time (<80 ms).

Spinning at the magic angle also accomplishes line-width reduction arising by another mechanism. This is the chemical shift anisotropy (CSA), and is a reflection of the fact that in a solid, it is not possible to ignore the orientational dependance of nuclear shielding.

The chemical shift of a nucleus in a molecule depends on its orientation with respect to the external magnetic field. In the case of a powder sample consisting of many crystallites (micro-crystals) placed in a magnetic field, there will be a distribution of orientations of each nuclear environment, each orientation having a different chemical shift. The form of the resulting spectrum, acquired with high power decoupling, but not MAS, depends on the packing symmetry of the crystal and is called the powder pattern. The nuclear shielding also exhibits a $(3\cos^2\theta - 1)$ dependence on its orientation (θ) to the static magnetic field and so is reduced by spinning at the magic angle. Spinning at 10 kHz will significantly reduce the CSA, but some atoms e.g. carbonyl carbons have such a large CSA, that the effect is not entirely eliminated. The effect can be identified by sidebands symmetrically displaced from the isotropic resonance and separated by the spinning speed. The CSA increases linearly with magnetic field, and is therefore more difficult to spin out at higher fields.

We have seen previously (Section 4.2.4.1) that the sensitivity of a nucleus depends on its magnetogyric ratio (γ), and that it was possible to achieve huge sensitivity gains by transferring polarisation (magnetisation) from proton to carbon in, for example, the DEPT experiment. It is also possible to transfer polarisation from proton to carbon in the solid state – indeed the mechanism for doing so relies on the dipolar coupling we have been trying to eliminate. By *spin-locking* the magnetisation of both ^{13}C and ^{1}H nuclei using radiofrequency fields that satisfy the Hartmann-Hahn condition, cross polarisation occurs, i.e. the polarisation of the proton spins is transferred to the carbon spins. Alternatively speaking, magnetisation flows from proton to carbon, effectively increasing the ^{13}C sensitivity by a factor of four (γ_H/γ_C). The length of time that the spin-lock is applied is called the *contact time*, and typically values of the order of 1–5 ms are used. Not only does the cross polarisation increase the ^{13}C sensitivity, but it is possible to pulse at a rate which is governed by the proton T_1 not the carbon T_1. This is a real benefit, as carbon T_1s in the solid can be of the order of seconds to hundreds of seconds, whereas proton T_1s are usually much shorter.

Thus the basic experimental procedure for acquiring ^{13}C solid state NMR spectra is cross polarisation magic angle spinning or *CP-MAS*. This experimental technique produces solid state carbon spectra with line-widths that are comparable to the solution state.

The discussions so far have been mainly aimed at experiments designed to observe the ^{13}C nucleus. High resolution proton solid state spectroscopy on the other hand is still in its infancy, mainly due to the fact that the homonuclear dipolar interaction of the abundant ^{1}H spins is large compared with the chemical shift range of this nucleus. The technical problems to overcome are thus much more substantial than for the ^{13}C nucleus. We shall see at the end of the next section that significant advances have been made in observing high quality ^{1}H spectra in the solid state, mainly through the use of *inverse* detection methods.

4.5.3 Methods of assignment of solid state NMR spectra

4.5.3.1 One-dimensional editing methods

It is common practise to compare the solid state ^{13}C NMR spectrum with that obtained in the solution state (if available), as the techniques available to assign the latter are well established and reasonably robust (see Section 4.2.4.1). In addition to the basic 1D ^{13}C CP-MAS experiment, there are a number of editing methods available which yield information about the multiplicity of the ^{13}C resonances, thereby aiding assignment. Some older methods are still routinely employed. For example, interrupted decoupling [219] which suppresses all but quaternary and methyl carbons due to the introduction of a dipolar dephasing delay; CP with phase inversion, CPPI [220] differentiates methine from methylene carbons due to differences in the rate of polarisation transfer from their attached proton(s); CP with short contact time yielding preferential observation of protonated carbons. Various combinations of these and related methods have been used to generate sub-spectra for all four types of carbon. For instance, the method of Wu *et al.* [221] uses a combination of five editing experiments to generate the sub-spectra. Improvements to the elements of this editing scheme have been published recently [222, 223] and offer better quality differentiation of carbon types in a shorter time.

An alternative approach is to take advantage of the different time-dependant behaviour of heteronuclear J couplings, in a similar way to the attached proton test (APT) experiment used in solution state [21]. This has been demonstrated using a similar pulse sequence to the solution state, modified by the inclusion of frequency switched Lee–Goldberg decoupling (FSLG – see below) [224]. In cases of moderately fast spinning ($\geq$10 KHz), such decoupling may be unnecessary [225]. It has been shown recently that the filtration (using phase cycling) of multiple quantum coherence generated via heteronuclear J couplings can be used to allow differentiation of all carbon multiplicities [226].

4.5.3.2 Two-dimensional solid state methods

It is possible to generate more information rich spectra by the introduction of a second frequency dimension. Correlated two-dimensional experiments have recently been developed to extend the assignment process of solid state spectra. For instance, the solution state experiment INADEQUATE (Incredible natural abundance double quantum transfer experiment) which shows direct ^{13}C connectivities based on $^{1}J_{CC}$ (essentially a ^{13}C COSY) has been found applicability for the solid state at natural ^{13}C abundance [227].

It is also possible to correlate ^{13}C and proton frequencies in a two-dimensional sense, leading to the generation of solid state proton spectra – by using f1 projections or individual columns of the 2D data matrix [228]. The resolution of the proton dimension is usually improved by CRAMPS (combined rotation and multiple pulse) techniques or by a very fast spinning ($\geq$20 kHz). The transfer of magnetisation from carbon to proton can be mediated by dipolar coupling or, as demonstrated recently, by heteronuclear J coupling [229]. These experiments are possible due to incorporation of the frequency switched Lee–Goldberg (FSLG) decoupling sequence that removes the homonuclear proton–proton dipolar coupling [230, 231]. For correlation methods based on dipolar coupling [232], the strongest correlations are observed, as expected, from carbons to their directly bonded protons, although weaker correlations are seen between non-bonded inter- and intra-molecular ^{13}C–^{1}H pairs.

Other methods are based on scalar heteronuclear J couplings and yield exclusively direct ^{13}C–^{1}H connectivities mediated by both multiple [229, 233] and single quantum heteronuclear coherence [234]. It is possible to increase the proton resolution in such experiments using a combination of very high magnetic fields and very fast MAS (up to 35 kHz) [235]. The reduction of proton linewidth to under 60 Hz under moderate MAS and more modest fields (500 MHz, 12.5 kHz MAS) has been demonstrated using new constant-time CRAMPS pulse sequences [236]. The 2D heteronuclear correlation methods described not only yield high quality solid state proton spectra, but have the potential to allow accurate carbon *and* proton assignment in the solid state e.g. [237]. It is possible to further extend the dimensionality of solid state proton detection method described above. By projecting the ^{1}H–^{1}H plane of a 3D ^{13}C detected dataset, it is possible to obtain a 2D ^{1}H–^{1}H dipolar correlation map, whose cross peak intensity is a function of the inter-nuclear separation, in a similar way to the familiar solution state 2D NOESY [238]. The following section describes other methods of determining distances in the solid state.

4.5.4 *Distance measurements in the solid state – recoupling*

The physical basis behind the ability of SSNMR to measure distances can be readily seen in Equation 4.5, in which the inter-nuclear distance exhibits an inverse cubic relationship to the dipolar coupling between the nuclei. However, it is precisely the dipolar coupling between nuclei that contributes to the broadness of solid state NMR spectra, and that we have striven to remove by using CP-MAS. Pulse sequences need to be developed that allow the re-introduction of the dipolar coupling – the general name for this technique is recoupling. Most of these methods use combinations of rotor-synchronised radio-frequency pulses to remove the averaging effects of the magic angle spinning. There are two main classes of re-coupling – heteronuclear and homonuclear, that will be discussed separately.

4.5.4.1 *Heteronuclear recoupling*

The most common technique used for heteronuclear distance measurement is rotational-echo double resonance – REDOR [239, 240, 241]. It is now relatively commonplace to label proteins with the spin-1/2 isotopes ^{13}C and ^{15}N, and C–N distance constraints can be derived using REDOR [242, 243] and used to determine elements of secondary structure. More importantly for molecules of interest to the pharmaceutical industry is the measurement of distances between ^{13}C and the quadrupolar deuterium (^{2}H) nucleus [244]. Introduction of the ^{2}H isotope is relatively straightforward and the measurement of ^{2}H–^{13}C distances at natural ^{13}C abundance over several bonds can be used to provide torsional constraints [245]. The measurement of ^{19}F–^{13}C distances (up to 6.8 Å) using REDOR has been used to determine a conformational change in a protein biosynthetically labelled with ^{13}C-cysteine and fluorophenylalanine [246].

4.5.4.2 Homonuclear recoupling

Several techniques are possible for the measurement of solid state homonuclear distances: rotational resonance [247]; radio-frequency driven dipolar recoupling (RFDR) [248]; dipolar recoupling with a windowless sequence (DRAWS) [249] and recently, sequences designed using symmetry arguments e.g. C- and R-type [250, 251] that can also be used to accomplish heteronuclear recoupling. The technique of rotational resonance [247] re-introduces the dipolar interaction between two nuclei by spinning the sample at an integral multiple of the difference in isotropic shift between them. The homonuclear dipolar coupling (and hence distances) can be obtained from the resulting lineshapes. This widely used method normally involves site-specific double ^{13}C labelling e.g. [252, 253]. For example, the rotational resonance spectra of doubly ^{13}C labelled cimetidine (7 – ^{13}C2 and ^{13}C16) has shown good agreement between the derived C2–C16 distance, and that determined crystallographically in two polymorphs (A and M1) [254]. In the case of the non-crystallisable C polymorph, a limited number of compatible structures was postulated using the C2–C16 distance derived from rotational resonance measurements

RFDR methodology has been used to measure inter-fluorine distances of up to 12 Å (+/–2 Å) [255], this application is expected to be of great utility in the solid state NMR analysis of fluorine containing pharmaceuticals. RFDR can be used to provide distance measurements between a labelled nucleus and others of the same isotope at natural abundance. For instance, such measurements were made between a singly ^{13}C labelled guest molecule and the natural abundance carbons of the host in an organic clathrate [256].

4.5.5 Application of solid state NMR to pharmaceuticals

The high degree of structural information and the sensitivity of chemical shifts to conformation and/or form makes NMR spectroscopy an attractive method of analysing solid phase pharmaceutical materials. The nulcei of a given molecule usually have different chemical shifts if they occupy sites which are crystallographically distinct e.g. polymorphs. It is a generally non-destructive technique, although high spinning speeds can cause form changes [257], and the data can be used in a quantitative fashion provided various precautions are taken [258].

Solid state NMR can be used to observe drug resonances in the presence of excipients (formulated product) [259] and is particularly important in this respect when the responses from the excipients interferes or masks those from the drug substance in vibrational spectroscopies and XRPD. Solid state NMR can thus be used to verify the integrity of polymorphic form during the passage of a drug through the development process. For instance, in the latter stage of development it is possible for form changes to occur as a direct result of the procedures used in formulation [260] or for chemical reactions [261] or interactions [262] to occur between drug substance and excipients.

It is instructive to compare the ^{13}C spectra of a simple molecule obtained in the solution and solid state. Figure 4.11 shows three such spectra of paracetamol (structure as shown in Fig. 4.11). The high resolution proton decoupled ^{13}C spectrum of pure paracetamol dissolved in DMSO-d_6 shows one resonance for each chemically distinct carbon atom – in this case six signals (Fig. 4.11A). The positions 2 and 2' are chemically equivalent due to the rapid rotation of the benzene nucleus and so have identical chemical shifts. This is not the case in the solid state (Fig. 4.11B), which shows that the carbon atoms labelled 2 and 2' are now inequivalent, suggesting that the exchange between these positions (rotation of the benzene nucleus) is slow on the NMR time scale in the solid phase. There are eight distinct carbon resonances in this spectrum, one for each carbon atom in the molecule. The spectrum in Fig. 4.11C shows the CP-MAS spectrum of a formulation blend of paracetamol. The resonances of the drug molecule can easily be seen by comparison with Fig. 4.11B, the other resonances are from the various excipients used in the formulation, in this case mainly lactose and avicel.

The influence of polymorphism on solid state NMR can be illustrated by comparing the CP-MAS spectra of forms A–C of the histamine H2 antagonist cimetidine (7), shown in Fig. 4.12 [263]. The spectra of the three polymorphs are clearly different, showing differences in the number of resonances and their chemical shifts. The relatively simple spectrum of form A indicates there is one molecule in the asymmetric unit, whereas for polymorphs B and C the greater number of resonances suggests a more complex crystalographic arrangement.

Cimetidine, 7

The prediction of the number of molecules in the unit cell from the number of resonances observed for each chemically distinct carbon environment in a simple CP-MAS NMR experiment can be of great importance in the structural refinement of X-ray data e.g. [264]. However, when multiple ^{13}C resonances are observed for a single ^{13}C environment, it is necessary to distinguish between the cases of a single polymorph having unit cell occupancy greater than one and a mixture of phases/polymorphs. Due to efficient spin diffusion in solids, each polymorph is usually characterised by a single proton T_1, so that a mixture of phases should show multiple T_1s. Proton T_1s are usually determined by observing the ^{13}C spectrum using a modified CP sequence.

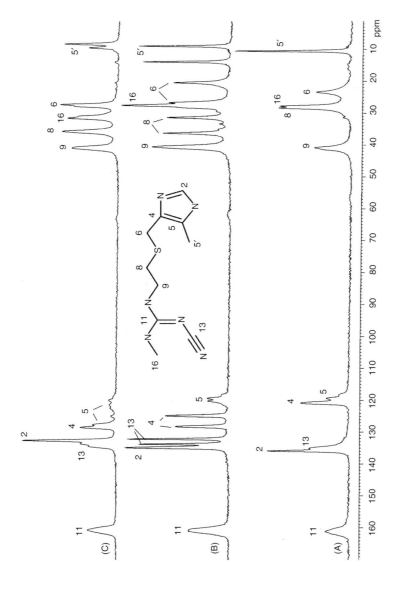

Fig. 4.11 Solid state NMR spectra of three polymorphs of Cimetidine. Spectra (A)–(C) represent forms designated (A)–(C) respectively. Spectra were acquired under standard CP-MAS conditions as described in [1] from which the resonance assignments shown are also taken. It is clear that the number of resonances observed in polymorph (B) is greater than the number of chemically distinct sites.

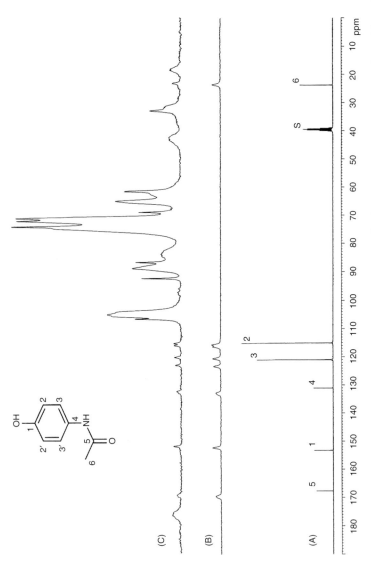

Fig. 4.12 Comparison of solution state and solid state NMR spectra of paracetamol: (A) the $^{13}C\{^1H\}$ solution state spectrum recorded on a 15 mg mL^{-1} solution of paracetamol in DMSO-d$_6$ 125 MHz, (B) the solid state spectrum obtained on pure paracetamol at 90.5 MHz, and (C) the solid state spectrum (90.5 MHz) of a crushed tablet of a paracetamol formulation. The solution state assignments are readily available from inverse correlation methods described in Section 4.2.4.1. The solid state isotropic chemical shifts are very similar to those obtained in the solution state, but note that the inequivalence of the *chemically equivalent* carbons e.g. 2, 2′ and 3, 3′ in the solid state spectrum. The drug resonances can easily be diffrerentiated from those of the excipients in spectrum (C). The resonances between 60 and 100 ppm in spectrum (C) originate from the cellulosic formulation ingredients.

The application of the resonance assignment methods discussed in Section 4.5.3 enable a direct comparison between the chemical shifs of each atom in each polymorph, allowing an insight into the structural differences between the polymorphs [265]. Structural information can also be obtained by the comparison of observed isotropic solid state chemical shifts with those calculated for particular molecular arrangements [266].

Acknowledgements

This chapter would not have been possible without the skill and insight of our coworkers, who contributed some of the (otherwise unpublished) examples in the text. These coworkers include Anthony Beck, Jamie Cotgreave, Zolani Dyosi, Unine Felix, Matthew Hallam, Richard Harris, Teyrnon Jones, David K O'Brien, Andrew Prescot, Jeff Richards and Pritpal Slaich.

References

1. Claridge, T.D.W. (1999) *High-Resolution NMR Techniques in Organic Chemistry*, Elsevier Science Ltd, Oxford.
2. Derome, A.E. (1987) *Modern NMR Techniques for Chemistry Research*, Pergamon Press, Oxford.
3. Akitt, J.W. & Mann, B.E. (2000) *NMR and Chemistry, An Introduction to Modern NMR Spectroscopy*, Stanley Thornes, Cheltenham.
4. Lindon, J.C., Nicholson, J.K., Holmes, E. & Everett, J.R. (2000) *Concepts Magn. Reson.*, **12**, 289–320.
5. Holzgrabe, U. (1999) *NMR Spectroscopy in Drug development and Analysis* (eds U. Holzgrabe, I. Wawer & B. Diehl), Wiley-VCH, Weinheim, pp. 118–133.
6. Stockman, B.J. (1998) *Prog. Nucl. Magn. Reson. Spectrosc.*, **33**, 109–151.
7. Szantay, Jr, C. & Demeter, A. (2000) *Identification and Determination of Impurities in Drugs* (ed. G. Gorog), Elsevier, Amsterdam, pp. 109–145.
8. Akita, E., Maeda, K. & Umezawa, H. (1964) *J. Antibiotics (Tokyo), Ser. A*, **17**, 200–217.
9. Akita, E., Maeda, K. & Umezawa, H. (1964) *J. Antibiotics (Tokyo) Ser. A*, **17**, 37–38.
10. Smith, R.J., Williams, D.H., Barna, J.C., *et al.* (1985) *J. Am. Chem. Soc.*, **107**, 2849–2857.
11. Neuhaus, D. & Williamson, M.P. (2000) *The Nuclear Overhauser Effect in Structural and Conformational Analysis*, Wiley-VCH, New York.
12. Spraul, M., personal communication.
13. Stewart, W.E. & Siddall, T.H. (1970) *Chem. Rev.*, **70**, 517–551.
14. Bothner-By, A.A., Stephens, R.L., Lee, J., Warren, C.D. & Jeanloz, R.W. (1984) *J. Am. Chem. Soc.*, **106**, 811–813.
15. Mansfield, P. & Chapman, B. (1986) *J. Magn. Reson.*, **66**, 573–576.
16. Hurd, R.E. (1990) *J. Magn. Reson.*, **87**, 422–428.
17. Norwood, T.J. (1994) *Chem. Rev.*, **23**, 59–66.
18. Parella, T. (1998) *Magn. Reson. Chem.*, **36**, 467–495.
19. Berger, S. (1997) *Prog. Nucl. Magn. Reson. Spectrosc.*, **30**, 137–156.
20. Keeler, J., Clowes, R.T., Davis, A.L. & Laue, E.D. (1994) *Methods Enzymol.*, **239**, 145–207.
21. Braun, S., Kalinowski, H.-O. & Berger, S. (1998) *150 and More Basic NMR Experiments: a Practical Course*, Wiley-VCH, Weinheim.

22. Wagner, R. & Berger, S. (1996) *J. Magn. Reson. Ser. A*, **123**, 119–121.
23. Stonehouse, J., Adell, P., Keeler, J. & Shaka, A.J. (1994) *J. Am. Chem. Soc.*, **116**, 6037–6038.
24. Stott, K., Stonehouse, J., Keeler, J., Hwang, T.-L. & Shaka, A.J. (1995) *J. Am. Chem. Soc.*, **117**, 4199–4200.
25. Stott, K., Keeler, J., Van, Q.N. & Shaka, A.J. (1997) *J. Magn. Reson.*, **125**, 302–324.
26. Parella, T., Sanchez-Ferrando, F. & Virgili, A. (1997) *J. Magn. Reson.*, **125**, 145–148.
27. Bauer, W., Soi, A. & Hirsch, A. (2000) *Magn. Reson. Chem.*, **38**, 500–503.
28. Piotto, M., Saudek, V. & Sklenar, V. (1992) *J. Biomol. NMR*, **2**, 661–665.
29. Ogg, R.J., Kingsley, P.B. & Taylor, J.S. (1994) *J. Magn. Reson. Ser. B*, **104**, 1–10.
30. Hwang, T.-L. & Shaka, A.J. (1995) *J. Magn. Reson. Ser. A*, **112**, 275–279.
31. Barjat, H., Morris, G.A., Smart, S., Swanson, A.G. & Williams, S.C. (1995) *J. Magn. Reson. Ser. B*, **108**, 170–172.
32. Pelta, M.D., Barjat, H., Morris, G.A., Davis, A.L. & Hammond, S.J. (1998) *Magn. Reson. Chem.*, **36**, 706–714.
33. Wu, D., Chen, A. & Johnson, Jr, C.S. (1996) *J. Magn. Reson. Ser. A*, **121**, 88–91.
34. Gozansky, E. & Gorenstein, D.G. (1996) *J. Magn. Reson. Ser. B*, **111**, 94–96.
35. Barjat, H., Morris, G.A. & Swanson, A.G. (1998) *J. Magn. Reson.*, **131**, 131–138.
36. Johnson, Jr, C.S. (1999) *Prog. Nucl. Magn. Reson. Spectrosc.*, **34**, 203–256.
37. Gounarides, J.S., Chen, A. & Shapiro, M.J. (1999) *J. Chromatogr. B: Biomed. Sci. Appl.*, **725**, 79–90.
38. Berger, S., Braun, S. & Kalinowski, H.-O. (1997) *NMR spectroscopy of the Non-Metallic Elements*, John Wiley & Sons, Chichester.
39. Gaillet, C., Lequart, C., Debeire, P. & Nuzillard, J.M. (1999) *J. Magn. Reson.*, **139**, 454–459.
40. Doddrell, D.M., Pegg, D.T. & Bendall, M.R. (1982) *J. Magn. Reson.*, **48**, 323–327.
41. Reynolds, W.F., Mclean, S. & Tay, L.-L., *et al*. (1997) *Magn. Reson. Chem.*, **35**, 455–462.
42. Willker, W., Leibfritz, D., Kerssebaum, R. & Bermel, W. (1993) *Magn. Reson. Chem.*, **31**, 287–292.
43. Zhu, G., Kong, X. & Sze, K. (1998) *J. Magn. Reson.*, **135**, 232–235.
44. Spitzer, T., Sefler, A.M. & Rutkowske, R. (2001) *Magn. Reson. Chem.*, **39**, 539–542.
45. Russell, D.J., Hadden, C.E., Martin, G.E., Gibson, A.A., Zens, A.P. & Carolan, J.J. (2000) *J. Nat. Prod.*, **2063**, 1047–1049.
46. Subramanian, R., Lam, R.R. & Webb, A.G. (1998) *J. Magn. Reson.*, **133**, 227–231.
47. Behnia, B. & Webb, A.G. (1998) *Anal. Chem.*, **70**, 5326–5331.
48. Bax, A. & Summers, M.F. (1986) *J. Am. Chem. Soc.*, **108**, 2093–2094.
49. Wagner, R. & Berger, S. (1998) *Magn. Reson. Chem.*, **36**, S44–S46.
50. Hadden, C.E., Martin, G.E. & Krishnamurthy, V.V. (1990) *J. Magn. Reson.*, **140**, 274–280.
51. Hadden, C.E., Martin, G.E. & Krishnamurthy, V.V. (2000) *Magn. Reson. Chem.*, **38**, 143–147.
52. Mallory, F.B. (1973) *J. Am. Chem. Soc.*, **95**, 7747–7752.
53. Gakh, Y.G., Gakh, A.A. & Gronenborn, A.M. (2000) *Magn. Reson. Chem.*, **38**, 551–558.
54. Everett, T.S. (1995) *ACS Monogr.*, **187**, 1037–1086.
55. Wray, V. (1983) *Annu. Rep. NMR Spectrosc.*, **14**, 1–406.
56. Russell, D.J., Hadden, C.E., Martin, G.E. & Krishnamurthy, K. (2002) *Magn. Reson. Chem.*, **40**, 207–210.
57. Quin, L.D. & Verkade, J.G. (eds) (1994) *Phosphorus-31 NMR Spectral Properties in Compound characterization and Structural Analysis*, VCH, New York.
58. Murray, M. (1994) in *Phosphorus-31 NMR Spectral Properties in Compound Characterization and Structural Analysis* (eds L.D. Quin & J.G. Verkade), VCH, New York, pp. 347–360.
59. Kuma, A. & Boykin, D.W. (1993) *J. Mol. Struct.*, **296**, 95–101.
60. Cerioni, G. & Plumitallo, A. (1993) *Magn. Reson. Chem.*, **31**, 320–321.
61. Cerioni, G., Plumitallo, A., Mocci, F., Rappoport, Z. & Rubin, M.B. (2001) *J. Chem. Soc., Perkin. Trans.*, **2**, 774–777.
62. Boykin, D.W. (1992) *Spectrosc. Lett.*, **25**, 1199–1205.

63. Boykin, D.W. (1995) *Stud. Nat. Prod. Chem.*, **17**, 549–600.
64. Boykin, D.W (ed.) (1991) *17O NMR Spectroscopy in Organic Chemistry*, CRC Press Inc., Boca Raton.
65. Crouch, R.C., Llanos, W., Mehr, K.G., Hadden, C.E., Russell, D.J. & Martin, G.E. (2001) *Magn. Reson. Chem.*, **39**, 555–558.
66. Crouch, R.C. & Martin, G.E. (1995) *J. Heterocycl. Chem.*, **32**, 1665–1669.
67. Martin, G.E., Crouch, R.C. & Andrews, C.W. (1995) *J. Heterocycl. Chem.*, **32**, 1759–1766.
68. Martin, G.E., Crouch, R.C., Sharaf, M.H.M. & Schiff, Jr, P.L. (1996) *J. Nat. Prod.*, **59**, 2–4.
69. Martin, G.E. & Crouch, R.C. (1995) *J. Heterocycl. Chem.*, **32**, 1839–1842.
70. Marek, R., Humpa, O., Dostal, J., Slavik, J. & Sklenar, V. (1999) *Magn. Reson. Chem.*, **37**, 195–202.
71. McDonnell, P.A., Gauthier, A.D. & Ferro, M.P. (1998) *Magn. Reson. Chem.*, **36**, 35–38.
72. Hadden, C.E., Bowman, P.B., Duholke, W.H., *et al.* (2000) *J. Heterocycl. Chem.*, **37**, 1623–1627.
73. Bodenhausen, G. & Ernst, R.R. (1982) *J. Am. Chem. Soc.*, **104**, 1304.
74. Martin, G.E., Hadden, C.E., Crouch, R.C. & Krishnamurthy, V.V. (1999) *Magn. Reson. Chem.*, **37**, 517–528.
75. Martin, G.E. & Hadden, C.E. (2002) *Magn. Reson. Chem.*, **38**, 251–256.
76. Martin, G.E., Hadden, C.E., Blinn, J.R., Sharaf, M.H.M., Tackie, A.N. & Schiff, Jr, P.L. (1999) *Magn. Reson. Chem.*, **37**, 1–6.
77. Farley, K.A., Bowman, P.B., Brumfield, J.C., *et al.* (1998) *Magn. Reson. Chem.*, **36**, S11–S16.
78. Witanowski, M., Stefaniak, L. & Webb, G.A. (1986) *Annu. Rep. NMR Spectrosc.*, **18**, 1–761.
79. Witanowski, M., Stefaniak, L. & Webb, G.A. (1993) *Annu. Rep. NMR Spectrosc.*, **25**, 1–480.
80. Martin, G.E. & Hadden, C.E. (2000) *J. Nat. Prod.*, **63**, 543–585.
81. Pretsch, E., Clerc, T., Seibl, J. & Simon, W. (1989) *Tables of Spectral Data for Structure Determination of Organic Compounds*.
82. Pouchert, C.J. & Behnke, J. (1993) *The Aldrich Library of 13C and 1H FT NMR Spectra*, Aldrich Chemical Company, Inc., Milwaukee.
83. Fuerst, A. & Pretsch, E. (1990) *Anal. Chim. Acta*, **229**, 17–25.
84. Buergin Schaller, R. & Pretsch, E. (1994) *Anal. Chim. Acta*, **290**, 295–302.
85. Buergin Schaller, R., Arnold, C. & Pretsch, E. (1995) *Anal. Chim. Acta*, **312**, 95–105.
86. Kalchhauser, H. & Robien, W. (1985) *J. Chem. Inf. Comput. Sci.*, **25**, 103–108.
87. Barth, A. (1993) *J. Chem. Inf. Comput. Sci.*, **33**, 52–58.
88. Thiele, H., Paape, R., Maier, W. & Grzonka, M. (1995) *GIT Fachz Lab.*, **39**, 668–670.
89. Advanced Chemical Design Inc. http://www.acdlabs.com (2002) Ref Type: Electronic Citation.
90. Chemical Concepts. http://www.chemicalconcepts.com (2002) Ref Type: Electronic Citation.
91. BioRad-Sadtler. http://www.bio-rad.com (2002) Ref Type: Electronic Citation.
92. Hearmon, R.A., Cook, G.D., Allan, P. & Chippendale, A.M. (1996) *Magn. Reson. Chem.*, **34**, 815–819.
93. Masunov, A. (2001) *J. Chem. Inf. Comput. Sci.*, **41**, 1093–1095.
94. Thomas, S., Bruehl, I., Heilmann, D. & Kleinpeter, E. (1997) *J. Chem. Inf. Comput. Sci.*, **37**, 726–730.
95. Abraham, R.J. (1999) *Prog. Nucl. Magn. Reson. Spectrosc.*, **35**, 85–152.
96. Perch. http://www.uku.fi/perch.html (2002) Ref Type: Electronic Citation.
97. Abraham, R.J. & Reid, M. (2001) *J. Chem. Soc. Perkin. Trans.*, **2**, 1195–1204.
98. Abraham, R.J., Canton, M. & Griffiths, L. (2001) *Magn. Reson. Chem.*, **39**, 421–431.
99. Aires-de-Sousa, J., Hemmer, M.C. & Gasteiger, M. (2002) *Anal. Chem.*, **74**, 80–90.
100. Radomski, J.P., van Halbeek, H. & Meyer, B. (1994) *Nat. Struct. Biol.*, **1**, 217–218.
101. Farrant, R.D., Lindon, J.C., Rahr, E. & Sweatman, B.C. (1992) *J. Pharm. Biomed. Anal.*, **10**, 141–144.
102. Spraul, M., Neidig, P., Klauck, U., *et al.* (1994) *J. Pharm. Biomed. Anal.*, **12**, 1215–1225.
103. Holmes, E., Foxall, P.J.D., Nicholson, J.K., *et al.* (1994) *Anal. Biochem.*, **220**, 284–296.
104. Holmes, E., Foxall, P.J.D., Spraul, M., Farrant, R.D., Nicholson, J.K. & Lindon, J.C. (1997) *J. Pharm. Biomed. Anal.*, **15**, 1647–1659.

105. Holmes, E., Nicholls, A.W., Lindon, J.C., *et al.* (2000) *Chem. Res. Tech.*, **13**, 471–478.
106. Lindon, J.C., Holmes, E. & Nicholson, J.K. (2001) *Prog. Nucl. Magn. Reson. Spectrosc.*, **39**, 1–40.
107. Schroder, H., Neidig, P. & Rosse, G. (2000) *Angew. Chem. Int. Ed.*, **39**, 3816–3819.
108. Steinbeck, C. (2001) *Curr. Opin. Drug Discov. Dev.*, **4**, 338–342.
109. Jaspars, M. (1999) *Nat. Prod. Rep.*, **16**, 241–248.
110. Munk, M.E. (1998) *J. Chem. Inf. Comput. Sci.*, **38**, 997–1009.
111. Schaller, R.B., Munk, M.E. & Pretsch, E. (1996) *J. Chem. Inf. Comput. Sci.*, **36**, 239–243.
112. Will, M., Fachinger, W. & Richert, J.R. (1996) *J. Chem. Inf. Comput. Sci.*, **36**, 221–227.
113. Meiler, J. & Will, M. (2001) *J. Chem. Inf. Comput. Sci.*, **41**, 1535–1546.
114. Lindel, T., Junker, J. & Koeck, M. (1999) *Eur. J. Org. Chem.*, 573–577.
115. Peng, C., Bodenhausen, G., Qiu, S., *et al.* (1998) *Magn. Reson. Chem.*, **36**, 267–268.
116. Steinbeck, C. (2001) *J. Chem. Inf. Comput. Sci.*, **41**, 1500–1507.
117. Kock, M., Junker, J. & Lindel, T. (1999) *Org. Lett.*, **1**, 2041–2044.
118. Blinov, K.A., Elyashberg, M.E., Molodtsov, S.G., Williams, A.J. & Martirosian, E.R. (2001) *Fresenius' J. Anal. Chem.*, **369**, 709–714.
119. Watanabe, N. & Niki, E. (1978) *Proc. Japan Acad. Ser. B*, **54**, 194–199.
120. Bayer, E., Albert, K., Nieder, M., Grom, E. & Keller, T. (1979) *J. Chrom.*, **186**, 497–507.
121. Buddrus, J. & Herzog, H. (1980) *Org. Magn. Reson.*, **13**, 153–155.
122. Buddrus, J., Herzog, H. & Cooper, J.W. (1981) *J. Magn. Reson.*, **43**, 453–459.
123. Bayer, E., Albert, K., Nieder, M., Grom, E., Wolff, G. & Rindlisbacher, M. (1982) *Anal. Chem.*, **54**, 1747–1750.
124. Albert, K., Nieder, M., Bayer, E. & Spraul, M. (1985) *J. Chrom.*, **346**, 17–24.
125. Laude, D.A. & Wilkins, C.L. (1987) *Anal. Chem.*, **59**, 546–551.
126. Albert, K., Kunst, M., Bayer, E., *et al.* (1989) *Anal. Chem.*, **61**, 772–775.
127. Hofmann, M., Spraul, M., Streck, R., Wilson, I.D. & Rapp, A. (1993) *LaborPraxis*, **17**, 36–41.
128. Liu, M., Mao, X.A., Ye, C., Huang, H., Nicholson, J.K. & Lindon, J.C. (1998) *J. Magn. Reson.*, **132**, 125–129.
129. Smallcombe, S.H., Patt, S.L. & Keifer, P.A. (1995) *J. Magn. Reson. Ser. A*, **117**, 295–303.
130. Parella, T., Adell, P., Sanchez-Ferrando, F. & Virgili, A. (1998) *Magn. Reson. Chem.*, **36**, 245–249.
131. Dalvit, C., Shapiro, G., Bohlen, J.M. & Parella, T. (1999) *Magn. Reson. Chem.*, **37**, 7–14.
132. Roberts, J.K. & Smith, R.J (1994) *J. Chrom. A*, **677**, 385–389.
133. Crowe, E.A., Roberts, J.K. & Smith, R.J. (1995) *Pharm. Sci.*, **1**, 103–105.
134. Sidelmann, U.G., Gavaghan, C., Carless, H.A.J., *et al.* (1995) *Anal. Chem.*, **67**, 4441–4445.
135. Sidelmann, U.G., Braumann, U., Hofmann, M., *et al.* (1997) *Anal. Chem.*, **69**, 607–612.
136. Griffiths, L. & Horton, R. (1998) *Magn. Reson. Chem.*, **36**, 104–109.
137. Wu, N., Webb, A., Peck, T.L. & Sweedler, J.V. (1995) *Anal. Chem.*, **67**, 3101–3107.
138. Behnke, B., Schlotterbeck, G., Tallarek, U., *et al.* (1996) *Anal. Chem.*, **68**, 1110–1115.
139. Lindon, J.C., Nicholson, J.K. & Wilson, I.D. (1996) in *Advances in Chromatography* (eds P.R. Brown & E. Grushka) Marcel Dekker, Inc., New York, pp. 315–382.
140. Albert, K. (1999) *J. Chromatogr. A*, **856**, 199–211.
141. Lindon, J.C., Nicholson, J.K. & Wilson, I.D. (2000) *J. Chrom. B: Biomed. Sci. Appl.*, **748**, 233–258.
142. Griffiths, L. (1995) *Anal. Chem.*, **67**, 4091–4095.
143. Peng, S.X. (2000) *Biomed. Chromatogr.*, **14**, 430–441.
144. Haw, J.F., Glass, T.E. & Dorn, H.C. (1982) *J. Magn. Reson.*, **49**, 22–31.
145. Godejohann, M., Preiss, A. & Muegge, C. (1998) *Anal. Chem.*, **70**, 590–595.
146. Godejohann, M., Preiss, A., Muegge, C. & Wuensch, G. (1997) *Anal. Chem.*, **69**, 3832–3837.
147. Mistry, N., Ismail, I.M., Smith, M.S., Nicholson, J.K. & Lindon, J.C. (1997) *J. Pharm. Biomed. Anal.*, **16**, 697–705.

148. Wilson, I.D., Griffiths, L., Lindon, J.C. & Nicholson, J.K. (2000) in *Identification and Determination of Impurities in Drugs* (ed. S. Gorog), Elsevier, Amsterdam, pp. 299–322.
149. Mccrossen, S.D., Bryant, D.K., Cook, B.R. & Richards, J.J. (1998) *J. Pharm. Biomed. Anal.*, **17**, 455–471.
150. Potts, B.C., Albizati, K.F., Johnson, M.O. & James, J.P. (1999) *Magn. Reson. Chem.*, **37**, 393–400.
151. Peng, S.X., Borah, B., Dobson, R.L., Liu, Y.D. & Pikul, S. (1999) *J. Pharm. Biomed. Anal.*, **20**, 75–89.
152. Chin, J., Fell, J.B., Jarosinski, M., Shapiro, M.J. & Wareing, J.R. (1998) *J. Org. Chem.*, **63**, 386–390.
153. Vogler, B., Klaiber, I., Roos, G., Walter, C.U., Hiller, W., Sandor, P. & Kraus, W. (1998) *J. Nat. Prod.*, **61**, 175–178.
154. Wolfender, J.L., Rodriguez, S. & Hostettmann, K. (1998) *J. Chromatogr. A*, **794**, 299–316.
155. Wolfender, J.L., Ndjoko, K. & Hostettmann, K. (1998) *Curr. Org. Chem.*, **2**, 575–596.
156. Cavin, A., Potterat, O., Wolfender, J.L., Hostettmann, K. & Dyatmyko, W. (1998) *J. Nat. Prod.*, **61**, 1497–1501.
157. Mutlib, A.E., Strupczewski, J.T. & Chesson, S.M. (1995) *Drug Metab. Dispos.*, **23**, 951–964.
158. Corcoran, O., Spraul, M., Hofmann, M., Ismail, I.M., Lindon, J.C. & Nicholson, J.K. (1997) *J. Pharm. Biomed. Anal.*, **16**, 481–489.
159. Ehlhardt, W.J., Woodland, J.M., Baughman, T.M., Vandenbranden, M. & Wrighton, S.A. (1998) *Drug Metab. Dispos.*, **26**, 42–51.
160. Nicholls, A.W., Lindon, J.C., Farrant, R.D., Shockcor, J.P., Wilson, I.D. & Nicholson, J.K. (1999) *J. Pharm. Biomed. Anal.*, **20**, 865–873.
161. Dachtler, M., Handel, H., Glaser, T., *et al.* (2000) *Magn. Reson. Chem.*, **38**, 951–956.
162. Lancelin, P. & Cleon, P. (1991) *Spectra 2000 [Deux Mille]*, **161**, 33–41.
163. Scarfe, G.B., Lindon, J.C., Nicholson, J.K., Wright, B., Clayton, E. & Wilson, I.D. (1999) *Drug Metab. Dispos.*, **27**, 1171–1178.
164. Scarfe, G.B., Clayton, E., Wilson, I.D. & Nicholson, J.K. (2000) *J. Chromatogr. B: Biomed. Sci. Appl.*, **748**, 311–319.
165. Scarfe, G.B., Wright, B., Clayton, E., *et al.* (1999) *Xenobiotica*, **29**, 77–91.
166. Farrant, R.D., Cupid, B.C., Nicholson, J.K. & Lindon, J.C. (1997) *J. Pharm. Biomed. Anal.*, **16**, 1–5.
167. Burton, K.I., Everett, J.R., Newman, M.J., Pullen, F.S., Richards, D.S. & Swanson, A.G. (1997) *J. Pharm. Biomed. Anal.*, **15**, 1903–1912.
168. Taylor, S.D., Wright, B., Clayton, E. & Wilson, I.D. (1998) *Rapid Commun. Mass Spectrom.*, **12**, 1732–1736.
169. Dear, G.J., Plumb, R.S., Sweatman, B.C., *et al.* (2000) *J. Chromatogr. B: Biomed. Sci. Appl.*, **748**, 281–293.
170. Lacey, M.E., Tan, Z.J., Webb, A.G. & Sweedler, J.V. (2001) *J. Chromatogr. A*, **922**, 139–149.
171. Albert, K., Braumann, U., Tseng, L.-H., *et al.* (1994) *Anal. Chem.*, **66**, 3042–3046.
172. Albert, K., Braumann, U., Streck, R., Spraul, M. & Ecker, R. (1995) *Fresenius' J. Anal. Chem.*, **352**, 521–528.
173. Braumann, U., Handel, H., Strohschein, S., *et al.* (1997) *J. Chromatogr. A*, **761**, 336–340.
174. Albert, K. (1997) *J. Chromatogr. A*, **785**, 65–83.
175. Wu, N., Peck, T.L., Webb, A.G., Magin, R.L. & Sweedler, J.V. (1995) *J. Am. Chem. Soc.*, **116**, 7929–7930.
176. Wu, N., Peck, T.L., Webb, A.G., Magin, R.L. & Sweedler, J.V. (1994) *Anal. Chem.*, **66**, 3849–3857.
177. Olson, D.L., Lacey, M.E. & Sweedler, J.V. (1998) *Anal. Chem.*, **70**, 645–650.
178. Olson, D.L., Lacey, M.E., Webb, A.G. & Sweedler, J.V. (1999) *Anal. Chem.*, **71**, 3070–3076.
179. Pusecker, K., Schewitz, J., Gfrorer, P., Tseng, L.H., Albert, K. & Bayer, E. (1998) *Anal. Chem.*, **70**, 3280–3285.
180. Schewitz, J., Gfrorer, P., Pusecker, K., *et al.* (1998) *Analyst*, **123**, 2835–2837.

181. Gfroerer, P., Schewitz, J., Pusecker, K., Tseng, L.H., Albert, K. & Bayer, E. (1999) *Electrophoresis*, **20**, 3–8.
182. Gfroerer, P., Tseng, L.H., Rapp, E., Albert, K. & Bayer, E. (2001) *Anal. Chem.*, **73**, 3234–3239.
183. Kautz, R.A., Lacey, M.E., Wolters, A.M., *et al.* (2001) *J. Am. Chem. Soc.*, **123**, 3159–3160.
184. Nyberg, N.T., Baumann, H. & Kenne, L. (2001) *Magn. Reson. Chem.*, **39**, 236–240.
185. Stockman, B.J. (2000) *Curr. Opin. Drug. Discov. Dev.*, **3**, 269–274.
186. Spraul, M., Hofmann, M., Ackermann, M., *et al.* (1997) *Anal. Commun.*, **34**, 339–341.
187. Lindon, J.C., Nicholson, J.K. & Everett, J.R. (1999) *Annu. Rep. NMR Spectrosc.*, **38**, 1–88.
188. Gard, D.R., Burquin, J.C. & Gard, J.K. (1992) *Anal. Chem.*, **64**, 557–561.
189. Fardella, G., Barbetti, P., Chiappini, I. & Grandolini, G. (1993) *Acta Technol. Legis. Med.*, **4**, 89–96.
190. Ozden, T., Senyuva, H. & Aksahin, I. (1999) *J. Pharm. Biomed. Anal.*, **21**, 467–472.
191. Goger, N.G., Parlatan, H.K., Basan, H., Berkkan, A. & Ozden, T. (1999) *J. Pharm. Biomed. Anal.*, **21**, 685–689.
192. Goger, N.G., Orbey, M.T., Ozden, T. & Aboul-Enein, H.Y. (1998) *Pharmazie*, **53**, 547–548.
193. Ozden, T., Ungormus, A., Tosun, A. & Ersan, S. (1997) *Spectrosc. Lett.*, **30**, 835–841.
194. Holzgrabe, U., Diehl, B.K. & Wawer, I. (1998) *J. Pharm. Biomed. Anal.*, **17**, 557–616.
195. Cookson, D.J. & Smith, B.E. (1982) *Anal. Chem.*, **54**, 2591–2593.
196. Griffiths, L. & Irving, A.M. (1998) *Analyst (Cambridge, UK)*, **123**, 1061–1068.
197. Maniara, G., Rajamoorthi, K., Rajan, S. & Stockton, G.W. (1998) *Anal. Chem.*, **70**, 4921–4928.
198. Lacroix, P.M., Dawson, B.A., Sears, R.W., Black, D.B., Cyr, T.D. & Ethier, J.C. (1998) *J. Pharm. Biomed. Anal.*, **18**, 383–402.
199. Griffiths, L. (2001) *Magn. Reson. Chem.*, **39**, 194–202.
200. Gerritz, S.W. & Sefler, A.M. (2001) *J. Comb. Chem.*, **2**, 39–41.
201. Fardella, G., Barbetti, P., Chiappini, I. & Grandolini, G. (1995) *Int. J. Pharm.*, **121**, 123–127.
202. Lankhorst, P.P., Poot, M.M. & de Lange, M.P.A. (1996) *Pharmacopeial Forum*, **22**, 2414–2422.
203. Hanna, G.M. & Lau-Cam, C.A. (2001) *Pharmazie*, **56**, 700–703.
204. Lindgren, B. & Martin, J.R. (2003) *Pharmeuropa*, **5**, 51–54.
205. Mistry, N., Ismail, I.M., Farrant, R.D., Liu, M., Nicholson, J.K. & Lindon, J.C. (1999) *J. Pharm. Biomed. Anal.*, **19**, 511–517.
206. Drew, M., Orton, E., Krolikowski, P., Salvino, J.M. & Kumar, N.V. (2000) *J. Comb. Chem.*, **2**, 8–9.
207. Corcoran, O., Lindon, J.C., Hall, R., Ismail, I.M. & Nicholson, J.K. (2001) *Analyst (Cambridge, UK)*, **126**, 2103–2106.
208. Hanna, G.M. & Evans, F.E. (2000) *J. Pharm. Biomed. Anal.*, **24**, 189–196.
209. Lacroix, P.M., Dawson, B.A., Sears, R.W. & Black, D.B. (1994) *Chirality*, **6**, 484–491.
210. Torrens, A., Castrillo, J.A., Frigola, J., Salgado, L. & Redondo, J. (1999) *Chirality*, **11**, 63–69.
211. Hanna, G.M. & Lau-Cam, C.A. (1995) *J. Pharm. Biomed. Anal.*, **13**, 1313–1319.
212. Branch, S.K. (1999) in *NMR Spectroscopy in Drug Development and Analysis* (eds U. Holzgrabe, I. Wawer & B.W.K. Diehl), Wiley-VCH, Weinheim, pp. 5–15.
213. Bauer, M., Bertario, A., Boccardi, G., Fontaine, X., Rao, R. & Verrier, D. (1998) *J. Pharm. Biomed. Anal.*, **17**, 419–425.
214. Bauer, J., Spanton, S., Henry, R., *et al.* (2001) *Pharm. Res.*, **18**, 859–866.
215. Bugay, D.E. (1993) *Pharm. Res.*, **10**, 317–327.
216. Jelinski, L.W. & Melchior, M.T. (2000) *Appl. Spectrosc. Rev.*, **35**, 25–93.
217. Fyfe, C.A. (1983) *Solid state NMR for Chemists*, CFC Press, Guelph.
218. Duer, M.J. (2002) *Solid State NMR Spectroscopy: Principles and Applications* (ed. M.J. Duer).
219. Opella, S.J. & Frey, M.H. (1979) *J. Am. Chem. Soc.*, **101**, 5854–5856.
220. Wu, X. & Zilm, K.W. (1993) *J. Magn. Reson. Ser. A*, **102**, 205–213.
221. Wu, X., Burns, S.T. & Zilm, K.W. (1994) *J. Magn. Reson. Ser. A*, **111**, 29–36.

222. Wu, X., Burns, S.T. & Zilm, K.W (2000) *J. Magn. Reson.*, **143**, 352–359.
223. Hu, J.Z., Harper, J.K., Taylor, C., Pugmire, R.J. & Grant, D.M. (2000) *J. Magn. Reson.*, **142**, 326–330.
224. Lesage, A., Steuernagel, S. & Emsley, L. (1998) *J. Am. Chem. Soc.*, **120**, 7095–7100.
225. De Vita, E. & Frydman, L. (2001) *J. Magn. Reson.*, **148**, 327–337.
226. Sakellariou, D., Lesage, A. & Emsley, L. (2001) *J. Magn. Reson.*, **151**, 40–47.
227. Lesage, A., Auger, C., Caldarelli, S. & Emsley, L. (1997) *J. Am. Chem. Soc.*, **119**, 7867–7868.
228. Vinogradov, E., Madhu, P.K. & Vega, S. (1999) *Chem. Phys. Lett.*, **314**, 443–450.
229. Lesage, A., Sakellariou, D., Steuernagel, S. & Emsley, L. (1998) *J. Am. Chem. Soc.*, **120**, 13194–13201.
230. Bielecki, A., Kolbert, A.C. & Levitt, M.H. (1998) *Chem. Phys. Lett.*, **155**, 341–346.
231. Levitt, M.H., Kolbert, A.C., Bielecki, A. & Ruben, D.J. (1993) *Solid State Nucl. Magn. Reson.*, **2**, 151–163.
232. Van Rossum, B.J., Foerster, H. & De Groot, H.J.M. (1997) *J. Magn. Reson.*, **124**, 516–519.
233. Saalwachter, K., Graf, R., Demco, D.E. & Spiess, H.W. (1999) *J. Magn. Reson.*, **139**, 287–301.
234. Lesage, A. & Emsley, L. (2001) *J. Magn. Reson.*, **148**, 449–454.
235. Saalwachter, K., Graf, R. & Spiess, H.W. (2001) *J. Magn. Reson.*, **148**, 398–418.
236. Lesage, A., Duma, L., Sakellariou, D. & Emsley, L. (2001) *J. Am. Chem. Soc.*, **123**, 5747–5752.
237. Lesage, A., Charmont, P., Steuernagel, S. & Emsley, L. (2000) *J. Am. Chem. Soc.*, **122**, 9739–9744.
238. Sakellariou, D., Lesage, A. & Emsley, L. (2001) *J. Am. Chem. Soc.*, **123**, 5604–5605.
239. Gullion, T. (1997) *Magn. Reson. Rev.*, **17**, 83–131.
240. Gullion, T. (1998) *Concepts Magn. Reson.*, **10**, 277–289.
241. Vogt, F.G., Mattingly, S.M., Gibson, J.M. & Mueller, K.T. (2000) *J. Magn. Reson.*, **147**, 26–35.
242. Arshava, B., Breslav, M., Antohi, O., *et al.* (1999) *Solid State Nucl. Magn. Reson.*, **14**, 117–136.
243. Macholl, S., Sack, I., Limbach, H.H., Pauli, J., Kelly, M. & Buntkowsky, G. (2000) *Magn. Reson. Chem.*, **38**, 596–603.
244. Gullion, T. (2000) *J. Magn. Reson.*, **146**, 220–222.
245. Sack, I. & Vega, S. (2000) *J. Magn. Reson.*, **145**, 52–61.
246. Murphy, O.J., Kovacs, F.A., Sicard, E.L. & Thompson, L.K. (2001) *Biochem.*, **40**, 1358–1366.
247. Raleigh, D.P., Levitt, M.H. & Griffin, R.G. (1988) *Chem. Phys. Lett.*, **146**, 71–76.
248. Bennett, A.E., Ok, J.H., Griffin, R.G. & Vega, S. (1992) *J. Chem. Phys.*, **96**, 8624–8627.
249. Gregory, D.M., Mitchell, D.J., Stringer, J.A., *et al.* (1995) *Chem. Phys. Lett.*, **246**, 654–663.
250. Lee, Y.K., Kurur, N.D., Helmle, M., Johannessen, O.G., Nielsen, N.C. & Levitt, M.H. (1995) *Chem. Phys. Lett.*, **242**, 304–309.
251. Zhao, X., Eden, M. & Levitt, M.H. (2001) *Chem. Phys. Lett.*, **342**, 353–361.
252. Isaac, B., Gallagher, G.J., Balazs, Y.S. & Thompson, L.K. (2002) *Biochem.*, **41**, 3025–3036.
253. McDermott, A.E., Creuzet, F., Gebhard, R., *et al.* (1994) *Biochem.*, **33**, 6129–6136.
254. Middleton, D.A., Le Duff, C.S., Peng, X., Reid, D.G. & Saunders, D. (2000) *J. Am. Chem. Soc.*, **122**, 1161–1170.
255. Lane Gilchrist, Jr, M., Monde, K., Tomita, Y., Iwashita, T., Nakanishi, K. & McDermott, A.E. (2001) *J. Magn. Reson.*, **152**, 1–6.
256. Zaborowski, E., Zimmermann, H. & Vega, S. (1999) *J. Magn. Reson.*, **136**, 47–53.
257. Zell, M.T., Padden, B.E., Grant, D.J.W., *et al.* (2000) *Tetrahedron*, **56**, 6603–6616.
258. Harris, R.K. (1985) *Analyst (London)*, **110**, 649–655.
259. Saindon, P.J., Cauchon, N.S., Sutton, P.A., Chang, C.J., Peck, G.E. & Byrn, S.R. (1993) *Pharm. Res.*, **10**, 197–203.
260. Brittain, H.G. & Fiese, E.F. (1999) *Drugs Pharm. Sci.*, **95**, 331–361.
261. Byrn, S.R., Xu, W. & Newman, A.W. (2001) *Adv. Drug Deliv. Rev.*, **48**, 115–136.
262. Lee, G.S.H., Taylor, R.C., Dawson, M., Kannangara, G.S.K. & Wilson, M.A. (2000) *Solid State Nucl. Magn. Res.*, **16**, 225–237.
263. Middleton, D.A., Le Duff, C.S., Berst, F. & Reid, D.G. (1997) *J. Pharm. Sci.*, **86**, 1400–1402.

264. Harris, R.K., Yeung, R.R., Lamont, R.B., Lancaster, R.W., Lynn, S.M. & Staniforth, S.E. (1997) *J. Chem. Soc. Perkin Trans.*, **2**, 2653–2659.
265. Apperley, D.C., Fletton, R.A., Harris, R.K., Lancaster, R.W., Tavener, S. & Threlfall, T.L. (1999) *J. Pharm. Sci.*, **88**, 1275–1280.
266. Smith, E.D.L., Hammond, R.B., Jones, M.J., *et al.* (2001) *J. Phys. Chem. B*, **105**, 5818–5826.
267. King, R.W. & Williams, K.R. (1990) *J. Chem. Ed.*, **67**, A100–A105.
268. Homans, S.W. (1989) *A Dictionary of Concepts in NMR*, OUP, Oxford.

5 Mass spectrometry in pharmaceutical analysis
Neville Haskins

5.1 Introduction

The mass spectrometer is now widely accepted as a crucial analytical tool for organic molecules in the pharmaceutical industry. Although usually treated as a spectroscopic technique, it does not rely on the interaction with electromagnetic radiation (light, infrared, etc.) for the analysis. Rather it is a microchemical technique relying on the production of characteristic ions in the gas phase, followed by the separation and acquisition of those ions. By its operation, it destroys the sample unlike other techniques, such as nuclear magnetic resonance (NMR), infrared (IR) and Raman/UV spectroscopies. Nonetheless, mass spectrometry is so sensitive that molecular weight and structural information can be provided on very small samples (attomolar (10^{-15} molar) quantities).

Mass spectrometry (MS) was introduced by J.J. Thomson and F.W. Aston in 1912 [1]. Until the 1940s it was used primarily for analysing elements and their isotopes, and gaseous species. The use for organic analysis in the petroleum industry started in 1942 [2], but its general use for organic structural analyses dates from the 1950s [3]. Over the next ten years, complex, double focusing spectrometers (e.g. AEI MS-9, Consolidated CEC 21-103) with a mass range of 500–800 Da but with sufficient resolving power to carry out accurate mass measurement were installed in larger pharmaceutical companies. In the last 50 years, a series of revolutions in the type and application of mass spectrometers, have brought the technique into every aspect of pharmaceutical research and development (Table 5.1).

Mass Spectrometry is used in a variety of ways in chemical development. It can be interfaced with a wide variety of separation techniques (such as gas liquid chromatography (GLC), high performance liquid chromatography (HPLC) or capillary electrophorosis (CE)), and give on-line analysis of components as they elute from the column. Its sensitivity allows detection, molecular-weight determination and structural elucidation of minor components eluting from a column. It is a quantitative technique and so it can be used as an assay tool, including on-line monitoring for the optimisation of yields and minimisation of impurities.

There is a variety of mass spectrometers available to the analyst. Each type has unique attributes which make it more suitable for certain operations. This review will briefly cover the major types of mass analysers suitable for applications in the Chemical Development Laboratory and their mode of operation.

Table 5.1 Notable developments in MS with application to chemical development

Year	Development	Authors
1942	First commercial instrument for organic analysis	Meyerson, 2
1953	Invention of the quadrupole mass spectrometer	Paul & Steinwedel, 4
1954	High resolution analysis of organic molecules	Beynon, 3
1955	Time of flight mass spectrometer	Wiley & McLaren, 7
1958	Direct insertion probe	Reed, 5
1958	GC/MS coupling	Gohlke, 6
1965	Chemical ionisation	Field & Munson, 8
1966	Jet separator	Ryhage, 9
1967	On-line computerised data processing systems	Bowen, 10
1974	FT-ICR/MS	Comisarow & Marshall, 11
1974	LC/MS coupling with direct introduction	Arpino, 12
1975	APCI	Carroll et al., 13
1978	Triple quadrupole	Yost & Enke, 14
1980	Fast atom bombardment ionisation	Barber et al., 15
1983	Thermospray	Blackley and Vestal, 16
1983	Commercial quadrupole ion trap mass spectrometer	Stafford et al., 17
1985	Electrospray	Yamashita & Fenn, 18
1987	MALDI	Karas et al., 19
1989	oa-ToF	Dawson & Guilhaus, 20
1994	Nanospray	Wilm & Mann, 21
1996	Quadrupole-oa ToF	Morris et al., 22
1999	Multiplexed electrospray source	de Biasi et al., 23

The second part of this review will cover typical applications and the use of various mass analysers for those applications.

5.2 Mass spectrometry

Mass spectrometry is a non-reversible process by which electrically neutral molecules are ionised to become positively or negatively charged, and subsequently separated in the gas phase according to their mass to charge ratios (m/z) where m is the mass of the ion and z is the charge. The charge z is made up of two components q and e. The component q is the number of charges and e is the electronic charge 1.6×10^{-19} coulomb. In addition to molecular weight, structural information can be obtained from molecular fragmentation to provide a fingerprint of the analyte molecule. When the charge is one, the mass is given in Daltons or atomic mass units (Da or u). The mass of the Dalton is defined as 1/12th mass of ^{12}C atom. Following mass separation, ions are measured by a detector which converts the impact of the ion into an electrical signal. The data is collected and analysed by a data system, usually to produce a bar graph mass spectrum where the abscissa is m/z and the ordinate is the relative abundance of the ions at each m/z value, the most intense being specified at 100%.

MASS SPECTROMETRY

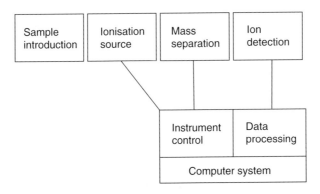

Fig. 5.1 The layout of the modern computer controlled mass spectrometer.

All mass spectrometers comprise a series of components to effect this (Fig. 5.1). The separation of ions in the gas phase requires the analyser to be held under high vacuum (better than 10^{-5} Torr) to reduce ion/gas molecule collisions. Thus a *sample introduction system* is required which maintains this. This introduction system can maintain the sample in an electrically neutral state, but often *ionisation* is effected during the sampling process. If the sample is neutral then ionisation takes place in an *ionisation source*. The mixture of ions formed is electrically accelerated into the *mass analyser*. This is the heart of the spectrometer and uses a variety of electrical and/or magnetic fields to separate ions and present them to a *detector* capable of amplifying the weak electrical signals generated by the ions to allow the response to be collected by a further analytical tool, usually a computerised *data system*. The computer also controls the operation of the mass analyser and often will control the sample introduction system as well.

5.2.1 Sample introduction systems

The sample introduction system used will depend on the type of sample and whether a chromatographic separation is required. The usual technique for separated and relatively pure samples is the direct insertion probe which carries a small amount of solid sample through an air lock into the high vacuum of the mass spectrometer. The sample would then be vapourised by applying gentle heat, and the vapours ionised in an *ionisation source*. Samples already in the gas phase, such as vapours or the eluent from gas chromatography, can be introduced directly into the ionisation source at low flows (~1 mL min^{-1}). Usually two ionisation techniques are used for these samples, *electron ionisation* (EI) or *chemical ionisation* (CI).

The traditional direct insertion probe (Fig. 5.2) comprises a wand with a holder at the tip for the sample. Modern probes can be heated by a coil around the tip, and may also be water cooled. Thus the temperature of the tip can be

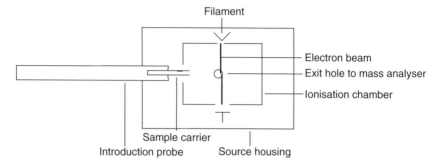

Fig. 5.2 The use of an insertion probe with an EI source.

controlled by balancing the heating rate against the cooling flow. The sample holder is normally a glass crucible. The holder can be removed and strongly heated in a Bunsen flame to burn off residue or be disposable to minimise sample carry over following the analysis. This technique is best suited for analysing relatively pure samples derived from synthesis or chromatographic purification.

An alternative probe is a wand with a desorption coil tip. This comprises a small coil of suitable wire, normally platinum. The sample is added to the coil as a solution in solvent (1–2 µL) and allowed to dry. The coil is then inserted directly into the source. Pickups can be used to allow the coil to be heated to desorb the sample from the coil. Again this technique is best suited to relatively pure samples.

Probably the most common separation systems used in the laboratory today require the sample to be in solution (e.g. HPLC, CE). The solvent may be aqueous or solvent based. However, one mL of such solution yields far too much vapour (1–2 L) to be accommodated by a mass spectrometer's vacuum system. Thus the aim of a sample introduction system for such solutions would require the sample to be ionised and the solvent to be separated from these sample ions. In addition the interface must maintain the integrity of the chromatography. The chromatographic separation must be maintained as well as allowing sufficient analyte through to generate a mass spectrum. A number of methods have been developed to do this, but the two main techniques used today are electrospray and atmospheric pressure chemical ionisation (APCI for short). These are described below under ionisation techniques.

5.2.2 Ionisation techniques

5.2.2.1 Electron ionisation

Electron ionisation (EI) (sometimes known as electron impact ionisation) is the traditional technique used from the earliest days of organic mass spectrometry. This works by passing a beam of high energy electrons through a cloud of sample vapour (Fig. 5.2). Traditionally electrons with energies of 70 eV are used, but the spectra can be optimised by altering the energy of the incident beam.

Ionisation is caused by close interaction of a high energy electron with a sample molecule which can displace an electron from the molecule. This generates a radical having a single electrical charge. Such molecules can be quite unstable, and fragment into smaller more stable portions of the molecule by elimination of neutral fragments or a radical. Reducing the energy of the incident beam (<20 eV) can create molecular ions but reduces the degree of fragmentation as a little excess energy is transferred. However, such low energy electrons are less efficient and the overall sensitivity is much reduced.

The electron ionisation process

$$M + e^{-*} \rightarrow M^{\cdot+} + 2e^{-} \quad \text{formation of the molecular ion}$$
$$M^{\cdot+} \rightarrow A^{+} + B^{\cdot} \quad \text{elimination of a radical B}$$
$$M^{\cdot+} \rightarrow C^{\cdot+} + D \quad \text{elimination of a neutral molecule D}$$

An EI spectrum comprises a mixture of ions of types A and C which may contain some molecular ion $M^{\cdot+}$ (Fig. 5.3). To complicate matters, the ions A and C may also fragment further to produce smaller ions. In these processes, the ion which fragments is known as the *precursor* or *parent* ion, whilst the smaller ions formed are known as the *product* ions. Understanding the relationship between precursor and product ions is at the heart of mass spectral interpretation and deducing the structure of the original molecule. A simple example is shown (Fig. 5.4).

EI spectra using 70 eV electrons are very reproducible and form the basis of many libraries of mass spectra available as an option with most data systems. There are some drawbacks. The technique is not suitable for involatile molecules, or compounds which are thermally unstable. Often the molecular ion is very weak or not detected. However, it is an excellent technique to analyse the eluent from Gas Chromatographs and many commercial GC/MS systems utilising EI are available.

5.2.2.2 Chemical ionisation
Chemical ionisation (CI) was introduced by Field and Munson in 1965 [8]. Like EI, the technique requires the sample to be in the gas phase. In positive ion mode the source contains a higher pressure reaction chamber in which a reagent gas is ionised by an electron beam or by an electrical discharge (Townsend discharge). Unlike EI, collisions then occur between the molecular ion and further gas molecules to form stable protonated species. Sample is introduced and sample molecules interact with these charged gas molecules. Several reactions can occur of which the most usual is proton transfer. This takes place if the sample has a higher proton affinity than the gas molecule. Loose complexes can also form between sample molecules and the reagent gas. Some fragmentation can occur, predominantly loss of small neutral molecules such as water. A commonly used reagent gas is ammonia.

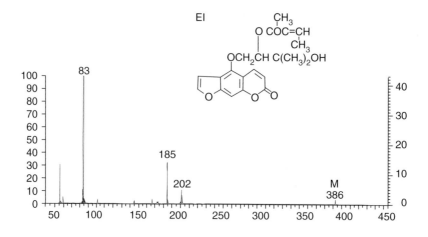

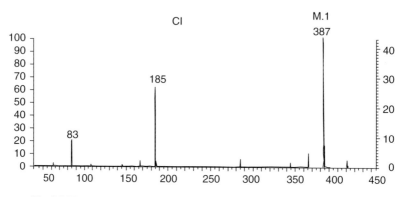

Fig. 5.3 Electron ionisation and methane chemical ionisation spectra of ostruthol.

The positive ion chemical ionisation process using ammonia as reagent gas

$NH_3 + e^-$	$\rightarrow NH_3^{\cdot+}$	Initial ionisation of reagent gas
$NH_3^{\cdot+} + NH_3$	$\rightarrow NH_4^+ + NH_2^{\cdot}$	Formation of ammonium reagent
$NH_2^{\cdot} + NH_2^{\cdot}$	$\rightarrow N_2H_4$	Loss of imine radical
$M + NH_4^+$	$\rightarrow M + H^+ + NH_3$	Ionisation of sample by proton transfer
$M + NH_4^+$	$\rightarrow M + NH_4^+$	Formation of a charged sample complex
$M + H^+$	$\rightarrow A + H^+ + C$	Possible fragmentation

Negative ions can also be formed using CI processes. The source is similar, although the polarities are reversed to expel negative rather than positive ions. Two types of reaction can occur. The high pressure of a modulating

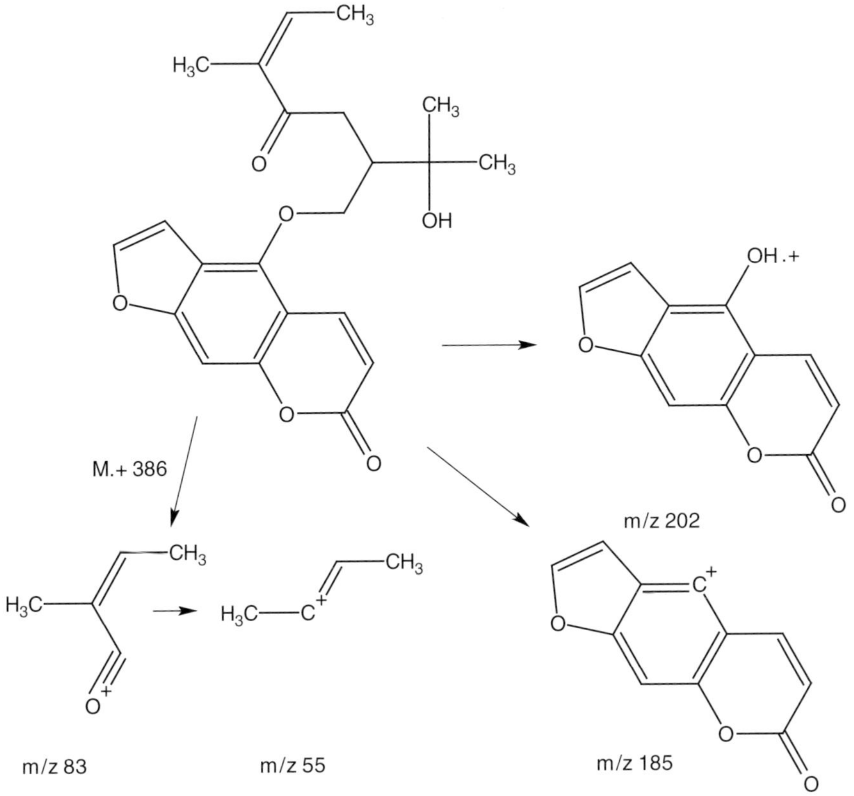

Fig. 5.4 The fragmentation of ostruthol following EI.

gas (such as nitrogen or methane) leads to multiple collisions with the electron beam yielding a population of *thermal* (i.e. low energy electrons <20 eV). These electrons can be *captured* by sample molecules. Unlike EI, in this case the molecule gains an extra electron to form a radical ion M^-. Fragmentation can occur yielding negatively charged ions.

Use of suitable reagent gases can yield negatively charged ions which can interact with the sample in analogous manner to positive ion. For instance using ammonia in negative ion mode gives a population of iminium ion NH_2^-. This can extract a proton from the sample molecule to yield a deprotonated species, e.g.

$$M + NH_2^- \rightarrow [M-H]^- + NH_3$$

Using a species such as ammonia, positive/negative switching can be used. In such cases alternative scans detect either positive or negative ions. A useful attribute of this experiment is that the difference between positively and negatively charged ions of the same major structure is two mass units.

Like EI, CI is often coupled to a gas chromatograph. As a lower energy process than EI the protonated species are often more stable giving intact molecular ions. This can be very useful when uncertainty exists about the molecular weight, especially as the protonated molecule is often accompanied by the corresponding cluster ion. Thus when using positive ion ammonia CI, the presence of two ions 17 u apart is probably due to $[M+H]^+$ and $M+NH_4^+$.

Separation is more usually carried out in the liquid, rather than gas phase. Neither EI nor CI can deal very effectively with samples in solution and a number of techniques to overcome this have arisen over the last few years. The first widely used method was thermospray [16]. However, this has been largely superseded by *atmospheric pressure chemical ionisation* (APCI). The other commonly applied technique is *electrospray*. These techniques are described in detail below.

5.2.2.3 *Electrospray*

Like APCI, electrospray acts as sample introduction interface as well as the means of producing the sample ions. In its simplest form eluent from a pumped system (syringe or HPLC, $1-10\,\mu L\,min^{-1}$) is fed along a narrow bore tube with a high voltage applied to the tip [18]. This generates a charged meniscus, as the solution emerges from the tube, and this charged meniscus is pulled towards a counter electrode, breaking into small droplets. The solvent evapourates and the droplets shrink until the density of charge is too high at which point repelling coulombic forces cause the droplet to break up into smaller droplets and charged ions. This charged stream is pulled towards a sampling cone, and passes through a two stage separator to reduce the solvent vapour pressure and allow the charged species through a pin hole into the mass spectrometer. This process can be aided by applying drying gas or some heat to aid evaporation of the droplet. A characteristic of such ionisation is that when a molecule has many sites capable of protonation multiple charged species can be formed. This has revolutionised the analysis of peptides and proteins, because the addition of multiple protons has brought the ions observed into the mass range of most mass spectrometers, which separate ions by their mass/charge ratio. Obviously a larger charge yields a lower m/z. The ionisation sites are rarely saturated giving an *envelope* of ions with different charges (Fig. 5.4). Simple mathematical manipulation can recreate the *real* molecular mass. A second characteristic of the electrospray source is that the signal intensity is due to the concentration of analyte in the solution and not due to the total amount of sample injected into the source, unlike most other sources. Thus the low flow rates into electrospray sources means that the effluent from traditional HPLC columns at flows of $1\,mL\,min^{-1}$ require dividing to reduce the flow into the mass spectrometer to a suitable level. This allows the bulk of the sample to be collected or analysed by an alternative technique.

5.2.2.4 Nanospray
The fact that the signal intensity is concentration dependent using electrospray means that very low flow rates can be used. This is used to advantage by miniaturised systems such as *nanospray* [21]. In this source a drawn glass capillary with a very narrow tip is used. The tip is plated to provide a high voltage and generate the electrospray. Usually the capillary is filled with ~1 µL of solution but the flow is so low that this will generate a signal for one hour or more. This allows off-line analysis by collecting a fraction of the eluent from an HPLC column and injecting a small volume into the capillary. The long time taken to consume the sample (opposed to the short time it takes a chromatographic peak to elute) allows further mass spectrometric experiments to be carried out. This kind of technology can also be used with very low flow separation systems such as capillary electrophoresis, capillary HPLC and chip style separators.

5.2.2.5 Atmospheric pressure chemical ionisation (APCI)
Atmospheric pressure chemical ionisation (APCI) uses gas phase ion/molecule reactions in an analogous manner to CI. APCI is applied to polar molecules up to 2000 Da. The analyte in solution (0.1–1 mL min^{-1}) from a syringe, or as HPLC eluent, is introduced into a nebuliser which creates a mist of droplets by a stream of nitrogen gas. The mist passes through a heated tube which vapourises the solvent. Ionization takes place using a corona discharge, i.e. a high voltage developed at the tip of a sharp needle electrode. This forms a primary ionisation of gas molecules (e.g. N_2^{+}) whose ions react with mobile phase to form CI reagents such as NH_4^{+}. These reagents react in turn with sample molecules.

All the atmospheric pressure sources use ions from HPLC eluent which are already formed in solution. For example, an acidic mobile phase in HPLC will generate protonated sample. Such ions will pass into the mass analyser and will be separated in the usual manner.

The main issue with atmospheric pressure sources is efficient sampling of the molecular ion species which must be transferred from normal atmospheric pressure to a vacuum chamber. This is effected by passing the ion stream through a pin hole and a series of focusing lenses into a two-stage pumped separator which removes the bulk of the neutral molecules. However, the nature of this transfer process makes the inlet prone to contamination, and the use of inorganic and non-volatile buffers in the HPLC separation will rapidly block these systems. Volatile buffers, such as ammonium formate, ammonium acetate and small concentrations of trifluoroacetic acid are best suited for LC/MS procedures.

5.2.3 Analysers

5.2.3.1 Sectors
The first mass spectrograph was based on a combined electrostatic and magnetic sector [2]. This was developed over the years so that modern instruments give

high field (allowing masses up to 3000 Da to pass), high scan rate (<1 s) and good resolving power (20 000). Improvements in resolving power for time of flight (ToF) and ion cyclotron resonance mass spectrometers since the 1990s has made sector instruments redundant and many pharmaceutical companies will have only one or two instruments where they used to have ten or twenty. Most texts on mass spectrometry – still discuss sector instruments but their use for chemical development has been largely superseded by ToF instruments.

5.2.3.2 Quadrupole analysers

The quadrupole was developed by Paul [4] and consists of four parallel circular, or ideally hyperbolic, rods (Fig. 5.5). Applying an oscillating radiofrequency (RF) field to rods carrying a fixed potential (DC) so that alternate rods have opposite polarity causes an ion entering the field to resonate with the field. Ions are injected into the mass filter along the z-axis and are acted upon by the variable field in the x,y plane. The complex equations concerning the motion of the ions can be solved so that for a given set of voltages (U, V), RF (w) and rod separation, a stable condition arises such that ions may be transmitted. Usually the voltages are adjusted so that only ions of the same nominal mass

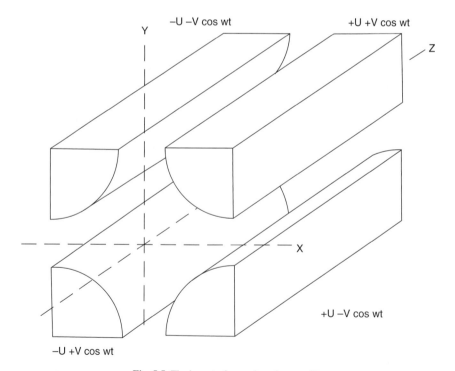

Fig. 5.5 The layout of a quadrupole mass filter.

are transferred. Other ions will undergo more violent oscillation and will be lost either by hitting the rods or escaping between them. By scanning the amplitude of V, U while keeping the ratio of V/U constant, ions of sequential mass are focused, thus generating a spectrum. Applying RF voltage only transmits all ions. Correct spacing and absolutely parallel rods are essential for the filter to work.

Quadrupole spectrometers are commonly available because they are relatively cheap to manufacture, and are compact. These spectrometers having a relatively short flight path (~30 cm), do not need such a high vacuum as sector spectrometers, and found early application for GC/MS and CI, and latterly for LC/MS. The transmission of ions is directly proportional to the applied voltage making the plot of m/z against V linear, unlike sectors where m/z is proportional to B^2, the magnetic field, generating an exponential curve. This makes for easier mass calibration. The instruments are robust, and portable systems for field monitoring have been developed.

The principal drawback for quadrupoles are their relatively poor resolving power and limited mass range (<3000 Da). However, in combination with electrospray ionisation they are capable of generating spectra of multiply charged ions and hence can be used for monitoring much heavier molecules (20 000 Da). They have a good dynamic range and are suitable for quantitative analyses. This has been a major application area for quadrupoles since the 1970s.

5.2.3.3 *Quadrupole ion traps*

More commonly called an ion trap, it has been developed from the original instrument [4] by Stafford *et al.* [17]. It is closely related to the quadrupole and can be described by considering one pair of rods forming the ring electrode, whilst the second pair have been formed into the circular end caps (Fig. 5.6). Like the quadrupole the ions are introduced and fall into a combined RF and electric field. With a broad range of RFs applied to the ring electrode, ions remain trapped within the cell. By applying a variable RF voltage to the end caps the stability of the ions is affected and by scanning these frequencies, ions of a certain m/z ratio are sequentially forced out through the end cap, to be detected by the normal dynode/electron multiplier.

Sophisticated control of the RFs applied can allow selected ions to be destabilised. This allows the trap to be used for multiple fragmentation experiments. In the first stage all ions are expelled apart from a selected precursor ion (P1). This is stored in a trap containing some helium, and fragmentation is induced. The ions derived from P1 can then be scanned out. However, a particular fragment ion can be selected as a further precursor ion P2 by scanning out the rest and the process repeated. Pragmatically five or six stages of fragmentation are about the limit, before the signal strength becomes too low to be detected.

Like their cousins the quadrupoles, ion traps are limited in mass range. They also suffer from charging effects. If too many ions are injected into the trap then coulombic repulsion destabilises the ion trajectories and mass resolution is poor. Most traps

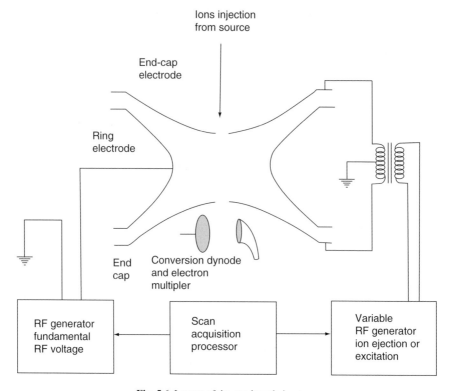

Fig. 5.6 Layout of the quadrupole ion trap.

have electronics to prevent this but it does mean that traps have a somewhat limited dynamic range. Their major advantages are the ability to interface with a wide range of separation systems and their small size making very compact systems.

5.2.3.4 Time of flight analysers
A recent review covers the renaissance of the ToF mass spectrometer [24]. The ToF spectrometer operates on the principle that ions of different mass accelerated to a uniform kinetic energy have different velocities, and hence different ToF over a given distance.

$$E_{kinetic} = qV = mv^2/2$$

where $q = ze$, $V =$ accelerating voltage, $m =$ mass and $v =$ velocity.

$$t = d/v$$

where $t =$ time of flight, $d =$ distance travelled and $v =$ velocity.

Solving for v between the two equations yields

$$m/q = 2Vt^2/d^2$$

i.e. m/z is proportional to t^2.

The ions need to be pulsed into the flight tube and accelerated with equal kinetic energy. This was difficult with older instruments but the development of MALDI [19], a pulsed ionisation technique, lead to the resurgence of ToF spectrometers in the 1980s for use with such sources. Because the mass range is theoretically limitless very high mass ions can be analysed making the technique suitable for the study of large proteins and polymers.

There were many problems with these early ToFs when using continuous ionisation sources, such as electrospray, APCI, EI. These concerned the gated sampling, deceleration and re-acceleration of the ion beam in a pulsed manner. Ions exiting continuous sources have a spread of kinetic energy and despite the electronics the beam in the ToF also had energy spread giving poor resolving power and limited application. This problem was overcome by allowing the focused ion beam from the source to pass the entrance slit to the ToF orthogonally, then pulsing a packet of ions at right angles into the ToF analyser (Fig. 5.7). Although the ions have momentum in their original direction they had no velocity in the orthogonal direction. Orthogonal velocity is independent of the axial beam, so acceleration orthogonally is not affected by residual energy spread in the primary beam. There remains a small drift due to the residual momentum from the primary beam but this is compensated by offsetting the detector. Development of fast detectors and clocks, improved focusing within the ToF and stable electronics has meant a substantial improvement in resolving power up to 20000 giving the ToF the ability to provide accurate mass measurement working with conventional continuous sources.

A major benefit of the ToF is its use with hyphenated separation systems. A problem with scanning instruments is that as the scan takes place the concentration of ions in the source will change. This is especially true with systems connected to chromatographs where the concentration can change rapidly over several orders of magnitude as peak elutes. This distorts the spectra obtained, with spectra from the leading edge being *enriched* with low mass ions, whilst those from the trailing edge have enhanced high mass ions (assuming conventional high to low mass scans) (Fig. 5.8). The ToF samples a complete packet of all ions instantaneously and so the distortion does not occur although the concentration changes (Fig. 5.9). Spectra are more representative of the relative concentrations of ions formed in the source.

5.2.3.5 Ion cyclotron resonance mass spectrometers
More commonly called FT (Fourier Transform)-ICRMS or even (wrongly) FT-MS, due to the high level computing deconvolution required to process

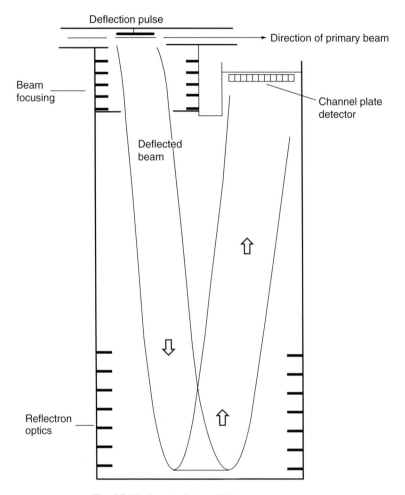

Fig. 5.7 The layout of the oa-ToF mass spectrometer.

the signal. Ion beams entering a magnetic field are deflected and this is the principal used in the first magnetic sector spectrometers. If an ion enters an intense magnetic field the deflection is so great that it becomes trapped into a circular orbit. This is the principle of the Penning or ion cyclotron trap (Fig. 5.10). The equations for this motion are:

$$\text{Centripetal force } F1 = vBq$$

where v = velocity, q = charge (ez) and B = magnetic field strength.

$$\text{Centrifugal force } F2 = mv^2/r$$

where m = mass, v = velocity and r = radius of the trajectory.

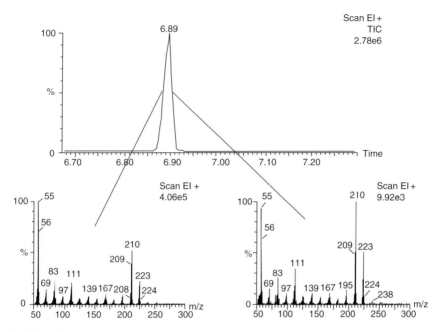

Fig. 5.8 Difference in spectra obtained across a chromatographic peak using a scanning spectrometer.

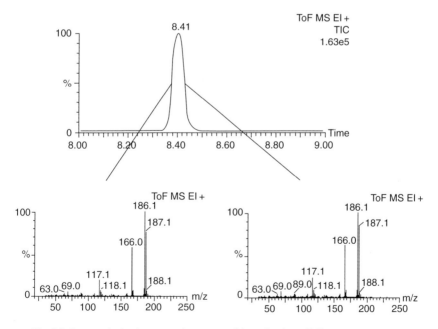

Fig. 5.9 Spectra obtained across a chromatographic peak using a ToF mass spectrometer.

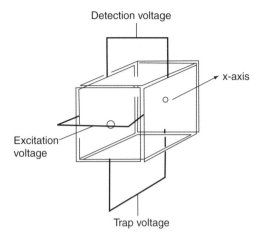

Fig. 5.10 The layout of the ICR mass spectrometer.

The ion enters a stable trajectory when $F1$ is balanced by $F2$, i.e. $qvB=mv^2/r$, with a frequency $\nu=v/2\pi r$. The angular velocity $\omega=2\pi\nu=v/r=Bq/m$. Thus the frequency and angular velocity depend on the ratio q/m and are independent of the velocity. However, the radius of the trajectory is dependent on the velocity and if the radius becomes too great then the ion is lost from the trap.

Practically the ion cyclotron resonance mass spectrometer comprises a box in an intense magnetic field as found in the centre of a superconducting magnet similar to those used by NMR spectrometers. The ion beam is gated into the trap along the x-axis. Currently (in 2002) field strengths of 3.0–9.4 Tesla are used. Stable trajectories are generated by applying cyclotron frequencies of 1.65 MHz (28 u ions) to 11.5 kHz (4000 u). The detection system uses FT analysis [11]. The ions are simultaneously excited by applying a pulse (1 μs) covering the whole frequency range. This increases the radius of the trajectories. As the ions relax coming close to the detection plates energy is detected as a complex pulse. This signal is then resolved through FT into a plot of intensity vs frequency, which itself is translated into a plot of intensity vs m/z. This requires a lot of computer processing and this is probably the major limitation on the availability of these instruments until recently.

The development of stable frequency generators have enabled very high resolving power to be claimed for this type of spectrometer. However accurate mass measurement usually requires the addition of an internal standard. This is caused by the problem of coulombic repulsion between ions in the orbit; increasing the number of ions within the trap causes slight deviations in the orbital trajectory which is reflected as a small mass shift.

In an analogous manner to quadrupole ion traps, adjusting the power and frequency of the applied RF voltage allows the expulsion of selected ions from the trap. This allows the selection of a precursor ion which can then fragment. The product ions can then be measured.

5.2.3.6 Hybrid instruments

The mass spectrometer is itself a separation tool. As described above quadrupole ion traps and FT-ICR/MS can both be used to select a particular ionic species and then study its fragmentation. However, other instruments can be designed which incorporate two or more mass analysers. Probably the two most commonly encountered are the triple quadrupole and the quadrupole-ToF spectrometer.

Triple quadrupole The triple quadrupole comprises two conventional quadrupole analysers separated by a third quadrupole acting as an ion focusing device and collision cell. A variety of scan combinations can be set to allow experiments to probe the structure of compounds, especially in combination with chromatographic separation. These modes are summarised in Table 5.2, and described below.

Product ion scan. Q1 is set to pass the selected precursor ion. Fragmentation occurs in the collision cell *Q2* and the emerging ions are separated by scanning Q3.

Precursor ion scan. Q1 is scanning normally, but ions are fragmented in Q2. Q3 is set to pass a selected product ion. The effect is that only compounds giving rise to that product ion will be detected in the spectrum from Q1. For example, if one wished to see all dimethyl amines giving rise to m/z 58, then Q3 would be set to pass m/z 58 only. The spectrum from Q1 would show only the protonated molecules due to the presence of dimethyl amines.

Constant neutral loss scan. Q1 and Q3 are both scanning, but the voltage applied to Q3 is offset by an amount representing loss of a neutral molecule. For example, if one scanned Q3 with an offset of 44 u (CO_2) then the scan from Q1 would show carboxylic acids present in the sample.

Fixed scans both Q1 and Q3. Q1 is set to pass a particular precursor ion whilst Q3 passes a significant fragment ion. This scan is a very sensitive means to detect a particular compound and can be used for assay work. For

Table 5.2 The various scan modes available using a triple quadrupole mass spectrometer

Q1	Q2	Effect of the scan
Scan	RF only	Conventional mass spectrum
RF only	Scan	Conventional mass spectrum
Fixed V	Scan	Select precursor ion to generate a product ion spectrum
Scan	Fixed V	Select product ion to generate a precursor ion spectrum
Scan	Scan ΔM	Generate spectra of compounds with the same neutral loss ΔM
Fixed V1	Fixed V2	Monitors a compound having a precursor ion M1 giving a product M2

example, setting Q1 to pass m/z 90 and Q3 to pass m/z 72 would form the basis of a sensitive detector for dimethylethanolamine. It would monitor the loss of neutral water from the protonated molecule.

Quadrupole-ToF The quadrupole-ToF is similar in layout to the triple quadrupole but Q3 is replaced by an orthogonal ToF mass spectrometer. Effectively two scan modes are available.
  Conventional spectrum. Q1 is set to RF only to pass all ions. The ToF scans as normal to produce a mass spectrum.
  Product ion scan. Q1 is set to select a precursor ion. This fragments in the Q2 collision cell. The product ions are sampled and separated by the ToF. The reason for this instrument is that the high resolving power of the ToF is not compromised. Accurate masses can be obtained for the precursor ions and also for the product ions. This makes the quadrupole-ToF a very powerful tool for structural elucidation and confirmation.

5.2.4 Ion detection systems

Apart from the FT-ICR/MS most spectrometers use a device which converts the ion beam into a usable signal by destruction of the ions. A number of means have been developed to do this but virtually all modern instruments use a conversion dynode and an electron multiplier to accomplish this. Faster signal processing required by ToF spectrometers, use array detectors.

5.2.4.1 Conversion dynode and electron multiplier
The ion beam leaving the mass analyser is deflected onto a plate. Impacting on the plate causes emission of secondary particles which include positive or negative ions, electrons and neutrals. These secondary particles are repelled into the mouth of an electron multiplier. These generally are of two types. In the first, called the continuous dynode type, the cathode is formed as a layer of semi-conducting material on the inner surface of a horn. Particles impacting this surface release a shower of electrons which impact further down the horn releasing more electrons. Thus a cascade of electrons builds up which generates a measurable current across the multiplier. The multiplying factor for a new multiplier can be as high as 10^7, but this performance falls off over two to three years. Array detectors are similar with many parallel cylinders. Again the electron cascade is generated by cascade through the channels which are coated with semi-conductor material.

  Miniaturisation of the array detector results in microchannelplate detectors (MCP). These can be quite large having many channels. They exhibit a fast response time (nanoseconds) and are used in oa-ToF instruments. However, they are easily saturated and also exhibit a dead time as the channels recover from a signal. This reduces the dynamic range to about 10^4. Careful choice of sampling rates and selection of lower intensity signals resolves these issues.

5.2.4.2 Conversion dynode and photon multipliers

This type of detector is made up of two components, a phosphorescent screen and a photomultiplier. After ion impact with the conversion dynode the secondary electrons given off are accelerated towards the screen and are converted into photons. The screen is earthed to prevent a build-up of charge on its surface. The photons enter a photomultiplier behind the screen and cascade to generate a detectable current. The advantage of this detector is its long lifetime compared to electron multipliers but the overall dynamic range is slightly reduced (10^5).

5.2.5 Data acquisition and processing

These days it is almost inconceivable to consider mass spectrometers without their attached computer. The first data acquisition systems were described in the late 1960s [10]. These early systems acquired the signal from the multiplier, processed this through an analogue/digital converter (ADC), and converted the time/intensity output to a mass/intensity plot. Within a short time the use of a comparable digital/analogue converter (DAC) allowed some control of the spectrometer functions enabling the setting of scan range, selected ion monitoring and switching valves.

Today, the use of sophisticated and large capacity computers with faster and more wide ranging ADCs and DACs allows complete control of all the mass spectrometers' functions, acquisition of data in several forms, and powerful post acquisition processing to automatically search the enormous data files for relevant information. A high degree of automation is possible including control of any hyphenated separation technique. The mass spectrometer operator can load the samples and then process the information at a comfortable work station away from the noisy laboratory as the instrument carries out the analyses. The scope of the computer's abilities will be exemplified by the examples given as applications.

5.3 Strategies for structural elucidation

At the start, the organic mass spectrometer was primarily a tool for analysing unknown compounds and providing clues to their structures [3]. It has its strengths and weaknesses like all spectroscopic techniques but it should be an important tool in conjunction with NMR and optical methods to generate structures. Unfortunately many analysts regard the mass spectrometer simply as a means of providing molecular weight information only, perhaps because of a perceived *difficulty* in interpreting mass spectra. This is often because of the misapprehension that mass spectra follow similar rigid rules to other spectroscopic techniques which do allow almost complete interpretation of the spectra. They do to some extent but the ions undergo gas phase chemical reactions to provide the spectral information, and we all know that chemistry is not always predictable.

Structural elucidation is required in all areas of the drug development process. In chemical development it is needed to identify impurities which may arise in the final drug substance from the synthesis, especially when route optimisation is being studied. It is especially important to identify unknowns if the chemical substance gives rise to unforeseen problems in toxicology due to minor impurities. Structural elucidation is similarly required to identify degradation products formed during stability testing of both drug substance and formulated products (pharmaceutical development). These are often due to oxidation, hydrolysis or interaction with formulation excipients. Thus addition of oxygen (+16 Da) or H_2O might be expected. Drug metabolism is a special case of degradation when the products formed may be due to enzymatic processes, such as addition of oxygen, or glucuronic acid or sulphate. The products may be quite different from those found by chemical degradation. When dealing with natural products obtained using biotechnology, the identification of unknown compounds may be essential to monitor the stability of the fermentation broths.

Mass spectrometry is ideally suited to this type of work and in some cases can be used on-line to give real time analyses.

5.3.1 Determination of molecular weight

This sounds so simple and often is the only piece of information required from a mass spectrum. However, the process is full of pitfalls to catch the unwary. In order to examine intact molecules, the amount of energy acquired during the ionisation process has to be kept well below that needed to break up the molecule. EI is a high energy process and often does not yield molecular ions. Softer ionisation techniques are needed but these involve other reagents, and reactions between samples and other compounds present may lead to ambiguous results. For example, CI may generate an unstable cluster ion which then decomposes. An example is the loss of water from an ammonia cluster which means that the ion observed is apparently at the correct nominal molecular weight.

$$M \text{ (m u)} + NH_4^+ \rightarrow (M + NH_4^+) \text{ (m + 18 u)} \rightarrow [(M - H_2O) + NH_4^+] \text{ (m u)} + H_2O$$

The differences in potential energy between the various species formed during CI can also yield odd effects. The protonated molecule formed by ammonia chemical ionisation of acetyl salicylic acid is so energetic that it decomposes losing water, in some sources, whilst the ammoniated cluster ion is stable. Thus the spectrum shows ions at m/z 163 ($M + H^+ - H_2O$) and 198 ($M + NH_4^+$), but no ion at the expected mass m/z 181 ($M + H^+$). Similar effects can be seen using APCI and thermospray, both *hot* sources where thermal energy can be absorbed to generate thermal decomposition of the intact molecule. Electrospray is a low energy process and usually yields a protonated molecular species in positive ion mode provided the analyte has a higher proton affinity than the reagent. Alternatively it may yield a deprotonated molecule in negative ion mode. However, because it is a lower energy process, fewer types of molecule will ionise using electrospray than by APCI.

It is important to be aware of the possibility of these reactions taking place, but in general, use of APCI with alternate scanning in positive and negative ion modes will generate suitable ions for molecular weight determination. This technique is used on open access mass spectrometer systems now often available to chemists to monitor reactions and products as a molecular weight check [25]. The presence of cluster ions and especially ions showing clustering with sodium ($M+Na^+$) ($m+23$ u) and potassium ($M+K^+$) ($m+39$ u) should be noted as open access systems are often contaminated with traces of these ions because they are used to monitor relatively crude mixtures. The use of APCI is likely to produce ions from more types of molecules than electrospray.

Salts are also observed. The presence of *preformed ions in solution*, i.e. cations and anions, perhaps generated in acidic HPLC effluent, will give rise to the cation or anion being directly observed without further manipulation. The analysis of an extract of *Coptis japonica* L. showed a protonated molecular ion due to canadine but the molecular cation is due to berberine, a naturally occurring quaternary ammonium salt (Fig. 5.11).

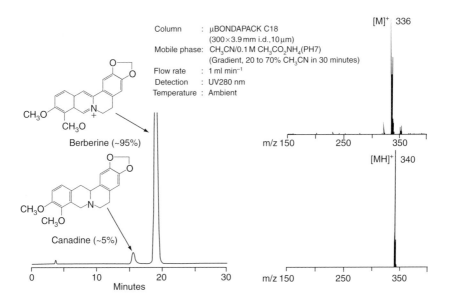

Fig. 5.11 Analysis of an extract of *Coptis japonica* showing protonated molecule for canadine and the cation for berberine.

The use of the mass spectrometer as a molecular weight detector has additional applications. The mass spectrometer can be used to monitor the effluent from a preparative column. When material with the required mass is detected a signal can be sent to a divert valve to allow collection of a fraction containing the desired material. When the ion intensity drops below a threshold the effluent is rediverted to waste. This procedure can be repeated following successive injections, thus providing an automated means of preparing an isolated fraction.

Attempts have also been made to use the mass spectrometer to monitor a synthesis. As the ions due to the reagents fall in intensity and those due to the product rise, the mass spectrometer can monitor these. When the optimum product/reagent ratio is reached the computer can quench the reaction or initiate a further stage. Although a desirable use the application is very difficult as the concentration of reagents and products in the raw reaction mixture is far too high for the mass spectrometer. An intermediate dilution step is required but this dilution, maybe $1:10^6$, has to be carried out quantitatively.

A major use of electrospray/nanospray is to obtain spectra of large molecules and determine their molecular weight. Probably the most used option is the determination of the mass of peptides and proteins. This can be accomplished because the prevalence of well separated amine and amide moieties in the structures of peptides allows the addition of more than one proton, generating multiply charged ions. The number of added protons can vary between individual molecules generating a spectrum containing many protonated molecular ions having varying numbers of protons (Fig. 5.10). Thus the apparent mass of any single peak is given by $(M+n)/n$ where n is the number of added protons and M is the peptide true mass. So for adjacent ions where the difference in the degree of protonation is 1, then the next low field peak is $(M+(n+1))/n+1$. If the mass of the higher ion is m_1 and the lower ion is m_2 then by solving the equations $n+1 = m_2/m_1 - m_2$. Knowing the value of n one calculates the true molecular weight. Most data systems carry out this deconvolution using software yielding a *spectrum* showing an apparent molecular ion (Fig. 5.12).

Nanospray requires very small quantities of material. Spectra of peptides can be generated on attomolar amounts, such as found from spots extracted from a two-dimensional plate analysis. Bio-informatics programs have been written which can take the peptide molecular weights obtained and match these with databases to provide an *identification* of the protein visible on the original 2D plate. This technique is used to support proteomics research, especially when seeking to identify any proteins which have dramatically changed in concentration following treatment with a drug.

5.3.2 *Collisionally induced decomposition and MS^n*

The generation of an ion related to the intact molecule is useful, but in itself tells us little about the structure. Many isomeric forms can exist for any formula

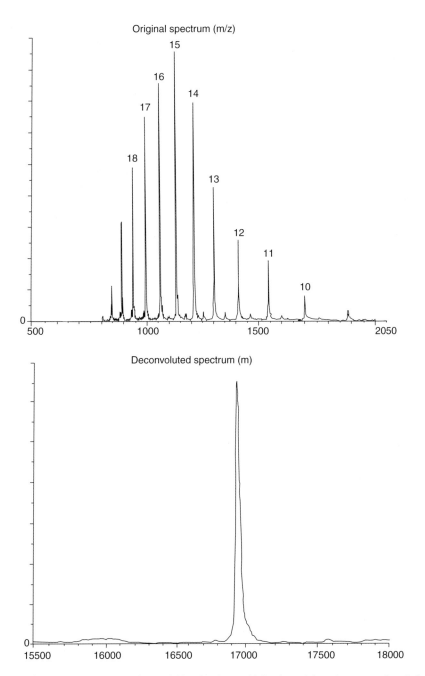

Fig. 5.12 The mass spectrum of myoglobin showing multiply charged ions (upper trace) and the deconvoluted spectrum showing the *molecular weight* (lower trace).

and so far all we know is that we have material of the appropriate molecular weight in our mixture. In order to study the structure of this ion we need to look at the various components making up the ion by fragmenting the molecule. This is accomplished by using a collision gas in a collision cell. The most useful technique for deducing the structure is to initially separate the precursor ion using a preliminary mass spectral separation in a multi-stage instrument (ion trap, FT-ICR/MS, triple quadrupole or quadrupole-ToF), but progress can be made by using the high pressure in an APCI or electrospray source to induce fragmentation by applying energy as a high voltage. This will produce fragment ions but unless the target precursor ion is the dominant ion in the spectrum at that point, the product ion spectrum will be contaminated with ions from other species present. It is also difficult to control the amount of fragmentation. Different product ion spectra can be generated by applying thermal energy in a thermospray source (Fig. 5.13) [26]. At low temperatures (110°C) the spectrum of I is dominated by the protonated molecular ion (m/z 285 u). As the temperature increases the fragment ion II (m/z 204) grows in relative abundance. At higher temperatures (>300°C) intermolecular reactions with acetate in the eluent lead to the formation of higher molecular weight species such as III.

Pre-separation of a precursor ion ensures that the product ions observed arise from the required precursor. The ion trap and ICR/MS use an input of RF voltage to generate fragmentation and often this gives predominantly one or two product ions due to a tight energy absorption. More traditional CID, using a separate collision cell, generates far more product ions and this yields more structural information. However, using a collision cell can also yield further stages of

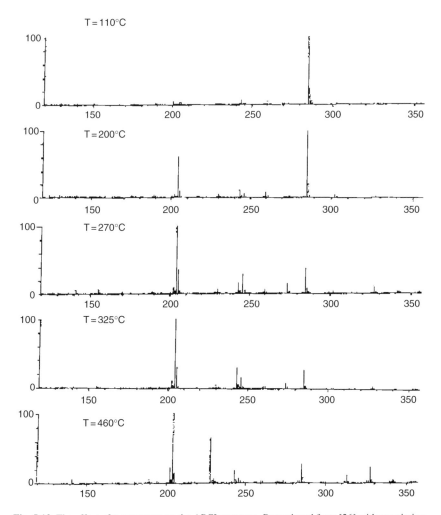

Fig. 5.13 The effect of temperature on the APCI spectrum. Reproduced from [26] with permission.

decomposition of the product ions to yield smaller products. The use of traps has the advantage that a product ion can be selected as the next precursor ion and a cascade of fragmentation can be followed. This is not so with traditional collision cell instruments but the energy used to decompose the ion can be adjusted to vary the degree of fragmentation. In practice this means that a higher energy will yield smaller product ions derived from more extensive fragmentation.

This effect is illustrated by considering the spectra (Fig. 5.14) of SB-243213 (IV) [27]. The primary spectrum obtained using a quadrupole-ToF instrument selecting the protonated molecular ion (m/z 429) with Q1 shows the ion m/z 429

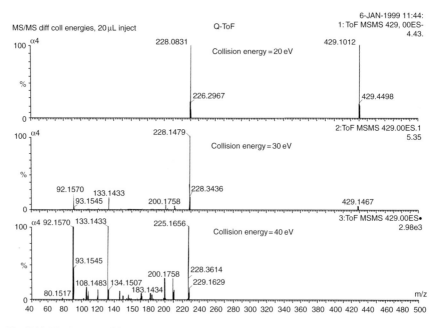

Fig. 5.14 The increase of fragmentation with increasing collision energy. Reproduced from [27] with permission.

and a fragment ion m/z 228 using a low collision energy of 20 eV. Increasing the collision energy to 30 eV almost completely fragments the protonated molecule, and generates some additional fragment ions m/z 200, 133 and 92. Increasing to 40 eV removes m/z 429 entirely and m/z 228 is no longer the base peak.

IV

A similar effect can be generated by manipulating the cone voltage in the atmospheric pressure source. Using a triple quadrupole instrument fitted with an identical source to the quadrupole-ToF, but scanning a normal spectrum, spectra were obtained for SB-243213 at various cone voltages (Fig. 5.15).

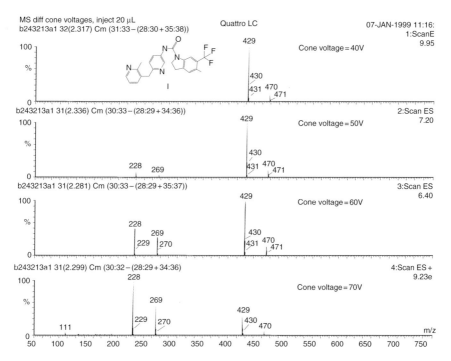

Fig. 5.15 The increase in fragmentation with increasing cone voltage. Reproduced from [27] with permission.

Below a cone voltage of 50V, the spectra are dominated by the protonated molecular ion. An adduct ion formed from interaction with acetonitrile used in the LC eluent is also seen (m/z 470). Increasing the voltage to 60V generates the fragment ions 228 and 269, whilst at 70V the fragment ion m/z 228 dominates the spectrum. These spectra additionally illustrate the problem of cone voltage fragmentation showing the presence of contaminating ions derived from interaction between the analyte and the eluent. Using multi-stage mass spectrometry allows selection of the precursor ion producing product ion spectra with little or no contamination.

5.3.3 Accurate mass measurement

All mass spectrometers can be tuned to yield accurate masses but generally unless pure analytes can be guaranteed it is better to use an instrument with good resolving power, such as a traditional electrostatic/magnetic, oa-ToF or FT-ICR/MS. With the relative cheapness of oa-ToFs and their ease of use, these are probably the modern instrument of choice for the average laboratory to acquire accurate masses. There is no reason why structural analyses should not be carried out at high resolution yielding accurate masses not only on the molecular species but also on fragment ions.

Unlike trapped ion spectrometers or sectors, oa-ToFs also have the advantage that they scan rapidly. This makes them ideal for carrying out accurate mass measurement with on-line separation systems such as GCs or HPLC [28]. A number of commercial instruments have been launched to do this. Accurate mass measurement requires good calibration and the stability of ToFs is such that an accurate mass can be obtained without adding calibrant. However, an electrospray system has been developed which allows stream switching between the LC effluent and a suitable calibration solution [29]. This has the advantage that the calibration sample need not be compatible with the analyte as it is being analysed independently.

Using MS/MS in a quadrupole-ToF instrument, the sample can be used to provide self calibration. The collision energy in the collision cell may be adjusted to ensure sufficient precursor ion is observed to allow it to be used as a lock mass (i.e. the calibration is adjusted to fit this ion of known accurate mass) other ions can then be mass measured. Often the mass of a low mass ion is also known, and improved accuracy can be gained by using that as a second calibration point and extrapolating between the two. Not only does this give an accurate mass for all the product ions in the spectrum but two other useful attributes arise. Firstly, the *difference* in mass between two ions gives the accurate mass of a lost neutral, such as 27.9949 (CO), aiding structural analysis. Secondly, because the accurate mass of the precursor ion is known, and hence potential formulae for that ion, the choice of atoms for any proposed product ion formulae are limited. Often within a 10 ppm window only one formula will fit, thus precision need not be as great as regular scans. This is useful because the collision cell will affect the dynamics of any ions passing through, yielding small deviations in mass measurement.

5.4 Structural confirmation

Mass spectra are rich in information about the structure of the analyte molecule but the total interpretation of the spectrum is usually too difficult to accomplish. However, the spectrum itself is often a unique fingerprint for the molecule and comparison of spectra from authentic samples against unknowns often gives sufficient information to provide an identification. In conjunction with a robust chromatographic separation yielding the correct retention time, this is sufficient to confirm the presence of the known substance.

Structural confirmation is essential for chemical development analyses to assure that the proposed compound is correct. For example if an impurity peak in a UV monitored trace increases it is essential to confirm that it is due to the proposed compound and not a co-eluting impurity. An important application is the monitoring of process indicative impurities in patent litigation cases. Extraction of suspect formulations and analysis of the extract is a powerful approach to confirm the presence of significant compounds often at ppb levels. Many other problems require the use of structural confirmation. A related strategy to this is the measurement of peak purity. Although UV is a commonly used detector for HPLC, co-eluting compounds of similar structure will be *lost*. Mass spectrometry can give assurance, by comparing the spectrum of authentic material against the spectrum obtained for an eluting peak, that no other material is present under the *peak*.

5.4.1 Library searches

Most manufacturers offer spectral libraries with their data system. These are usually based on the NIST compilation of EI spectra of many thousands of compounds [30]. This is useful for matching spectra from GC/EI/MS systems but not for LC/MS which do not give EI spectra. There are no official libraries of electrospray or APCI spectra due to the vagaries of these spectra dependent on the instrument used, the exact parameters of the source and the applied voltages in the transfer region. However, the library software can be used to compile one's own library and this can help to confirm identity. MS/MS data obtained from multi-sector instruments also require careful monitoring of the collision cell pressure and applied voltage, but provided these are controlled, reproducible MS/MS spectra can be obtained. These can then be compared with spectra obtained from authenticated materials. If an eluting peak from a chromatograph is monitored, then *background subtraction* is required. This is accomplished using subtraction software provided by the manufacturer and usually entails summing the spectra across the peak and then subtracting summed spectra from immediately before and after the peak.

Library programs also contain subtraction routines which allow the raw spectrum to be processed by subtracting the spectrum of an authentic standard from the test spectrum. The remaining *spectrum* can then be subjected to further library processing to identify the extra component.

When dealing with a peak eluting from a chromatograph, spectra can be compared across the peak. If the spectrum from the leading edge of the peak differs from that obtained in the tail it is probable that more than one component is present. However care must be taken. Firstly, the background spectrum must be eliminated. The background spectrum is due to ionised eluent, buffer, column bleed, etc. Once this is done then one should also confirm that *second* components are not due to reactions in the source between the analyte and any reagent, especially if the concentration is high. Also one should monitor the profile for major ions ensuring that none are saturating the detector system (usually denoted by a flat top to the mass chromatogram peak). If peaks saturate then the relative abundances of minor ions will be distorted. If all these tests

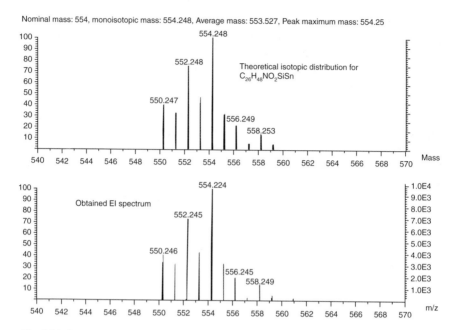

Fig. 5.16 Comparison of the theoretical and experimental isotopic distribution for $C_{26}H_{48}NO_2SiSn$.

are done and the spectrum of a second component still persists then it is likely due to a co-eluting component.

Most elements occur naturally as a mixture of isotopes, differing from one another by the number of neutrons present in the nucleus. Natural carbon comprises a mixture of mainly ^{12}C and ^{13}C (98.9 and 1.1% respectively) with a trace of the radioactive isotope ^{14}C. Chlorine has isotopes ^{35}Cl (75.77%) and ^{37}Cl (24.23%). Thus any mass spectrum will demonstrate a number of molecular ions due to the isotopomers present. Most data systems have programs which allow the input of a molecular formula which generates a theoretical isotopic distribution. This can then be compared with the actual spectrum obtained (Fig. 5.16). It may be necessary to add or subtract a proton from the inputted formula, hydrogen contains 0.015% deuterium.

5.4.2 Using MS^n techniques

The detection of significant components from GC/MS or LC/MS analyses is often helped by constructing *mass chromatograms*. The normal chromatogram is the so-called *total ion chromatogram* (TIC) which is obtained by summing the intensities of all the ions in each sequential mass spectrum and plotting these. The result approximates to a UV chromatogram. However, in a busy chromatogram it is often difficult to accurately assign each peak. This can be aided by drawing the chromatogram due to a significant ion or ions in the target analyte. Such traces plotting the intensity of selected ions are called *mass chromatograms* and are available with all mass spectrometers. Such traces are an aid to identify the possible presence of a compound but selecting few ions to represent a component may not be sufficient, especially when analysing mixtures containing isomeric compounds, and identification of a possible target peak should be confirmed by extracting the full spectrum from the TIC, using the mass chromatogram to isolate the required scans.

The above approach is not a particularly sensitive technique for confirming the presence of a particular component in a mixture and MS/MS approaches give increased specificity of detection. By selecting a suitable precursor ion at the expected retention time a product ion spectrum is obtained. This can be compared with authentic material to confirm the presence of the targeted analyte. Although this is less sensitive than a full spectrum the reduction in noise is so great that the overall limit of detection is improved due to the improved signal/noise ratio. The most sensitive technique is to select the precursor ion in MS1 and then monitor the product ion in MS2. This eliminates any chemical noise almost completely and often allows the limit of detection to be lowered to ppt (parts per 10^{12}) levels.

5.4.3 Process monitoring

A very specific application for structural confirmation and detection is to monitor the appearance of product and/or disappearance of reactants during a chemical process. Mass spectrometers are being used to monitor the headspace over fermentation broths to indicate the completion of particular phases of the process. This is relatively easy due to the gas sampling that goes on. However, monitoring liquid phase reactions is not so easy due to the presence of involatile materials in suspension and the very high concentrations of reactants. Untreated, this would rapidly destroy any atmospheric pressure source. Sampling has to include a substantial dilution phase, perhaps $1:10^6$, and filtering of any insoluble material. Some attempts are being made to do this but to date there are no commercial systems available. Off-line monitoring is available and although not so immediate has been used to monitor reactions successfully.

5.5 Quantitation

The development of chromatography mass spectrometry rapidly led to the use of these systems for quantitation. Mass spectrometers are specific detectors and the substantial reduction in noise allows sensitive detection and quantitation of analytes even in complex matrices such as biofluids or formulated tablets. A good review of *Quantitative Mass Spectrometry* was published in 1978, covering methodology and application [31].

The development of quantitative assays for specific compounds in formulated products is a useful tool in pharmaceutical development. For example during stability studies tracking the increasing concentration of a degradant over time to establish the rate of degradation may be required. A second example is the measurement of a specific *impurity* namely penicillin in active substance isolated from a fermentation broth which was treated with penicillin during work up. However, the main use of quantitative assays remains drug metabolism studies, in which measuring the concentration of active substance in biofluids (blood, plasma, urine) is essential to allow the determination of the pharmacokinetics of the compound after dosage. A technique used in this context is *stable isotope dilution analysis* which improves accuracy and precision.

5.5.1 Development of an assay

A typical analysis will follow a series of stages, each of which may introduce errors. These are as follows:

1. Take a known weight/volume of sample.
2. Extract the sample to recover as much of the analyte as possible with suitable solvent.
3. Dry the extract to remove excess solvent and then dilute to a known volume.
4. Take an aliquot of this solution for injection into a GC/MS, LC/MS, etc.
5. Carry out the mass spectral analysis and calculate the results.

Usually stage 1 requires the use of either a pipette to measure a known volume or weighing an aliquot. When dealing with sticky plasma or blood it can be difficult to ensure an accurate volume is taken. In addition the sample should be homogeneous which may require mixing or macerating before sampling. One commonly used technique to overcome the sampling difficulties is to add a known amount of an internal standard which is a substance with different physical properties, usually mass, but identical chemical properties so that it is extracted to the same extent as the analyte.

Stage 2, the extraction process, has to be reproducible from aliquot to aliquot. Thus no matter what concentration of analyte is present in the sample the degree of extraction should remain the same. If a high degree of recovery is obtained (>80%) then this is not usually a problem but for low recoveries of material reproducibility can be a major problem. In these cases the use of an internal standard is essential.

Drying the extract also has its drawbacks. The material of the vessels used is important. The glassware should be inert and not absorb the extract. An example involved the analysis of contaminated groundwater. When extracted immediately after sampling, a wide range of significant components were observed, but at low concentration. Water stored for 48 h in glass containers when extracted appeared to contain far fewer components; presumably the rest had either degraded or were adsorbed onto the glass bottle. It is best, when drying down, not to take the extract to complete dryness.

The extract should then be redissolved in a known volume of the injection solvent. Again there is a potential error here. If the analyte is in trace of the extraction solvent, say 0.1 mL, then what volume would be given by adding 1.0 mL of injection of the solvent. It may be 1.0, or 1.1 mL. Again the use of internal standards removes the necessity to measure the final volume because the ratio of analyte to standard remains constant, no matter what the final volume of the aliquot may be.

Most of these problems can be alleviated by adding an internal standard to the initial sample. The errors which arise from the addition of internal standard involve making up the original internal standard (weighing and diluting) and adding an aliquot to the sample. Provided care is taken in the withdrawal of the aliquot using validated dispensing pipettors these errors can be kept small.

5.5.2 The calibration process

Most quantitative analyses depend on measuring a response against a known value of analyte. In mass spectrometric terms this entails measuring the intensity of the signal obtained for the analyte against a known concentration of it. This is called the *calibration curve* although normally the curve is a straight line. In simple terms a series of samples of the analyte are prepared at varying concentrations covering the expected sample concentrations. These are then injected, and selected ion monitoring used to record the intensity of the area under the mass chromatogram peak.

Some assays not requiring high precision may not require addition of an internal standard. In these cases the analyte is extracted, and a known proportion of the extract injected. The amount of analyte in the aliquot can be estimated by comparison with the calibration curve. Such assays can be used for simple procedures requiring identification of analyte above a certain concentration where absolute concentration measurement is not required. Most assays will require internal standard to be added.

When using an internal standard the same amount of the standard is added to each analyte solution. Selected ion recording is used to measure the peak areas for both standard and analyte, and the ratio of these areas is plotted against the concentration of the analyte. Statistical procedures are used to calculate the equation of the regression line and the coefficient of variation due to the sampling errors.

To analyse *real* samples the same amount of standard is added and the ratio of peak areas determined. Simple application of the regression equation will yield the concentration of analyte in the sample.

In order to improve the accuracy and precision of the measurement it is important that the known samples of analyte used for the calibration curve are made up in the same matrix as the real samples, i.e. blank plasma, urine, etc. For quality control a series of samples undergoing analysis should also contain some standards of known concentration (~10%) and also some *blind* samples where the concentration is known but not by the operative carrying out the assay.

5.5.3 Use of stable isotopically labelled substance

Assays requiring an internal calibration depend on the near identical behaviour of the calibration compound and the analyte. This is easily achieved by using an analogue of the analyte which has some protons in its structure replaced by deuterons. Provided that the deuterons are in such positions that they are stable to the analytical procedure and not back exchanged for proton, and do not affect the physical properties of the molecule (for instance changing the proton affinity) then the substitution of three or more protons will shift the deuterium labelled response clear from the isotopic peaks of the analyte due to natural ^{13}C and ^{2}H contributions. Such compounds act as excellent internal standards. Suitable labelling also enables the deuterium labelled compound to act as a carrier to improve recovery of the analyte at low levels (the so called stable isotope carrier effect [32]).

5.5.4 Use of double labelling

The ability of the mass spectrometer to distinguish between various entities differing only in the degree of isotopic labelling allows drug bioavailability studies to be carried out with single dosage regimes. In these studies the total amount of drug which reaches the circulation after administration by various routes is determined (usually comparison between an intravenous bolus and oral administration of a formulated product). A second isotopically labelled

Va $R_1 = H$, $R_2 = H$
Vb $R_1 = {}^2H$, $R_2 = H$
Vc $R_1 = {}^2H$, $R_2 = {}^2H$

compound is used as an internal standard for both analogues. An example was a bioavailability study for SC-27166 (V) [33].

In this work SC-27166 (Va) was administered orally and d_4-SC-27166 (Vb) intravenously to two baboons. D_6-SC-27166 (Vc) was used as the internal standard to quantify the concentration of Va and Vb in plasma and urine. Comparison of the area under the plasma concentration vs time curve showed that oral bioavailability was approximately 76%.

5.6 Using the data system

5.6.1 Automation

Automation of mass spectrometric analyses is now well established. Using the computer to both control the inlet system, especially HPLC or gas chromatographs, and the mass spectrometer allows a series of samples to be loaded into an autosampler connected to the chromatograph and analysed sequentially. In early applications [34] the chromatograph controller tended to drive the mass spectrometer by simply initiating data collection scans. The need to process many samples in pharmacokinetic studies led to more sophisticated interaction between very large autosamplers (up to 1000 samples) and fast LC/MS systems taking perhaps two or three minutes to complete an LC separation which effectively only separated endogenous substances (lipids, small proteins, etc.) from the analyte.

The need for such high throughput in chemical development did not arise for a while. The advent of combinatorial chemistry technology and the processing of hundreds of samples a day, needed a significant increase in throughput. Applying fast chromatographic separations aided turnover, but parallel analysis offered an immediate improvement by linking up to eight chromatographs to one mass spectrometer [23] (Fig. 5.17). Using electrospray, sources switching between eight sprays takes place every 0.2s so that each spray is sampled every 1.6s (Fig. 5.18). Separation of the sprays is generated by a rotating shield because the electrosprays are required to spray continuously to obtain stability. Samples are analysed from 96 well plates, and the data system has a virtual representation of this. Making a mouse-click over a virtual well brings up the data for that sample. This data is pre-processed by the computer which compares the expected result with the actual obtained, highlighting problem samples.

Automation has also allowed the use of open access mass spectrometry for chemical analysis. Initially developed for single analysis of chemical samples with no prior separation, the system simply requires sample details to be logged onto a computer. A choice of positive and/or negative ion APCI (or electrospray depending on the source available) is allowed. More recently LC/MS is available with perhaps a choice of two separation systems. The user is then directed to place the sample in a particular location in an autosampler tray.

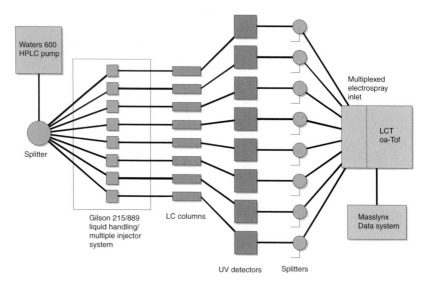

Fig. 5.17 Multiple inlet electrospray source. The rotating shield steps across the sprays allowing a single spray to be sample at any time. Diagram by permission of Micromass UK.

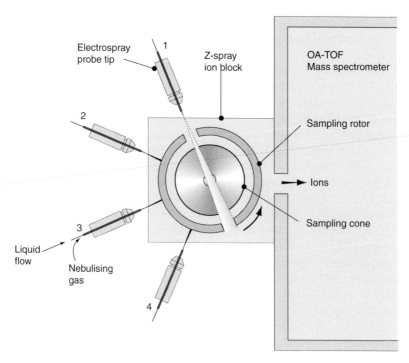

Fig. 5.18 The multiple sample interface (this version for 4 sprays) *MUX*™. Figure supplied by Micromass (UK) Ltd.

A click on OK initiates the analysis. After data acquisition the computer scans the chromatogram and produces summed spectra for any chromatographic peaks above a set threshold. These data can then be transmitted electronically to the user's personal work station. Most manufacturers offer packages to carry this out.

5.6.2 Data processing

Automated data processing is gradually being introduced but currently is not robust. Programs have been developed to predict the fragmentation likely to be generated from an input structure [35]. However, no programs have been fully developed to allow the interpretation of an unknown spectrum with structure prediction.

Automation has been introduced into the analysis of peptides and proteins, especially in the study of proteomes. Extracts of proteins from cells are separated using two dimensional gel techniques, usually a chromatographic and an electrophoretic separation. Protein *spots* are identified and cut out. After tryptic digestion the mixture of peptide fragments is analysed by mass spectrometry, usually either by MALDI-ToF or by nanoelectrospray analysis. The pattern of masses obtained from the peptides is then compared with a database showing the patterns obtained from many proteins. Matches are compared and identifications are proposed. Several manufacturers offer complete systems which carry out this analysis automatically.

5.6.3 Data mining

The old miners panning for gold were hard pushed to separate the small nuggets from the vastly greater amounts of mud, pebbles and rock. So it is with isolating significant data from large data sets, such as files of LC/MS data. The sensitivity of the mass spectrometer has allowed the detection of components within complex mixtures at very low levels. These components, which may have significance, would often not appear as discrete peaks in the chromatographic trace above the background noise. Extracts from complex matrices such as formulated product, bio-fluids or waste streams would contain many non-relevant impurity peaks. There is a need to develop computerised systems which can process a file of data and pull out the interesting spectra whilst ignoring the noise. Such systems are being developed and the application of statistical procedures such as principal components analysis enabling small differences between a sample, and a control to be highlighted appears to hold considerable promise. Other systems using a simpler forward search procedure are available for metabolite identification [36]. The application of these programs is at an early stage, but doubtless will be a major part of the future of mass spectrometry and its application in chemical development.

References

1. Thomson, J.J. (1913) *Rays of Positive Electricity and their Application to Chemical Analysis*, Longmans Green, London.
2. Meyerson, S. (1984) *Retrospective Lectures*, 32nd ASMS Conference, San Antonio, 64.
3. Beynon, J.H. (1954) *Nature*, **174**, 735.
4. Paul, W. & Steinwedel, H.S. (1953) *Z. Naturforsch.*, **8a**, 448.
5. Reed, R.I. (1958) *J. Chem. Soc.*, 3432.
6. Gohlke, R.S. (1958) *Anal. Chem.*, **34**, 1332.
7. Wiley, W.L. & McLaren, I.H. (1955) *Rev. Sci. Instrum.*, **16**, 1150.
8. Field, F.H. & Munson, M.S.B. (1965) *J. Am. Chem. Soc.*, **87**, 3289.
9. Ryhage, R. (1966) *Arkiv. Kemi.*, **26**, 305.
10. Bowen, H.C., et al. (1967) *J. Scient. Instrum.*, **44**, 343.
11. Comisarov, M.B. & Marshall, A.G. (1974) *Chem. Phys. Lett.*, **25**, 282.
12. Arpino, P.J., Baldwin, M.A. & McLafferty, F.W. (1974) *Biomed. Mass Spectrom.*, **1**, 80.
13. Carroll, D.I., et al. (1975) *Anal. Chem.*, **47**, 2369.
14. Yost, R.A. & Enke, C.G. (1978) *J. Am. Chem. Soc.*, **100**, 2274.
15. Barber, M., et al. (1982) *J. Chem. Soc. Chem. Comm.*, **16**, 936.
16. Blackley, C.R. & Vestal, M.L. (1983) *Anal. Chem.*, **55**, 2280.
17. Stafford, G.C., et al. (1984) *Int. J. Mass Spectrom. Ion Proc.*, **60**, 85.
18. Yamashita, M. & Fenn, J.B. (1984) *J. Chem. Phys.*, **80**, 4451.
19. Karas, M., et al. (1987) *Int. J. Mass Spectrom. Ion Proc.*, **78**, 53.
20. Dawson, J.H.J. & Guilhaus, M. (1989) *Rapid Comm. Mass Spectrom.*, **3**, 155.
21. Wilm, M. & Mann, M. (1994) Proc. 42nd ASMS Conference, Chicago, 774.
22. Morris, H.R., et al. (1996) *Rapid Comm. Mass Spectrom.*, **10**, 889.
23. de Biasi, V., et al. (1999) *Rapid Comm. Mass Spectrom.*, **13**, 1165.
24. Guilhaus, M., Selby, D. & Mlynski, V. (2000) *Mass Spec. Reviews*, **19**, 65.
25. Pullen, F.S., et al. (1995) *J. Am. Soc. Mass Spectrom.*, **6**, 394.
26. Haskins, N.J. & New, A. (1988) *Biomed. Environ. Mass Spectrom.*, **17**, 169.
27. Wolff, J.-C., et al. (1999) *Rapid Comm. Mass Spectrom.*, **13**, 1797.
28. Cottee, F., et al. (2000) *Eur. J. Mass Spectrom.*, **6**, 219.
29. Eckers, C., et al. (2000) *Anal. Chem.*, **72**, 3683.
30. NBS, EPA/NIH (1983) Mass Spectral Data Base, National Bureau of Standards, Faithersburg MO.
31. Millard, B.J. (1978) *Quantitative Mass Spectrometry*, Heyden & Son, London.
32. Haskins, N.J., et al. (1978) *Biomed. Mass Spectrom.*, **5**, 423.
33. Haskins, N.J., et al. (1979) Stable Isotopes, *Proceedings of the Third International Conferences* (eds E.R. Klein & P.D. Klein), Academic Press Inc., New York, pp. 485–491.
34. Haskins, N.J., et al. (1978) *Quantitative Mass Spectrometry in Life Sciences II* (eds A.P. de Leenheer, R.R. Roncucci & C. van Petegham), Elsevier, Amsterdam, pp. 287–293.
35. Mistrik, R., et al. (1999) 47th ASMS Conference, Dallas, ThPA 007.
36. Metabolynx (2000) Micromass UK, Ltd.

6 Vibrational spectroscopy in pharmaceutical analysis

Clare L. Anderton

6.1 Introduction

Vibrational spectroscopy provides a means of obtaining information on the molecular structure of a compound and its local environment in the gaseous, liquid and solid states. The vibration of molecules can be studied by both infrared (IR) and Raman spectroscopies. Although these techniques are related to each other, the two types of spectra are not exact duplicates and each has inherent advantages and disadvantages. IR and Raman spectroscopies are in many ways complementary techniques, primarily since the different origins of the spectra and selection rules mean that vibrational modes which give rise to strong bands in one spectrum will have weak bands in the other, and vice versa. In addition, the practicalities of the techniques mean that each lends itself to specific applications.

This chapter gives a brief overview of IR and Raman spectroscopy, with examples from a number of application areas, demonstrating the usefulness of both techniques in the pharmaceutical laboratory. In general, these examples will be grouped by application, rather than by technique, since the two should ideally be used together, in conjunction with other analytical techniques, in a problem-driven manner.

6.1.1 *Molecular motion*

The energy of a molecule consists of translational, electronic, rotational and vibrational energy. Electronic transitions are associated with the ultraviolet and visible regions of the electromagnetic spectrum, and rotational transitions in the microwave region. Molecular vibrations are observed within the mid-IR region of the spectrum (400–4000 cm^{-1}).

The frequency of a particular molecular vibrational mode is dependent on its structure. The approximate frequency of vibration will be determined by the bonding between atoms and so, for organic molecules, is characteristic of the functional groups e.g. the C=O vibration is observed in the region 1900–1550 cm^{-1}. The exact frequency will vary depending on the precise nature of the bonding and so will be affected by the neighbouring functional groups, hydrogen bonding, isomerism and steric effects, e.g. amides, ketones and aldehydes may be distinguished by the exact position of the carbonyl stretching frequency.

So, the position of the bands in a vibrational spectrum gives information both on the functional groups present and the nature of the bonding within that functional group. Tables of the vibrational frequencies associated with functional groups exist that allow assignment of observed bands [1–6]. Libraries of reference spectra are also available [7–9].

6.1.2 Infrared spectroscopy

Infrared absorption offers a direct means of investigating molecular vibration. In the spectrometer, the molecule is irradiated with the full range of IR frequencies but can only absorb energy at the frequencies that match the vibrational frequency of the molecule. However, not all vibrational modes will absorb as this is governed by the selection rules which state that in order to absorb IR radiation, a molecular vibration must cause a change in the dipole moment of the molecule. The intensity of the absorption band is proportional to the square of the change in dipole moment. Thus, vibrations involving polar bonds such as O–H, C=O and N–H will, in general, be strong in the IR spectrum.

6.1.2.1 Instrumentation for infrared spectroscopy

Most modern IR spectrometers are Fourier transform (FT) instruments based on a Michelson interferometer. FT instruments offer significant benefits in analysis time, throughput and wavenumber reproducibility over grating spectrometers. FT theory and instrument design is beyond the scope of this chapter and is covered in depth in the literature [10].

6.1.2.2 Sample preparation for infrared spectroscopy

The advent of FT spectrometers offers a variety of sampling methodologies based on transmission or reflectance [11]. The classical method of preparing a sample suitable for IR transmission measurements is an alkali halide disk. A small quantity of the analyte is mixed and ground with KBr (or KCl) typically in a 1:100 ratio by weight, and compressed with ca. 10 000 psi pressure in a die. This gives a glass pellet, in which the analyte is dispersed. A significant disadvantage of this technique is that the high pressure may cause alteration of the sample through solid-state transformation or halide exchange, and water absorption may also be a problem.

A less aggressive sample preparation method is a mull in which the sample is dispersed in a liquid matrix, most commonly used is the mineral oil Nujol but other liquids may also be used (e.g. fluorolube and hexachlorobutadiene). The sample is ground and mixed with the oil (Nujol) and then *sandwiched* between two IR transparent (metal halide) plates. Although the sample is less likely to be altered using this technique, the oil has strong absorbance bands which obscure some regions of the spectrum (by using alternatives to Nujol different regions will be obscured). This method is most suited to qualitative identification of the sample.

Diffuse reflectance or DRIFTS (diffuse reflectance infrared Fourier-transform spectroscopy) allows the sample to be analysed neat, or diluted in a non-absorbing matrix (e.g. KCl or KBr at 1–5% w/w analyte). DRIFTS also may be used to obtain the spectrum of a solute in a volatile solvent by evaporating the solution onto KBr. When the IR radiation interacts with the powdered sample it will be absorbed, reflected and diffracted. The radiation which has been diffusely reflected contains vibrational information on the molecule. This technique allows non-destructive testing of neat materials and is suited to quantitative analysis, although care must be taken to ensure that a consistent particle size is used.

Attenuated total reflectance (ATR) may also be used to collect IR spectra. The sample is placed in intimate contact with an IR transmitting crystal which has a high refractive index (e.g. Ge or ZnSe). For solid samples, pressure from a top plate is used to give good contact between the sample and crystal. The IR beam is directed through the crystal and penetrates the sample, obtaining spectral information on the surface of the sample. Since the IR beam penetration is dependent on its wavelength with longer wavelengths penetrating more deeply (effectively increasing the pathlength), there will be some perturbation of the spectrum with bands at low wavenumber being more intense than those at high wavenumbers. An advantage of this technique is that little sample preparation is required.

FT-IR spectrometers may be interfaced with an optical microscope allowing analysis of very small sample quantities and simultaneous chemical and physical investigation of the sample [12]. The spectrum may be collected from an area as small as 10×10 microns, the effective diffraction limit of the IR radiation. These instruments may be used to collect the spectrum in reflectance, transmission or ATR modes; the choice will be determined by the sample thickness and transmission properties.

Mid-IR spectroscopy is generally unsuitable for non-invasive analysis of samples in glass or plastic containers since these materials absorb strongly in this region of the spectrum. The use of fibre optics for IR measurements is also limited, and is restricted to short fibre lengths but *remote* sampling may be achieved through lightguides, i.e. tubes with mirrors at joints, although the number of joints is limited and the system rapidly becomes inefficient.

6.1.3 Raman spectroscopy

The origin of the Raman spectrum is markedly different to that of the IR spectrum. The Raman effect arises from the inelastic scattering of a monochromatic light source, usually a laser with a wavelength of between 200 and 1400 nm (ultraviolet, visible or near-IR). Figure 6.1 shows a schematic representation of the Raman effect. As the incident photons of the laser beam interact with the

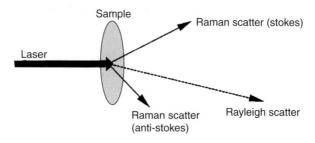

Fig. 6.1 Schematic of the Raman scattering effect. Redrawn from Petty et al. [14].

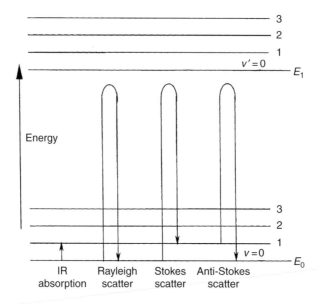

Fig. 6.2 Energy level diagram showing Rayleigh, Stokes and anti-Stokes scatter in comparison with IR absorption.

sample, the majority will be scattered with no change in energy giving rise to Rayleigh scatter. Approximately, 10^{-5} of the incident photons will be scattered with either a loss or gain in energy [13]. If the molecule is in the ground state, the photon will lose energy to the molecule (Stokes scatter) but if the molecule already exists in an excited state, the photon will gain energy from the molecule (anti-Stokes scatter) (Fig. 6.2). At room temperature, most molecules will be in the ground state the Stokes lines will be more intense. In solid and liquid samples, the quanta of energy lost or gained will equate to vibrational modes. The absolute energy of the photons will be dependent on the wavelength of the laser so as to obtain the frequency of the vibrational modes and produce a trace comparable to an IR spectrum, the difference between the laser frequency and the Stokes scatter in wavenumbers is used as the abscissa (the Rayleigh line will be at $0\,cm^{-1}$).

The observation of vibrational modes in the Raman spectrum is governed by selection rules and is determined by the change in polarisability associated with the vibration. The more non-polar bonds, such as C–C, C=C, will give rise to the strongest bands in the Raman spectrum. The complementary nature of IR and Raman spectroscopy is demonstrated in Fig. 6.3; it can be seen that bands which are strong in one spectrum are weak in the other thus showing modes attributable to all the functional groups of the molecule. It should be noted that obtaining the both Raman and IR spectra does not necessarily show all of the vibrational modes since it is possible for some modes to cause neither a change in dipole moment nor polarisability. However, for drug molecules which are highly asymmetric, it is unlikely that any mode will be both IR and Raman inactive.

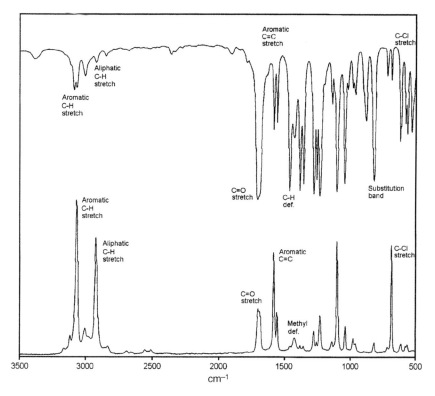

Fig. 6.3 The IR (top trace) and Raman (bottom trace) spectra of 2,5-dichloroacetophenone, some bands are assigned to group frequencies. Reproduced with permission from Hendra, P., Jones, C. and Warnes, G., *Fourier Transform Raman Spectroscopy* [15]; published by Ellis Horward, 1991.

As for IR spectra, the frequency of the vibrational modes is characteristic of functional groups and so the molecular structure can be deduced from interpretation of the spectrum. Unfortunately, the availability of comprehensive correlation tables for Raman, lags behind that of IR. Tables that correlate both IR and Raman frequencies are available [4, 6].

6.1.3.1 Instrumentation for Raman spectroscopy

Thorough descriptions of instrumentation can be found elsewhere [15–17] and only brief details are given here. Instrumentation for Raman spectroscopy falls into two distinct categories: FT instruments and dispersive instruments. Both types of instrument have useful application in the pharmaceutical industry and a comparison is summarised in Table 6.1.

Fourier transform Raman spectroscopy (FT-Raman) was developed following the observation that it was possible to record near-IR excited Raman spectra using a Michelson interferometer [19]. Significant developments in instrumentation did not occur until the late 1980s when FT-Raman instruments became commercially available [15, 20]. It was the introduction of FT-Raman instruments that heralded the use of Raman as a routine analytical tool through their ease-of-use, accuracy, high performance and the opportunity to analyse samples non-invasively [21]. These instruments also overcome what is probably the greatest (or certainly the most quoted) disadvantage of Raman spectroscopy, namely fluorescence. This is caused by electronic excitation of either the molecule of interest or an impurity by the laser. Since the fluorescent emission is far stronger than the Raman signal and has a much broader bandwidth it *swamps* the spectrum. FT-Raman instruments universally use near-IR excitation at 1064 nm and few organic molecules have an electronic transition at this frequency, so the chances of fluorescence are, therefore, greatly reduced. FT-Raman instruments may be a stand-alone instrument, or an accessory to a standard FT-IR bench. These instruments typically provide the *workhorse* Raman instrument in an analytical laboratory and are commonly used for applications such as routine chemical identification.

Table 6.1 Comparison of the main features of dispersive and FT-Raman spectroscopy. Adapted from Frank [18]

Dispersive Raman	FT-Raman
Applicable to >80% samples (unsuitable for fluorescent samples)	Applicable to >90% samples (unsuitable for highly coloured samples)
Potential interference from glass	Little interference from glass
Response function changes drastically across spectrum	Response function uniform across spectrum
Utilises low laser powers (<100 mW)	Utilises high laser powers (<500 mW)
High sensitivity	Low sensitivity
Microscopy provides very good spectra with short integration times (e.g. <30 s)	Microscopy requires long integration times (e.g. 5–10 min)
Fibre optic probes possible	Limited fibre optic probe options

Dispersive instruments are the *classical* instrumentation. The scattered radiation is dispersed using a grating and subsequently detected. In the past, their use was limited to specialist applications or academia but through developments in solid-state lasers with far-red emission which reduce fluorescence [22], high efficiency optics (e.g. holographic gratings and notch filters [23]) and multichannel detectors (such as the charge coupled device [24]), these instruments have become more compact, reliable and robust [16, 25]. As such their use has been extended to the non-expert and have been applied to a number of applications. The use of shorter wavelength visible and far-red lasers for excitation of the Raman spectrum leads to some distinct advantages over FT instruments in applications such as microscopy where spot sizes as small as 1 micron can be achieved and remote sampling using fibre optics becomes a viable option.

6.1.3.2 Sample preparation for Raman spectroscopy
Raman spectra are obtained by the scattering of light and so very little, in some cases no, sample preparation is required. The only requirement is that the sample is *visible* to the laser beam. It is this feature that can offer significant advantages over IR spectroscopy in pharmaceutical analysis. Spectra may be obtained of solids or liquids in vials (either clear or coloured), through some blister packs or through plastic bags since these are transparent to the laser beam. Most commonly a solid or liquid sample will be held in a colourless glass or quartz vial. Some glasses may fluoresce or give broad bands in the spectrum; this is dependent on the composition of the glass and will vary between manufacturers. The spectrum is obtained only from the area in which the laser is focused and so a representative spectrum may not be obtained if the sample is heterogeneous. To overcome this problem, the sample may be rotated in the beam and an *average* spectrum obtained.

As with IR, Raman spectrometers of both types may be coupled to an optical microscope to obtain spectra of small amounts of material. Using a dispersive instrument, the laser and scattered light are in the visible region and so conventional optics may be used and much smaller spot sizes achieved (as small as 1 micron) [26]. When using microsampling with Raman spectroscopy, it is possible that the spectrum may be altered by polarisation of the scattered light. This is illustrated in Fig. 6.4 which shows the spectra obtained from a single crystal; the crystal was rotated by 90° between each acquisition. The relative intensities of the band are altered depending on the orientation of the crystallographic axes with respect to the direction and polarisation of both the excitation and scattered light [27]. For this reason, care should be taken when taking measurements of this kind, either to consider these effects when interpreting spectra or to minimise the effect by scrambling the polarisation of the excitation source and averaging spectra collected from a number of crystals. For FT-Raman, orientation effects are less of a problem, firstly, because the larger sampling volume is likely to contain crystals in many orientations

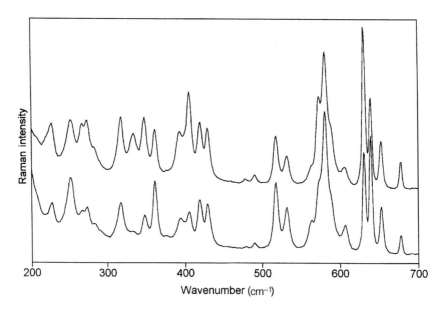

Fig. 6.4 Raman spectra of a single crystal showing the effects of polarisation on the relative intensity of the vibrational bands. The spectra were acquired using a Raman microscope and the sample was rotated 90° between measurements.

and, secondly, reflection of the excitation and scattered light within the sample will scramble the polarisation [27].

Another advantage of using visible light in dispersive instruments is that fibre optics may be used to link the spectrometer with a sampling probe. These types of instrument have many applications in online monitoring of chemical processes (Section 6.4.2).

6.1.4 Applications

The applications of vibrational spectroscopy within the pharmaceutical industry are numerous and diverse. The techniques are applied from early discovery, throughout development, and to the quality control of final drug substance and product in a manufacturing environment.

Applications may be grouped into the following categories, naturally there are no distinct boundaries between these, and applications such as quantitative analysis apply throughout:

- Structural investigations, using the molecular origin of the spectra to provide information which is often in support of nuclear magnetic resonance (NMR) or mass spectrometry studies.
- Solid-state analysis – this is probably the most common application of vibrational spectroscopy and includes both drug substance and product.

- Monitoring structural changes and identifying components in situ, from analysing single crystals on a hot-stage microscope to monitoring chemical processes on a plant scale.

In the following sections, examples in each of these areas will be given.

6.2 Vibrational spectroscopy to investigate molecular structure

6.2.1 Introduction

Many samples are analysed by vibrational spectroscopy purely to obtain structural information. This may be either to identify an unknown or, more commonly, to offer complementary and additional information to that obtained by NMR and/or mass spectrometry studies.

A *practical and pragmatic* approach to interpreting vibrational spectra has been proposed by Chalmers and Dent [11]. This approach considers factors which may alter data, such as sample preparation, and offers a suggested order in which to systematically interpret the spectrum. More detailed interpretation can be achieved using correlation tables which are based on extensive review of literature examples [1–4, 6].

In this section, some common structural problems are described where IR and Raman spectroscopy have particular utility in understanding the molecular structure. Also, some selected examples of specialist applications are described, which, although by no means exhaustive, give an indication of the breadth of the use of vibrational spectroscopy within the pharmaceutical industry.

6.2.2 Techniques in structural elucidation and specialist applications

6.2.2.1 Tautomerism

Vibrational spectroscopy, in particular IR, can provide a rapid and sensitive method for studying tautomerism in the solid state. Systems involving keto/enol tautomerism may be distinguished by the presence of a C=O or O–H stretch for the keto and enol forms, respectively. Amido forms are readily distinguishable from the imido by the presence of C=O and N–H vibrations. Although the bands are weaker, thio/thiol systems may also be investigated. An example of changes in tautomerism is given in Section 6.3.2.

6.2.2.2 Hydrogen bonding

The vibrational bands associated with O–H, N–H and C=O bonds are usually strong in the IR spectrum making this an ideal technique for studying hydrogen bonding. The frequency of vibration will be altered when hydrogen bonds are formed due to changes in the electronic density within the bond [2]. Amide groups are commonly found in drug molecules and the position of the

N–H and C=O stretching modes in the IR spectrum may be used to infer the inter- and intra-molecular hydrogen bonding pattern [4], for example between polymorphs [28, 29]. IR may also be used to study the hydrogen bonding within hydrates as described in Section 6.3.2.

6.2.2.3 Chirality
In general, the vibrational spectra of enantiomers and racemic mixtures will be identical. Exceptions arise where there are differences in the hydrogen bonding pattern between the two enantiomers which may cause differences in the spectra, or where a racemate crystallises as a racemic crystal (i.e. each unit cell contains one molecule of each enantiomer, rather than a two-phase mixture of the components).

However, it is possible to distinguish enantiomers using vibrational optical activity (VOA). VOA may be measured using IR radiation to probe the vibrational transitions, this is called vibrational circular dichroism (VCD). The VCD spectrum is the spectral difference between the spectra acquired using right- and left-handed circularly polarised light. Alternatively, VOA may be measured using Raman scattered light; the Raman optical activity (ROA) is the intensity difference between the scattering measured with right- or left-handed circularly polarised incident or scattered light. In both cases, the band positions will be identical to those observed in the normal spectrum, but the band intensity in the individual spectra will vary depending on the polarisation of the vibrational mode with respect to the incident light. Thus, in the different spectra, bands will have a sign (either positive or negative) associated with them and the VCD or ROA spectra of the two enantiomers will be mirror images of each other. A more thorough explanation of these techniques and the instrumentation required are available in the literature [30].

An example of an ROA spectrum is shown in Fig. 6.5 [31]. It can be seen that each band in the Raman spectrum has an associated feature in the ROA spectrum (the ROA of the opposite enantiomer would be the same but opposite in sign) and that the ROA signal is approximately a quarter of that of the Raman.

Many studies using VOA have been conducted on proteins but its application to small organic molecules is developing. Two applications which may have relevance to the pharmaceutical industry are the possibility of determining the absolute configuration of a molecule thus offering an alternative to obtaining single crystal X-ray diffraction data, but currently this is not possible if the compound contains a heavy atom or is a salt. Secondly, the determination of enantiomeric excess which, although much less accurate than chromatography, has the advantage that no separation is required.

6.2.2.4 Contaminant analysis
Spectroscopic microscopy may be used to non-destructively identify contaminants, such as specks or fibres, which are embedded in a product, thus aiding

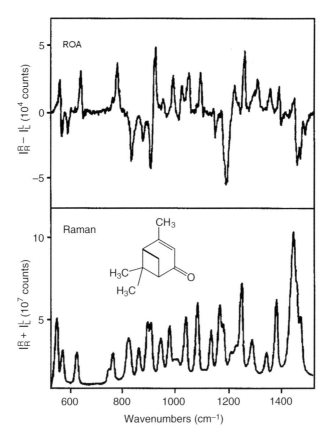

Fig. 6.5 Raman and ROA spectra of verbenone. Reproduced with permission from L.A. Nafie, G.-S. Yu and T.B. Freedman, *Raman Optical Activity of Biological Molecules* [31]; published by Elsevier, 1995.

investigations into the origin of the contaminant [32]. Raman microscopy is most suited to this application due to the small sample size required (1–2 micron) and the requirement for little sample preparation [13].

6.2.2.5 Products from combinatorial chemistry

The use of combinatorial chemistry where a library of compounds is prepared through a series of synthetic reactions, often on a solid support such as a bead, has had a huge impact on the way drug candidates are prepared. The rapid characterisation of the resulting compounds which by definition have extensive molecular variation, has proved challenging and usually requires cleaving the compound from the solid support. IR and Raman spectroscopy, however, offer the opportunity of ascertaining the structure of the compound without cleavage.

Table 6.2 Characteristic vibrational frequencies of IR and Raman tags used in combinatorial synthesis. Adapted from Rahman et al. [33]

Tag	Characteristic band position cm^{-1}
(benzonitrile tag, –C≡N)	2231 (IR and Raman)
(2,6-di-tBu phenol tag, –O–H)	3624 (IR only)
(terminal alkyne tag)	2120 (IR and Raman)

IR spectra of compounds on beads can be readily obtained using ATR-IR, the beads are simply compressed against the crystal and the spectrum obtained. Spectra also may be obtained from single beads using either IR or Raman microspectrometry [33]. This study demonstrated that by incorporating IR and Raman specific tags into the synthesis (i.e. functional groups known to give distinct bands) the molecular structure of the product may be inferred. An example of this is shown in Table 6.2 and Fig. 6.6; the tags were chosen to introduce bands in the 4000–3500 and 2800–1800 cm^{-1} regions, which are clear of bands due to the polymer bead. By determining the presence and/or absence of the tag bands for each bead and with reference to the synthesis steps followed, the complete structure of the compound may be deduced.

6.2.2.6 Theoretical calculations

Theoretical calculation of the frequencies of the normal modes of vibration is possible and has been demonstrated for moderately complex molecules such as acetazolamide [34]. This facilitates assignment of the observed spectral bands beyond what may be possible by comparison with correlation tables and *manual* interpretation. The ability to predict the visual appearance of IR and Raman spectra is, however, more challenging since the inclusion of selection rules to ascertain which bands will be IR or Raman active and the calculation of their intensity must be considered.

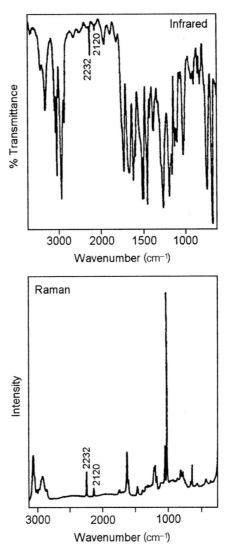

Fig. 6.6 IR and Raman spectra of a bead from a combinatorial library. Both spectra were obtained using microspectrometry of a single bead. Redrawn from Rahman *et al.* [33].

IR and Raman may be used to confirm results obtained from structure prediction software. In a study on eniluracil [35], the crystal structure was predicted from the X-ray powder diffraction pattern. Two feasible crystal structures were identified in which the hydrogen bonding pattern was very similar with only the orientation of the non-bonded carbonyl differing between the two. The IR and Raman spectra of eniluracil support the presence of such a hydrogen bonding pattern with two distinct C=O stretches at 1701 and 1664 cm^{-1} corresponding to the bonded and non-bonded carbonyl, respectively. Only the Raman spectrum shows the acetylene stretch (2115 cm^{-1}) demonstrating the advantage of using both techniques. Although the vibrational spectra offer strong evidence that one of these structures is correct, it is unable to distinguish between them as there are no other differences between them which may be predicted from inspection of the crystal structures.

6.2.3 Routine chemical identification

IR spectroscopy is often the method of choice for chemical identification on specifications for release of an active pharmaceutical ingredient and is included in all the major pharmacopoeias. This method is simple, reliable and ensures correct chemical identification when compared with the spectrum of a suitable reference standard. However, the sample preparation required means that carrying out these tests can become time-consuming and there is little opportunity for automation; in addition, poor sample preparation can lead to erroneous results.

Raman spectroscopy is included in the US Pharmacopoeia [36] but to date it does not appear in any monographs, however, it may offer a viable alternative to the established IR identification test. The main advantage in using Raman is the requirement for little (or no) sample preparation with the opportunity to analyse samples within glass vials thus facilitating automation and allowing much more rapid throughput of samples. The use of Raman for this application has been demonstrated using both FT and dispersive instruments [18, 37]. McCreery and co-workers [37] also assessed the use of spectral libraries with Raman spectra and included a balanced discussion on the relative merits of this method, and the robustness and sensitivity of the Raman technique.

The feasibility of using Raman as a non-destructive quality control method for drug product has been demonstrated for intact gel capsules, for which a spectrum could be obtained even from the capsule inside a blister pack [38], since the gel capsule and blister are transparent to the wavelength of the laser excitation.

6.3 Vibrational spectroscopy of polymorphs, hydrates and solvates

6.3.1 Introduction

Polymorphism is a solid-state phenomenon and is the ability of an element or compound to exist in more than one crystalline form. These forms may differ in the packing arrangement of the molecules within the crystal, or the conformation of the molecule, or a combination of both. Crystals may also have a different shape or morphology (habit), yet have the same crystal structure and thus be of the same solid-state form. The study of polymorphism within the pharmaceutical industry is usually expanded to include the study of solvates and hydrates. These forms are sometimes referred to as pseudopolymorphs but there is a growing dislike of this term since its definition can be ambiguous. It has been estimated that approximately one-third of organic molecules exhibit polymorphism, whilst a further third form solvates (or hydrates) [39]. Whilst some polymorphic compounds may have only two forms, it is not unknown for a compound to have numerous polymorphs and pseudopolymorphs (for example, cortisone acetate has seven forms [40] whilst sulfathiazole forms over 100 solvates [41]). The term *solid-state form* is used here to cover both polymorphs and the so-called pseudopolymorphs. In addition to the crystalline phase, an understanding of the amorphous (non-crystalline) phase is also of importance.

In polymorphic systems, the electrostatic micro-environment of the functional groups may vary between forms, for example through differences in hydrogen bonding (inter- or intra-molecular), and the steric forces within the molecule may also alter due to conformational or crystal-packing changes. Therefore, the frequencies of some, or all, of the vibrational modes will probably differ between solid-state forms. In addition, changes in symmetry, due to conformational or packing effects, can alter the appearance of the spectrum by causing band splitting or relative intensity changes. Band splitting may also be due to an asymmetric unit cell containing more than one molecule. Vibrational spectroscopy can be of great use in studying the amorphous phase since even though there is no long-range order in the material, it will still give a characteristic spectral *fingerprint* and reveal chemical information. Therefore, IR and Raman spectroscopy may be used to identify and distinguish solid-state forms using the spectra as a *fingerprint* and through assignment of the spectra, a structural basis for the differences between the forms may be made.

The fundamental crystal structure of a solid has a major bearing on its physical and chemical properties. Solid-state forms may exhibit variation in, for example, colour, stability, processability, dissolution and bioavailabilty ranging from the subtle to the severe [42]. It is for this reason that an understanding of the solid-state behaviour of a drug molecule, including the identification of all possible forms, and thorough characterisation of the observed forms are of prime importance within the pharmaceutical industry. The financial impact of problems associated with polymorphism was highlighted recently when the manufacturing process for ritonavir (an anti-retroviral drug produced by Abbott) failed to give the correct form, and in fact produced a more stable form. This new form was not suitable for the existing capsule formulation due to differences in solubility and led to the temporary change to a liquid formulation [43]. Demonstrated control of the polymorphic purity of the drug substance, and sometimes of the active component of the formulation, through all stages of drug development is a regulatory requirement. A flowchart describing the characterisation requirements has been devised [44, 45].

A detailed account of polymorphism and its relevance in the pharmaceutical industry is given elsewhere in this volume and in the literature [42, 46, 47]. This section will focus on the use of vibrational spectroscopy as a technique for solid-state analysis. However, it should be noted that these techniques must be used as an integral part of a multidisciplinary approach to solid-state characterisation since various physical analytical techniques offer complimentary information when compared to each other. The most suitable technique will depend on the compound, and the objectives and requirements of the analysis. Techniques commonly used in solid-state analysis include crystallographic methods (single crystal and powder diffraction), thermal methods (e.g. differential scanning calorimetry, thermogravimetry, solution calorimetry) and structural methods (IR, Raman and solid-state NMR spectroscopies). Comprehensive reviews on solid-state analysis using a wide variety of techniques are available in the literature [39, 42, 47–49].

6.3.2 Identifying structural differences between solid-state forms

6.3.2.1 Polymorphs

The differentiation of polymorphic forms by IR and Raman spectroscopy is demonstrated in Fig. 6.7 which shows the spectra for three polymorphic forms of cimetidine. These spectra clearly demonstrate how either IR or Raman spectroscopy may be used to *fingerprint* and identify the solid-state form of the drug substance. Care should be taken, as the spectral differences may sometimes be subtle [50] and it is possible that the manual processing involved in sample preparation, especially for IR analysis, *may* cause conversion of the solid-state form. In general, if the differences in the IR spectra are slight, then the Raman may show greater differentiation and vice versa. It is for these reasons that initial investigation of solid-state forms should always utilise a range of tests characterising different properties of the substance to confirm the presence of polymorphism or pseudopolymorphism [39, 49].

It is possible to progress the utilisation of spectroscopy beyond simple identification of solid-state forms to an understanding of their structural differences, and possibly to link to the physical and chemical behaviour of each form. In a study of acetohexamide, Takla and Dakas [51] demonstrated that in Form A, the drug was present as the enol form, whilst in Form B the keto form was present. Spectra of both forms show bands in the 1685–1690 cm^{-1} region due to the presence of an aryl ketone but a band attributable to a urea carbonyl is not observed for Form A due to enolisation. In the N–H stretching region of the spectrum, a shift is observed from 3310 cm^{-1} (Form A) to 3360 cm^{-1} (Form B) which may be caused by the extra degree of conjugation associated with the enolised system. The different structures lead to a change in the hydrogen bonding pattern between the two forms. In Form A, a sharp band at 3240 cm^{-1}, attributable to the O–H vibration, is characteristic of intra-molecular bonding

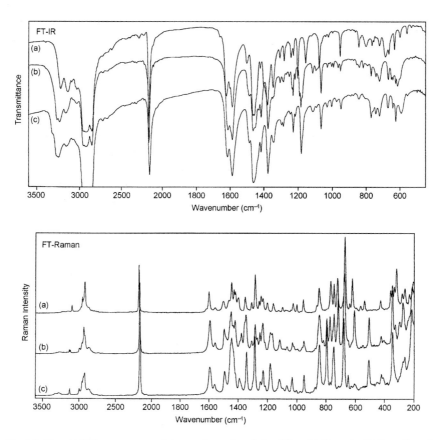

Fig. 6.7 IR and FT-Raman spectra of three polymorphs of cimetidine, (a) Form A, (b) Form B, and (c) Form C. Samples for IR were prepared as Nujol mulls.

between O–H and S=O. In Form B, there is inter-molecular bonding between the urea carbonyl of one molecule and the sulphonamide N–H of its neighbour characterised by a broad band at ca. 3100 cm^{-1} attributed to the sulphonamide N–H. In comparison, an IR and Raman study of spironolactone [52] showed no evidence of enolic tautomerism between the forms. In this study, the benefits of using both IR and Raman to characterise polymorphs are shown by the observation of the C–S stretching modes at 637 and 655 cm^{-1} which are not observed in the IR spectra. This study also highlights the care that must be taken with sample preparation to avoid conversion between forms; IR spectra recorded by DRIFTS and the Raman spectra show variation, whereas IR spectra recorded as potassium halide disk are identical.

In a study of dirithromycin, Stephenson *et al.* [53] demonstrate how the position of the carbonyl stretching mode of the lactone moiety may be used to determine the level of hydrogen bonding. From the crystal structures, it is known that in Form II, the lactone carbonyl is hydrogen bonded and in the IR spectrum the stretching mode appears at 1712 cm^{-1}. However, for the 1-propanol solvate in which the lactone C=O is not involved in hydrogen bonding, the band appears at 1735 cm^{-1}. This information was used to infer that in the amorphous form, which has a band at 1735 cm^{-1}, there is no hydrogen bonding as might be expected. In Form I, where bands at both 1712 and 1735 cm^{-1} are observed, it is likely that two molecules are present in the unit cell in which one is bonded and the other non-bonded. A similar analysis has been made of the IR and Raman spectra for a steroidal anaesthetic [54] which has a hydrogen-bonded carbonyl in Form II but not in Form I. It was also noted that, in the Raman spectrum, the ν(C=O) band Form II was much more intense than that in Form I resulting from proposed differences in the polarisability of bonded and non-bonded forms. The position of the N–H and O–H stretching modes in the IR spectra of hydrogen bonding molecules may be also used to determine the relative stability of polymorphs. According to one of Burger's rules, 'the highest frequency OH or NH stretching band will be associated with the form least stable at absolute zero' [55] since the highest frequency will be associated with the form containing the weakest hydrogen bonding. This approach has also been applied to Raman spectra [56].

The low frequency region of the vibrational spectrum of solids, between 10 and 150 cm^{-1}, contains bands attributable to lattice vibrations (or phonons) which are extremely sensitive to structural changes in the solid state. Because of the low frequency, this region of the spectrum is only readily obtained by Raman spectroscopy since the optics commonly used in IR spectrometers and materials used in sample preparation absorb strongly in this region. The lattice vibrations may be used to obtain, both to characterise and distinguish polymorphs, and to investigate the intermolecular interactions between the lattice and solvent [57].

6.3.2.2 *Hydrates and solvates*

Solvates are formed when spaces within the crystal lattice are occupied by solvent molecules. Due to its small size and prevalence, water is the most common solvent to be associated with the lattice and the term *hydrate* is then used. Unlike hydrates, solvates are rarely used as an active pharmaceutical ingredient. However, they may be formed as precursors to the required form and so their characterisation is often essential in order to understand the solid-state properties of the drug substance. The potential impact of changes in hydration state exists throughout the development process since these changes can impinge on the primary and secondary manufacture of the drug. Substances may hydrate or dehydrate in response to changes in processing conditions (e.g. a change in water content of the process solvent) or environmental conditions (e.g. changing the site of manufacture).

The location of water molecules within a crystal lattice, and their interaction with the drug substance molecules, can alter the properties of the substance. Hydrates are often classified as *isolated-site*, *channel* or *adventitious* [58]; these descriptions apply equally well to solvates. Isolated-site hydrates are formed when water is held within spaces in the crystal lattice; often there will be hydrogen bonding between the water and the drug molecule. It is possible that not all the sites will be occupied. In channel hydrates, long open channels exist within the crystal lattice into which water may enter and reside in sites stabilised by hydrogen bonding to the drug molecule or other water molecules. A common feature of channel hydrates is the ability to exist in a number of hydration states dependent on the relative humidity of the environment. In contrast, for adventitious hydrates there is no interaction between the water and the crystalline lattice; these forms are often described as *wet* and are not true hydrates. The ability of water to form hydrogen bonds to the drug molecules, to other water molecules or to the anion in salts is a major factor in the understanding and characterisation of hydrates.

The vibrations of the O–H functional group are strong in the IR spectrum and therefore IR spectroscopy is an invaluable technique for characterising hydrates. The exact position of the bands will depend on the extent of hydrogen bonding: the lower the ν(O–H) frequency, the stronger the hydrogen bonding. More than one bond may be observed if each water hydrogen is involved in different hydrogen bonds. In higher hydrates (i.e. greater than one), each water molecule may be involved in different hydrogen-bonding patterns giving multiple vibrational bands [58]. Water is a weak scatterer of light and so does not give a strong Raman spectrum [13]. However, the Raman spectra of hydrates will show variation due to the steric and polar influences of the water on the molecular and crystal structure.

In a study of a reductase inhibitor [59], IR spectra of a hydrate exposed to relative humidities of 3.5–84% all show a sharp ν(O–H) band at 3640 cm^{-1} indicating that the water molecules are bound within the lattice. This is contrary to the results from DSC which show no dehydration endotherm. These results in combination suggest a channel hydrate with channels in the crystal lattice which can take up varying levels of water molecules, but that within the channels the water is bound to other water molecules and/or the drug molecule via hydrogen bonding. The *continuous* nature of the hydration was confirmed by a variable temperature IR experiment which showed that as a hydrated sample was heated, the intensity of the 3640 cm^{-1} band reduced as the water level decreased.

Lattice vibrations observed in the Raman spectrum may also be used to characterise solvates. The spectrum of a chloroform solvate of griseofulvin and the anhydrous form shows good correlation [57] indicating that the structure of the solvate is similar to that of the parent molecule. The lattice vibration modes of the solvate have shifted to lower frequency (Fig. 6.8) which can be

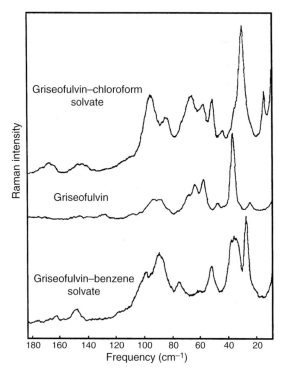

Fig. 6.8 Raman spectra of griseofulvin and its solvates with chloroform and benzene in the 20–180 cm^{-1} region. Reproduced with permission from J.C. Bellows, F.P. Chen and P.N. Prasad, *Determination of Drug Polymorphs by Laser Raman Spectroscopy* [57]; published by Marcel Dekker, Inc., 1977.

attributed to an expansion of the griseofulvin lattice to accommodate the chloroform molecules. The expansion reduces the interactions between the drug molecules thus decreasing the frequency of the lattice vibrations. In comparison, the spectrum of a benzene solvate shows no correlation to that of the anhydrate (Fig. 6.8) suggesting that the two structures are different.

6.3.2.3 Amorphous material

Amorphous materials have no long-range crystalline order but since they maintain a molecular structure they still give a vibrational spectrum which may be distinct from the crystalline material. In comparison, the X-ray powder diffraction pattern of an amorphous material contains only a broad signal which yields no structural information.

Figure 6.9 shows the IR and Raman spectra of two crystalline forms and amorphous indomethacin. The increased bandwidth in the spectrum of the amorphous material is a common consequence of the loss of order in the solid.

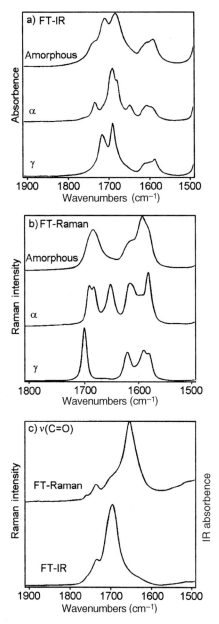

Fig. 6.9 (a) IR and (b) Raman spectra of γ-indomethacin, α-indomethacin and amorphous indomethacin showing the acid and benzoyl carbonyl vibrations, (c) the carbonyl stretching frequency region of acetic acid illustrating the asymmetric stretch observed in the IR and the symmetric stretch present in the FT-Raman spectrum. Redrawn from L.S. Taylor and C. Zografi [73].

Despite this, differences can still be observed allowing identification of amorphous material relative to the crystalline forms.

6.3.3 Quantifying solid-state forms in drug substance

If the polymorphic purity of the drug substance is known to significantly affect the drug's properties then it may be necessary to quantify the amount of an undesirable form in the required form [45]. Both IR and Raman may be used to quantitatively analyse mixtures of solid-state forms since the intensity of the vibrational bands is directly proportional to the concentration of the species. In general, a calibration curve based on the spectroscopic data is generated using mixtures of known composition [11]. The most likely source of error in any quantitative study will arise from inhomogeneity in the calibration samples. Insufficient mixing of the phases will lead to false results since only a small portion of the mixture will be analysed and this will not be a representative sample. This is of particular importance with Raman spectroscopy since only the relatively small quantity of sample within the laser focus will be analysed. The effects of inhomogeneity may be reduced by thorough mixing and averaging of several spectra for each sample (e.g. by spinning the sample during a Raman measurement). A second source of error is variation in particle size between components of the calibration mixture and/or subsequent samples and it is of particular importance when using diffuse reflectance IR measurements. Particle size also has an effect on the intensity of Raman bands [60] but if a relative intensity is measured then, assuming the particle size of all components is similar, the changes in intensity will be uniform across the spectrum and any variation in relative intensity will be minimal. Other variations between data such as artefacts (e.g. water vapour or CO_2 in the IR spectra, or room-light emission bands in the Raman), and changes in crystallinity may also cause errors.

The use of diffuse reflectance IR for quantifying the level of cefepime dihydrochloride dihydrate in batches of monohydrate has been demonstrated [61]. In order to obtain homogeneous mixtures of the components, a slurry method was developed since attempts at mechanical dry mixing were unsuccessful. The failure of dry mixing was attributed to differences in the morphology of the two forms since the monohydrate has flat, plate-like crystals whereas the dihydrate exists as long needles. By slurrying in a solvent, in which both forms are insoluble, a homogeneous mixture with good particle-size distribution was obtained. This mixing technique is only suitable for solid-state forms where there is no solvent-mediated conversion and so it is important to analyse a mixture before and after slurrying to ensure that no conversion has occurred. Quantification of the forms was based on bands in the O–H stretching regions of the spectrum: The monohydrate has a single ν(O–H) band at $3529\,cm^{-1}$ whereas the dihydrate has two bands at 3574 and $3432\,cm^{-1}$ corresponding to distinct vibrations of the two water molecules. To quantify the dihydrate content, the integrated intensity of the 3574 cm^{-1} band was used and the calibration was based on Beers law [11, 62]. The calibration was found to be valid over a range of 1.0–8.0% w/w dihydrate in monohydrate with a detection limit of 0.3%. Detailed methods for carrying out the calibration, validation, and the evaluation of errors in the measurement are given by Bugay et al. [61]. Quantitation of samples prepared as Nujol mulls is also possible [63] although with transmission techniques accurate measuring of the pathlength may be required [11].

Unlike IR spectroscopy, the intensity of Raman bands cannot be related to Beers law. Although the intensity is a function of the number of molecules present, the overall intensity will be dependent on the power of the excitation laser, the scattering cross section and the packing of the solid [11, 13]. The overall intensity may vary between subsequent measurements of the same sample as well as between samples and is on an absolute scale rather than a relative scale (such as absorbance). To obtain quantitative information from Raman spectra, it is necessary to measure the relative intensity of a band attributed to the species of interest – this may be relative to a band of another species (of which the concentration will vary) or to an internal standard (of which the concentration will be constant). In a study by Langkilde et al. [64], the ratio of distinct Raman bands for two forms was used for quantification. Large variations between measurements for the same sample were observed highlighting the problem of non-representative sampling of mixtures. This was minimised using a step-wise rotating sample holder.

The accuracy of quantification methods may be improved using multivariate methods since these have the advantage of analysing variation across the whole spectrum rather than the intensity of individual bands [11, 65]. The use of IR spectroscopy with factor analysis [66], partial least squares [67, 68] and neural networks [69], and FT-Raman with factor analysis [70] and principal component analysis [71] has been demonstrated.

Since spectral differences are often observed between crystalline and amorphous phases, IR and Raman may also be used to quantify the degree of crystallinity in a sample. Unlike powder diffraction methods in which the presence of amorphous material is based on the absence of a signal, in the IR or Raman spectrum a positive signal is often observed for amorphous material (Fig. 6.9). In the IR spectrum of crystalline cefazolin sodium pentahydrate, the amide II band is observed at 1542 cm^{-1} [72]. As the degree of crystallinity is reduced the intensity of this band is observed to decrease, and in amorphous material it is totally absent. The area of this band relative to that of the beta-lactam carbonyl band at 1760 cm^{-1} which is independent of crystallinity has been used to estimate the quantity of amorphous material. A linear calibration line from 0 to 100% amorphous was obtained, and results show good correlation with X-ray powder diffraction results. In the Raman spectrum of indomethacin, a distinct band at 1680 cm^{-1} is observed for the amorphous form and at 1698 cm^{-1} for the crystalline material; these bands are attributed to the benzoyl carbonyl [73]. The intensity ratio of these two bands showed linear correlation across the entire range of crystallinity with a proposed limit of detection of 1% amorphous in crystalline, and vice versa. As for mixtures of polymorphs, the largest source of error arose from inhomogeneous mixing of the sample.

6.3.4 Identifying the solid-state form in formulated product

The majority of drugs are formulated and delivered as solid dosage forms, typically tablets or capsules. It is recognised that drug substances can undergo chemical or solid-state form transitions during processing to form the drug product or during storage and, in cases where the solid-state form is known to have an effect on critical properties such as bioavailability or toxicity, it is required that the solid-state form present in the tablet is identified [44, 45]. In many formulations, the concentration of the active ingredient may be less than 1% and its identification is therefore challenging. Reference IR and Raman spectra for common excipients are available in the literature [9].

The use of mid-IR for analysis of drug substance in formulated product has been demonstrated using DRIFTS of ground tablets [74]. However, Raman spectroscopy offers several advantages over IR for this type of analysis. It is usually possible to analyse the tablets or capsule intact, since only Raman scattering at the surface is required, removing the need for sample preparation which might alter the sample. Probably the most advantageous factor is that most common excipients are non-aromatic, non-crystalline or hydrophilic and are therefore poor Raman scatterers compared to the drug molecules [75], thus making it easier to identify bands due to drug substance even if its concentration is relatively low. Functional groups such as carbonyls are not often found in excipients (commonly used exceptions, for example, are povidone and stearic acid) leaving this highly diagnostic region of the spectrum clear of interference from excipient bands. In general, the limit of detection for a drug substance will be dependent on the strength of the Raman spectrum of the active compared with that of the excipients. To obtain information on the polymorphic purity, the concentration of the active and the spectral contrast between the forms will determine the limit of detection of the undesired form. An excellent overview of the use of FT-Raman for analysing solid dosage forms is given by Taylor and Langkilde [76]. Potential issues such as non-representative sampling, fluorescence and sample heating are addressed and case studies covering a range of active concentration and dosage forms are presented. The ability of FT-Raman to distinguish between polymorphs in formulations is demonstrated by the spectra of tablets containing ranitidine hydrochloride from two manufacturers (Fig. 6.10a) [76]. It can clearly be seen that in one tablet the drug is present as Form I, and in the other Form II has been used. It is also possible to determine the polymorphic purity in the formulation. Tablets containing 1.25 and 2.5% w/w of Form II (23.75 and 22.5% Form I, respectively) were analysed; a band attributable to Form II is observed in both samples (Fig. 6.10b).

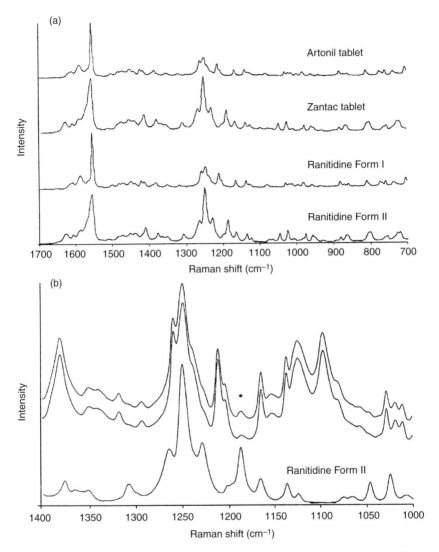

Fig. 6.10 (a) FT-Raman spectra of ranitidine hydrochloride tablets obtained from two difference manufacturers. Comparison with reference substances indicates that Aronil tablets contain the Form I polymorph whereas Zantac tablets are comprised of Form II. (b) Spectra of ranitidine hydrochloride tablets containing predominantly the Form I polymorph with Form II present as a polymorphic impurity. The top spectrum is from a 100 mg ranitidine hydrochloride tablet that has 10% w/w Form II ranitidine hydrochloride. The middle spectrum is from a tablet of the same strength but containing 5% w/w Form II, and the bottom spectrum is from pure Form II substance. The peak which shows the presence of Form II is marked*. Both figures reproduced from *Evaluation of Solid-State Forms Present in Tablets by Raman Spectroscopy*, L.S. Taylor and F.W. Langkilde, *J. Pharm. Sci.*, © 2000. John Wiley & Sons. Reprinted with permission of Wiley-Liss Inc., a subsidiary of John Wiley & Sons, Inc.

A more detailed examination of formulations may be achieved using vibrational spectroscopy–microscopy mapping systems. These systems allow the spectrum to be obtained from a small area of the sample (ca. 10 micron diameter for IR, and as little as 1 micron for Raman); the microscope stage is then moved and so spectra are sequentially obtained from different areas of the sample. High quality motorised stages are required for this application to give good accuracy and precision during data collection. Calculating a numerical value for a spectroscopic parameter, such as band area or ratio, allows a two-dimensional representation showing the distribution of each component. Obtaining data in this way can be time-consuming with data collection time taking upto several days. Data collection can be quicker using an imaging system utilising global illumination of the sample and imaging onto an array detector [77]. These instruments are still in their infancy with application to drug product analysis due primarily to issues with spectral resolution, which may not be adequate to distinguish solid-state forms, and the requirement for high-power diode lasers to give sufficient signal. However, their potential benefits are obvious. Raman mapping has been used to determine the distribution of ibuprofen in a polymer matrix [78], oestradiol distribution in a transdermal drug delivery device [79], and in combination with near IR microscopy to obtain a detailed analysis of the distribution of both drug substance and excipients [80].

6.4 Vibrational spectroscopy for in situ characterisation

6.4.1 *Investigating solid phase transformations*

Transitions from one solid phase to another can occur between polymorphs, solvates of different stoichiometry, an unsolvated form and a solvated form, or from amorphous to crystalline [42]. These changes may by induced by heat, stress, grinding or tabletting. Characterisation of *before and after* samples of either drug substance or product can be achieved by the methods already described (Section 6.3). This section focuses on using vibrational spectroscopy to monitor these changes in situ to give structural information on the transformation and phases not stable at ambient conditions.

6.4.1.1 *Variable temperature spectroscopy*
Variable temperature spectroscopy is a valuable tool for monitoring changes in the sample during heating and allows phases identified by thermal techniques, such as thermogravimetry and differential scanning calorimetry (DSC) to be identified, in particular those that are unstable at ambient conditions.

Variable temperature spectroscopy, both IR and Raman, can be achieved either by using a variable temperature cell or chamber in a standard spectrometer or by including a variable temperature stage when using an IR or Raman microscope. Variable temperature cells for FT-IR use DRIFTS or transmission (Nujol mull or disk) sampling however, the quality of the data acquired is often poor making interpretation of the raw data difficult. Gu [81] utilised factor analysis in order to determine the number of phases present as pentaerythritol tetrastearate was heated. Spectra were acquired sequentially as the sample was heated at a constant weight. Using factor analysis, two solid–solid phase transitions prior to the melt were identified; the data also indicated the presence of two conformational changes not observed by DSC prior to the first endothermic transition. FT-Raman spectroscopy at elevated temperatures using a heated cell was demonstrated by McMahon and co-workers in a study of carbamazepine dihydrates [82]. Changes in the low frequency region of the spectrum, which contains bands due to lattice vibrations, were observed as the temperature was increased, which allowed the identification of the phases indicated by DSC. To further correlate with DSC experiments, the FT-Raman was performed with the sample chamber both open and sealed. It was observed that the form produced prior to dehydration was different in each case.

A direct correlation between visual observation of changes within the sample and the characterising spectra of the material can be obtained using a spectroscopic microscope with a heated stage. Raman spectroscopy can be advantageous for this kind of study since no preparation of the sample is necessary. For IR microscopy using transmission, very thin sections of the solid are required which may necessitate *rolling* of the solid which may alter the solid-state form. In an extensive study, Szelagiewicz *et al.* [83] used a Raman microscope with a heated stage to investigate the behaviour of paracetamol and lufenuron. The spectra obtained for paracetamol at different temperatures are shown in Fig. 6.11. These spectra allow the identification of all the known forms of paracetamol. In Fig. 6.11a, spectra acquired from different areas of the crystal show some variation which is due to polarisation effects. In Fig. 6.11b, it can be seen that there are two differently shaped crystals: small prisms at the centre surrounded by larger crystals. However, the Raman spectra are essentially the same indicating that both crystal habits are associated with Form II. The authors also report shifts in band positions as a function of the temperature; this can be observed in Fig. 6.11c where the Form I spectra recorded at 160°C and room temperature show considerable temperature effects for some modes. However, it was still possible to identify both spectra as Form I since a continuous frequency shift was observed as the sample was cooled. Lufenuron is a racemic crystal, i.e. there is one molecule of each enantiomer in the unit cell. The thermal behaviour of the racemic crystal, which has up to five forms, was compared with that of the individual enantiomers and found to differ significantly. The forms generated were identified by their Raman spectra and X-ray powder diffraction pattern [83]. Variable temperature IR microscopy has also been used to investigate conformational isomerism in the solid state [84].

Randall *et al.* [85] carried out a variable temperature Raman study on two forms of a leukotrine antagonist using a heated stage on a Raman microscope. DSC

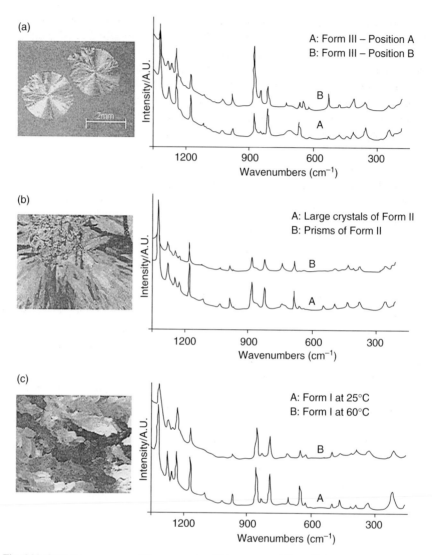

Fig. 6.11 Optical appearance and Raman spectra of (a) paracetamol Form III (b) paracetamol Form II, and (c) paracetamol Form I. Reproduced with permission from M. Szelagiewicz *et al.*, *In Situ Characterisation of Polymorphic Forms* [83]; published by Akadémiai Kiadó Rt. 1999.

and XRPD data indicated that the two *forms* were not true polymorphs and that Type I may be a less crystalline version of Form II. Results also showed that Type I converted to Type II with temperature cycling and high humidity. Differences in the Raman spectra of the two forms are most distinct in the C–H stretching region. Variable temperature studies showed that the spectrum of Type I (low crystallinity) was temperature independent (prior to melting) whereas that of Type II (high crystallinity) showed significant differences as the temperature was raised. The bands that showed variation were attributed to the aliphatic C–H and suggest that there is greater mobility of the methylene groups in Type I compared with Type II.

Low temperature IR or Raman may be used to aid interpretation of the spectrum since the bandwidth will be reduced thus resolving broad bands, or to obtain spectra of species unstable under ambient conditions [28, 86].

6.4.1.2 Combining spectroscopy with other solid-state techniques

The study and understanding of polymorphism is most effectively achieved through a multidisciplinary approach [39]. In order to correlate the data from techniques probing different properties of the compound of interest, it may be advantageous to combine two techniques in order to obtain data simultaneously. IR and Raman spectrometers may be coupled to analytical instruments and offer the advantage of obtaining structural information on-line during the experiment.

In thermogravimetic analysis (TG), the mass of a compound is measured whilst applying a heating ramp and is usually used for the characterisation of hydrates and solvates. Although this accurately determines the quantity of any volatile component, and the temperature at which it is liberated, there is no positive identification of the volatile species. By coupling an IR spectrometer to the off-gas of the TG instrument, identification of the volatile component is possible. Rodriguez and Bugay [87] demonstrate the application of TG-IR to the study of hydrates and solvates. They show how co-eluting solvents can be identified and use TG-IR to determine which solvate has been formed for solids crystallised from mixed solvent systems.

The availability of Raman spectrometers with fibre optic probes facilitates coupling with other analytical instruments. The use of Raman spectroscopy coupled with DSC and variable temperature/humidity X-ray powder diffraction [88] has been demonstrated.

6.4.2 Monitoring chemical reactions and processes

The ability to monitor the chemical composition of a reaction mixture in situ in real time allows both sensing and control of the process variables in order to provide more consistent product, improved efficiency and reduced costs. In situ monitoring also allows analysis of processes not amenable to sampling such as extremes of temperature and/or pressure, and toxic or air-sensitive reagents. Instruments designed for in situ measurement of IR and Raman spectra are available commercially. IR instruments typically use an ATR element at the end of a metal immersion probe (usually stainless steel or hastelloy to give good chemical compatibility). The probe is linked to the spectrometer by an arm that has joints to allow movement and positioning of the probe. The length of this arm is typically 1 m. Since the ATR technique collects the spectrum from the material in intimate contact with the element, these probes only collect spectra from the liquid phase of the reaction mixture (this, of course, may or may not be advantageous) and so IR monitoring can only be used to obtain information on dissolved or liquid components. Monitoring of aqueous reactions may also be problematic due to the intense absorption of the solvent. Dispersive Raman instruments may also be configured for process monitoring [16, 89]. Various probe configurations are available and these may either be immersed in the reaction mixture, or non-invasive sampling may be used with the probe sitting outside the vessel and the laser focused through the vessel wall. The use of visible and near-IR frequencies means that fibre optics may be used to link the probe to the spectrometer. Fibre lengths of upto 200 m are possible allowing the spectrometer to be remotely sited from the reaction vessel. Multichannel detectors allow the so-called multiplexing i.e. simultaneous measurement from more than one probe using the same spectrometer.

FT-IR monitoring has been used to monitor the initiation of Grignard reactions [90]. Sampling and off-line analysis of these reactions are problematic since the Grignard is moisture sensitive. On-line monitoring allows the point of initiation and the subsequent formation of the Grignard to be monitored via the R-Mg-X vibration and means that the halide addition to be controlled on a plant scale, thus avoiding potential safety issues.

The utility of IR monitoring in mechanistic studies through assignment of the observed bands is demonstrated in Fig. 6.12 which shows data obtained during the addition of $LiAlH_4$ to a Weinreb amide. As the reaction progresses, the depletion of the starting material is characterised by the carbamate and amide carbonyl and the appearance of a band at 1620 cm^{-1} which is consistent with a carbamate anion. This indicates that the amide is reduced with concurrent deprotonation of the carbamate N-H to give an intermediate metal-chelated anion.

Hydrogenation reactions can be difficult to monitor off-line and so an in situ method is advantageous. Vibrations due to the C=C bond are generally strong in the Raman spectrum so this is potentially a good technique for monitoring hydrogenations. Figure 6.13 shows the Raman spectra from the catalytic hydrogenation. An intense band due to the C=C stretch was observed at 1597 cm^{-1} superimposed on a broad feature at ca. 1600 cm^{-1} which is due to vibrations of the catalyst [91]. As the reaction progresses, the intensity of the ν(C=C) band decreases and may be used to determine the end-point of the reaction.

In Raman monitoring of a slurry, the spectrum will be dominated by bands due to the solid, since the scattering cross section of the solid particles exceeds that of the solvent and solute molecules. This offers the opportunity to monitor solvent-mediated form changes in situ without the need for sampling, isolation and analysis. In situ monitoring has many advantages for this type of study

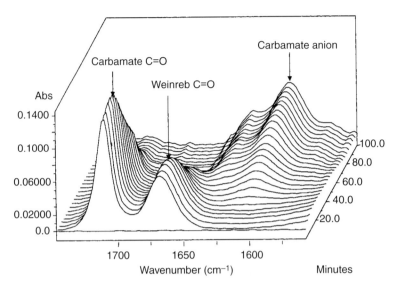

Fig. 6.12 In situ IR data acquired during reduction of a Weinreb amide. The loss of starting material and subsequent formation of a carbamate intermediate may be followed.

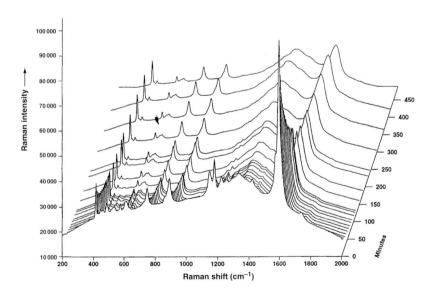

Fig. 6.13 In situ Raman data acquired during a catalytic hydrogenation. The progress of the reaction may be followed by the loss in intensity of the $\nu(C=C)$ band at 1597 cm^{-1}. A broad absorption at ca. 1600 cm^{-1} is also observed. This had a constant intensity throughout the reaction and is attributable to vibrations of the carbon-containing catalyst.

since the isolation and drying may cause a form change (e.g. desolvation) and short-lived species may be *missed* between sampling points. The polymorphic conversion of progesterone Form II to Form I has been monitored using such an instrument [92]. It was shown that good quality spectra suitable for form identification can be obtained from a slurry. The observation of the conversion was used to confirm the kinetic and thermodynamic relationship between the two forms, and the spectroscopic data used to quantify the concentrations of the forms to estimate the rate of conversion over a range of process temperatures.

6.4.3 Chromatography detection

IR or Raman spectroscopy may be used as the detection method for chromatography providing directly from components [11]. These techniques are not universally applicable and are not as widely used as, for example, mass spectrometry or NMR but may offer an advantage for some specialised applications.

Structural information on the eluent of liquid chromatography can be obtained using IR spectroscopy in one of two ways, either by flowing the column eluent through a flow cell of suitable pathlength and obtaining spectra continuously, or by depositing the eluent onto a substrate, removing the solvent, and then obtaining the spectrum using diffuse reflectance or transmission. The flow cell method is more appropriate for the detection and quantification of major components. The solvent-elimination method is more suited to the identification of minor constituents of mixtures since their concentration is too low in solution to be detected, but by acquiring the spectra off-line longer acquisition times allow better quality data to be obtained (commercially available devices are of this type). An extensive review of both methods and the practical application of these devices is given in the literature [93]. Raman spectroscopy has been demonstrated to be a viable detection method of capillary electrophoresis and micro-chromatography [94].

FT-IR instruments may also be coupled to GC systems in order to identify volatile mixtures. Spectra are either obtained in the gas phase using a flow cell, or the eluent is cryogenically frozen onto a surface suitable for IR analysis (e.g. a ZnSe window) [95]. It is also possible to add IR detection onto GC-MS systems. A hyphenated technique such as this, in which samples are analysed in series, is of use since multidimensional complementary information can be obtained from a single analysis although it should be noted that the IR is obtained of the intact molecule prior to MS analysis rather than fragments.

Raman and IR spectroscopy may be used to analyse spots on thin layer chromatography (TLC) plates. Unlike other methods, such as mass spectrometry or NMR, the spectrum may be obtained in situ without the need of scraping the spot from the plate and extracting the substance from the silica matrix. Dispersive Raman microscopy has been used to collect spectra from active drug substance on a plate [18, 88].

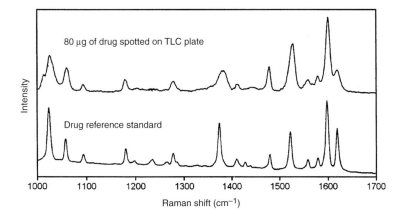

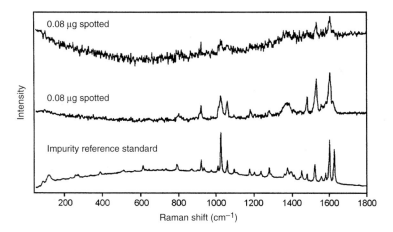

Fig. 6.14 Application of Raman microspectrometry to TLC drug spot identification (top trace) and impurity identification (bottom trace). Reproduced with permission from C.J. Frank, *Raman Spectroscopy for Identity Testing* [18]; published by SPIE 1999.

Figure 6.14 shows that good quality data can be obtained from a spot in as little as 10 s, the spectrum of the plate is subtracted to give the spectrum of the drug. Although the spectrum shows band shifts, broadening and relative intensity differences when compared to the reference material (due to interaction between the drug and the silica plate), it is still possible to make a positive identification. The identification of impurities which may be 100–1000 times less concentrated than the active is more challenging [18, 88], but Raman spectra may still be acquired at these low levels (Fig. 6.14). The spectra can be compared to reference material or interpreted to provide additional structural information. The detection and identification of impurities spotted on TLC plates using FT-IR has also been demonstrated [96]. The plate is moved under the IR beam using at computer controlled stage and diffuse reflectance IR measurements taken. Vibrational spectroscopy may also be used to investigate the molecular structure and conformation in stationary phases [97]. For example, Raman spectroscopy has been used to elucidate the conformation of alkyl side chains in octadecylsilane over a range of temperatures [98].

6.5 Summary

Vibrational spectroscopy continues to be used as a routine analytical tool for investigating and characterising drug substance throughout discovery, development and manufacture. Both IR and Raman can add valuable information to a wide variety of analytical problems, although the main area of application is in solid-state analysis. More esoteric techniques, such as microscopy and imaging, are being applied to niche applications providing information that is not attainable from other sources. The relatively new technique of Raman spectroscopy has become firmly established alongside traditional IR spectroscopy for structural investigation and characterisation studies within pharmaceutical analysis, and the advantages of non-invasive sampling and little sample preparation have lead to some novel applications. Instrument developments in both IR and Raman spectrometers are opening up new areas in process monitoring, formulation analysis, automation, rapid analysis and microanalysis.

Acknowledgements

Many thanks to Bob Lancaster (GlaxoSmithKline) for his valuable contribution of examples and data, and for his constructive review and comments prior to submission. Thanks also to William Martin and David Busby (GlaxoSmithKline) for providing data and examples.

References

1. Bellamy, L.J. (1975) *The Infrared Spectra of Complex Molecules Volume One*, 3rd edn, Chapman and Hall, London.
2. Bellamy, L.J. (1980) *The Infrared Spectra of Complex Molecules Volume Two*, 2nd edn, Chapman and Cole, London.
3. Colthup, N.B., Daly, L.H. & Wiberley, S.E. (1990) *Introduction to Infrared and Raman Spectroscopy*, 3rd edn, Academic Press, London.
4. Lin-Vien, D., Colthup, N.B., Fateley, W.G. & Grasselli, J.G. (1991) *The Handbook of Infrared and Raman Characteristic Frequencies of Organic Molecules*, Academic Press, San Diego.
5. Roeges, N.P.G. (1994) *A Guide to the Complete Interpretation of Infrared Spectra of Organic Structures*, John Wiley & Sons, Chichester.

6. Socrates, G. (2001) *Infrared and Raman Characteristic Group Frequencies: Tables and Charts*, 3rd edn, John Wiley & Sons, New York.
7. Pouchert, C.J. (1985) *The Aldrich Library of FT-IR Spectra*. Aldrich Chemical Company, Milwaukee.
8. Schrader, B. (1989) *Raman/Infrared Atlas of Organic Compounds*, 2nd edn, VCH, Weinheim/ New York.
9. Bugay, D.E. & Findlay, W.P. (1999) *Pharmaceutical Excipients*, Marcel Dekker, New York.
10. Griffiths, P.R. (1987) in *Laboratory Methods in Vibrational Spectroscopy*, 3rd edn (eds H.A. Willis, J.H. van der Maas & R.G.J. Millar), John Wiley & Sons, Chichester, pp. 121–143.
11. Chalmers, J.M. & Dent, G. (1997) *Industrial Analysis with Vibrational Spectroscopy*, The Royal Society of Chemistry, Cambridge.
12. Whitehouse, M.J. & Curry, C.J. (1987) in *Laboratory Methods in Vibrational Spectroscopy*, 3rd edn (eds H.A. Willis, J.H. van der Maas & R.G.J. Millar), John Wiley & Sons, Chichester, pp. 251–279.
13. Ferraro, J.R. & Nakamoto, K. (1994) *Introductory Raman Spectroscopy*, Academic Press, San Diego.
14. Petty, C.J., Bugay, D.E., Findlay, W.P. & Rodriguez, C. (1996) *Spectroscopy*, **11**, 41–45.
15. Hendra, P., Jones, C. & Warnes, G. (1991) *Fourier Transform Raman Spectroscopy*, Ellis Horwood, Chichester.
16. Pelletier, M.J. (1999) in *Analytical Applications of Raman Spectroscopy* (ed. M.J. Pelletier), Blackwell Science, Oxford, pp. 53–105.
17. Delhaye, M., Barbillat, J., Aubard, J., Bridoux, M. & Da Silva, E. (1996) in *Raman Microscopy Developments and Applications* (eds G. Turrell & J. Corset), Academic Press, London, pp. 51–173.
18. Frank, C.J. (1999) *Proc. SPIE-Int. Soc. Opt. Eng.*, **3608**, 30–36.
19. Chantry, G.W., Gebbie, H.A. & Hilsum, C. (1964) *Nature*, **203**, 1052–1053.
20. Chase, D.G. (1986) *J. Am. Chem. Soc.*, **108**, 7485–7491.
21. Hendra, P.J. (1993) *Vib. Spectrosc.*, **1993**, 5, 25–32.
22. Lewis, I.R., Daniel, Jr, N.W., Chaffin, N.C., Griffiths, P.R. & Tungol, M.W. (1985) *Spectrochim. Acta*, **51A**, 1985–2000.
23. Tedesco, J.M., Owen, H., Pallister, D.M. & Morris, M.D. (1993) *Anal. Chem.*, **65**, 441A–448A.
24. Harnley, J.M. & Fields, R.E. (1997) *Appl. Spectrosc.*, **51**, 334A–351A.
25. Adar, F. (2001) in *Handbook of Raman Spectroscopy: From the Research Laboratory to the Process Line* (eds I.R. Lewis & H.G.M. Edwards), Marcel Dekker, New York, pp. 11–40.
26. Baldwin, K.J., Batchelder, D.N. & Webster, S. (2001) in *Handbook of Raman Spectroscopy: From the Research Laboratory to the Process Line* (eds I.R. Lewis & H.G.M. Edwards), Marcel Dekker, New York, pp. 145–190.
27. Turrell, G. (1996) in *Raman Microscopy Developments and Applications* (eds G. Turrell & J. Corset), Academic Press, London, pp. 1–25.
28. Fletton, R.A., Lancaster, R.W., Harris, R.K., *et al.* (1986) *J. Chem. Soc. Perkin Trans.*, **II**, 1705–1709.
29. Wenslow, R.M., Baum, M.W., Ball, R.G., McCauley, J.A. & Varsolona, R.J. (2000) *J. Pharm. Sci.*, **89**, 1271–1285.
30. Nafie, L.A. (1997) *Ann. Rev. Phys. Chem.*, **48**, 357–386.
31. Nafie, L.A., Yu, G.-S. & Freedman, T.B. (1995) *Vib. Spectrosc.*, **8**, 231–239.
32. Clark, D. (2002) in *Handbook of Vibrational Spectroscopy Volume 5* (eds J.C. Chalmers & P.R. Griffiths), John Wiley & Sons, Chichester.
33. Rahman, S.S., Busby, D.J. & Lee, D.C. (1998) *J. Org. Chem.*, **63**, 6196–6199.
34. Camí, G.E., Chufán, E.E., Pedregosa, J.C. & Varetti, E.L. (2001) *J. Mol. Struc.*, **570**, 119–127.
35. Sacchetti, M., Varlashkin, P.G., Long, S.T. & Lancaster, R.W. (2001) *J. Pharm. Sci.*, **90**, 1049–1055.
36. The Board of Trustees, The United States Pharmacopoeia Convention Inc. (2000) *USP 24 The United States Pharmacopoeia 2000*. United States Pharmacopeial Convention Inc., Rockville, pp. 1992–1997.

37. McCreery, R.L., Horn, A.J., Spencer, J. & Jefferson, E. (1998) *J. Pharm. Sci.*, **87**, 1–8.
38. Niemczyk, T.M., Delgado-Lopez, M.M. & Allen, F.S. (1998) *Anal. Chem.*, **70**, 2762–2765.
39. Threlfall, T.L. (1995) *Analyst*, **120**, 2435–2460.
40. Harris, R.K., Kenwright, A.M., Say, B.J., *et al.* (1990) *Spectrochim. Acta*, **46A**, 927–935.
41. Bingham, A.L., Hughes, D.S., Hursthouse, M.B., Lancaster, R.W., Taverner, S. & Threlfall, T.L. (2001) *Chem. Commun.*, **7**, 603–604.
42. Byrn, S.R., Pfeiffer, R.R. & Stowell, J.G. (1999) *Solid-State Chemistry of Drugs*, SSCI Inc., West Lafayette.
43. Editorial comment (1998) *Pharmaceutical Journal*, **261**, 150.
44. Byrn, S., Pfeiffer, R., Ganey, M., Hoiberg, C. & Poochikian, G. (1995) *Pharm. Res.*, **12**, 945–954.
45. US Government. (1997) Q6A Specifications, Test Procedures and Acceptance Criteria for New Drug Substances and New Drug Products, Chemical Substances. *Federal Register*, **62**(227), 62890–62910.
46. Caira, M.R. (1998) *Topics in Current Chemistry*, **198**, 163–208.
47. Brittain, H.G. (ed.) (1999) *Polymorphism in Pharmaceutical Solids*, Marcel Dekker, New York.
48. Brittain, H.G. (1997) *J. Pharm. Sci.*, **86**, 405–412.
49. Bugay, D.E. (2001) *Adv. Drug Delivery Rev.*, **48**, 43–65.
50. Reutzel-Edens, S.M., Russell, V.A. & Yu, L. (2000) *J. Chem. Soc. Perkin Trans.*, **2**, 913–924.
51. Takla, P.G. & Dakas, C.J. (1989) *J. Pharm. Pharmacol.*, **41**, 227–230.
52. Neville, G.A., Beckstead, H.D. & Shurvell, H.F. (1992) *J. Pharm. Sci.*, **81**, 1141–1146.
53. Stephenson, G.A., Stowell, J.G., Toma, P.H., Dorman, D.E., Greene, J.R. & Byrn, S.R. (1994) *J. Am. Chem. Soc.*, **116**, 5766–5773.
54. Harris, R.K., Kenwright, A.M., Fletton, R.A. & Lancaster, R.W. (1998) *Spectrochim. Acta*, **54A**, 1837–1847.
55. Burger, A. & Ramberger, R. (1979) *Mikrochim. Acta*, **2**, 259–271.
56. Bugay, D.E. (1999) *Proc. SPIE-Int. Soc. Opt. Eng.*, **3608**, 56–63.
57. Bellows, J.C., Chen, F.P. & Prasad, P.N. (1977) *Drug Dev. Ind. Pharm.*, **3**, 451–458.
58. Morris, K.R. (1999) in *Polymorphism in Pharmaceutical Solids* (ed. H.G. Brittain), Marcel Dekker, New York, pp. 125–182.
59. Morris, K.R., Newman, A.W., Bugay, D.E., *et al.* (1994) *Int. J. Pharm.*, **108**, 195–206.
60. Pellow-Jarman, M.V., Hendra, P.J. & Lehnert, R.J. (1996) *Vib. Spectrosc.*, **12**, 257–261.
61. Bugay, D.E., Newman, A.W. & Findlay, W.P. (1996) *J. Pharm. Biomed. Anal.*, **15**, 49–61.
62. Griffiths, P.R. & de Haseth, J.A. (1986) *Fourier Transform Infrared Spectrometry*, John Wiley & Sons, New York.
63. Skrdla, P.J., Antonucci, V., Crocker, L.S., Wenslow, R.M., Wright, L. & Zhou, G. (2001) *J. Pharm Biomed. Anal.*, **25**, 731–739.
64. Langkilde, F.W., Sjöblom, J., Tekenbergs-Hjelte, L. & Mrak, J. (1997) *J. Pharm. Biomed. Anal.*, **15**, 687–696.
65. Adams, M.J. (1995) *Chemometrics in Analytical Spectroscopy*, The Royal Society of Chemistry, Cambridge.
66. Sarver, R.W., Meulman, P.A., Bowerman, D.K. & Havens, J.L. (1998) *Int. J. Pharm.*, **167**, 105–120.
67. Salari, A. & Young, R.E. (1998) *Int. J. Pharm.*, **163**, 157–166.
68. Lee, K.R., Zuber, G. & Katrincic, L. (2000) *Drug Dev. Ind. Pharm.*, **26**, 135–147.
69. Agatonovic-Kustrin, A., Tucker, I.G. & Schmierer, D. (1999) *Pharm. Res.*, **16**, 1477–1482.
70. Tudor, A.M., Chruch, S.J., Hendra, P.J., Davies, M.C. & Melia, C.D. (1993) *Pharm. Res.*, **10**, 1772–1776.
71. Jalsovszky, G., Egyed, O., Holly, S. & Hegedüs, B. (1995) *Appl. Spectrosc.*, **49**, 1142–1145.
72. Kamat, M.S., Osawa, T., DeAngelis, R.J., Koyama, U. & DeLuca, P.P. (1988) *Pharm. Res.*, **5**, 426–429.
73. Taylor, L.S. & Zografi, G. (1988) *Pharm. Res.*, **15**, 755–761.
74. Ryan, J.A., Compton, S.V., Brooks, M.A. & Compton, D.A.C. (1991) *J. Pharm. Biomed. Anal.*, **9**, 303–310.

75. Hendra, P.J. (1996) *Am. Lab.*, **28**, 17–24.
76. Taylor, L.S. & Langkilde, F.W. (2000) *J. Pharm. Sci.*, **89**, 1342–1353.
77. Barbillat, J. (1996) in *Raman Microscopy Developments and Applications* (eds G. Turrell & J. Corset), Academic Press, London, pp. 175–200.
78. Brietenbach, J., Schrof, W. & Neumann, J. (1996) *Pharm. Res.*, **16**, 1109–1113.
79. Armstrong, C.L., Edwards, H.G.M., Farwell, D.W. & Williams, A.C. (1996) *Vib. Spectrosc.*, **11**, 105–113.
80. Clarke, F.C., Jamieson, M.J., Clark, D.A., Hammond, S.V., Jee, R.D. & Moffat, A.C. *Anal. Chem.*, **73**, 2213–2220.
81. Gu, W. (1993) *Anal. Chem.*, **65**, 827–833.
82. McMahon, L.E., Timmins, P., Williams, A.C. & York, P.J. (1996) *J. Pharm. Sci.*, **85**, 1064–1069.
83. Szelagiewicz, M., Marcolli, C., Cianferani, S., *et al.* (1999) *Therm. Anal. Calorim.*, **57**, 23–43.
84. Wang, S.-L., Lin, S.-Y., Chen, T.-F. & Chuang, C.-H. (2001) *J. Pharm. Sci.*, **90**, 1034–1039.
85. Randall, C.S., DiNenno, B.K., Schultz, R.K., Dayter, L., Konieczny, M. & Wunder, S.L. (1995) *Int. J. Pharm.*, **120**, 235–245.
86. Sheppard, N. (1955) in *Molecular Spectroscopy* (ed. G. Sell), Institute of Petroleum, London, pp. 136–149.
87. Rodriguez, C. & Bugay, D.E. (1997) *J. Pharm. Sci.*, **86**, 263–266.
88. Frank, C.J. (1999) in *Analytical Applications of Raman Spectroscopy* (ed. M.J. Pelletier), Blackwell Science, Oxford, pp. 224–275.
89. Lewis, I.R. & Griffiths, P.R. (1996) *Appl. Spectrosc.*, **50**, 12A–30A.
90. am Ende, D.J., Clifford, P.J., DeAntonis, D.M., SantaMaria, C. & Brenek, S.J. (1999) *Org. Proc. R & D*, **3**, 319–329.
91. Dresselhaus, M.S., Dresselhaus, G., Pimenta, M.A. & Eklund, P.C. (1999) in *Analytical Applications of Raman Spectroscopy* (ed. M.J. Pelletier), Blackwell Science, Oxford, pp. 367–434.
92. Wang, F., Wachter, J.A., Antosz, F.J. & Berglund, K.A. (2000) *Org. Proc. R & D*, **4**, 391–395.
93. Somsen, G.W., Gooijer, C. & Brinkman, U.A.Th. (1999) *J. Chromatogr. A.*, **856**, 213–242.
94. Ruddick, A., Batchelder, D.N., Bartle, K.D., Gilby, A.C. & Pitt, G.D. (2000) *Appl. Spectrosc.*, **54**, 1857–1863.
95. Sasaki, T.A. & Wilkins, C.L. (1999) *J. Chromatogr. A*, **842**, 341–349.
96. Stahlmann, S. & Kovar, K.A. (1998) *J. Chromatogr. A*, **813**, 145–152.
97. Mulvaney, S.P. & Keating, C.D. (2000) *Anal. Chem.*, **72**, 145R–157R.
98. Ho, M. & Pemberton, J.E. (1998) *Anal. Chem.*, **70**, 4915–4920.

7 Solid-state analysis and polymorphism

Ulrich J. Griesser and Joseph G. Stowell

7.1 Introduction

The fact that the vast majority of drug products are solids represents a special challenge in pharmaceutical analysis. The main reason for the use of drug compounds in the solid state is that compounds generally have a higher chemical stability in the solid state than in solution. Moreover, solid dosage forms show a higher dosage precision, are easier to handle, and all factors result in a higher safety and reliability of the drug product. However, we often have to pay a high price for these advantages, because the solid state has many variables of which we must have knowledge, and specify, in order to avoid manufacturing problems, physical and chemical changes in drug products, or even liberation and bioavailability problems.

The variable properties of solids are connected with the ability of molecules to exist in different states of order, ranging from closely packed molecular crystals with a minimum free energy to metastable crystal phases and, finally, to the glassy state with the highest free energy. This phenomenon is commonly referred to as *polymorphism*. Lattice defects in crystals and particularly solvate formation add another level of complexity. Whether a solid in any metastable state can be handled and analysed is a kinetic issue, which again is affected by many factors (e.g. chemical impurities, solvent residues, moisture, and interactions with drug excipients).

The majority of drug compounds and excipients are known to exist in different solid-state forms (i.e. polymorphs, solvates, amorphous state), each form having different physical properties (e.g. melting point, solubility, density, and hardness). Some of these properties are more of analytical and theoretical interest (e.g. melting point, index of refraction, conductivity) and others can be considered as key properties (e.g. solubility) for drug formulation and the quality of drug products. Parallel to chemical purity, it is also important to know the physical purity of drug substances, even though drug monographs and pharmacopoeias grossly neglect the physical purity of drugs and excipients. It is obvious that in contrast to synthetic and opportunistic impurities that are detected through various forms of chromatography and spectroscopy, physical impurities are impossible to detect with any technique requiring the dissolution of the sample in a solvent. In order to test the chemical identity of drug compounds it is, on the other hand, dangerous using solid sampling techniques (e.g. by comparison of solid-state infrared (IR) spectra, a common method in pharmacopoeias) because dissimilar results may only be attributed to differences in the solid state but not differences of the chemical integrity.

The solid state becomes particularly crucial when we deal with poorly water-soluble drug compounds. In order to pass through biological membranes, the active ingredient must first be released from the solid drug product and dissolve in body fluids (liberation). This critical step strongly depends on the solid-state properties since the dissolution characteristics of a drug substance are a function of molecular, supramolecular, and particulate features. The thermodynamics of solubility (intrinsic solubility) depends mainly on the molecular (chemical) and supramolecular features. The particulate characteristics of the solid (e.g. particle size, habit) affect the dissolution rate (kinetics of solubility). Solubility is undoubtedly one of the key properties of a drug compound and it should be stressed that this property is not a quantity that exclusively depends on the molecular structure, such as, the acid/base dissociation constant (pK_a) and the lipophilicity (partition coefficient, $logP$). At present, we are not able to calculate and predict the solubility of a compound with a sufficient accuracy in contrast to the pure molecular quantities (pK_a, and $logP$) that, in most cases, can be predicted rather accurately. In order to successfully predict the intrinsic solubility of a compound, we must be able to predict the thermodynamic properties of the supramolecular state which, however, is variable (i.e. we have to deal with phenomena such as polymorphism, amorphicity, and solvate formation). Each individual crystal form exhibits different lattice energies and thus different intrinsic solubilities. However, the *ab initio* prediction of reliable solid-state properties is still unresolved. Hence, we have to measure the fundamental solid-state data experimentally and apply suitable solid-state analytical techniques. Having said this, it is clear that precise data and innovations in solid-state analysis become more and more important not only because we become increasingly aware of the impact on the solid state, on drug formulation and drug performance, but also in order to explore and understand the basic principles of the solid state and to be able to make accurate predictions.

Solid-state properties of drugs and excipients should be examined as early as possible in the development of drug products, typically in the preformulation stage [1, 2]. It is also important to realize that the assessment of the relevant solid-state properties requires a multidisciplinary approach using assembly of different analytical methods. Ideally, these methods cover different aspects of the solid state (chemical, structural, energetic, physical). The main goal of this chapter is to summarize and briefly discuss only the most important analytical techniques in the investigation of pharmaceutical solids as well as some of the prevailing solid-state phenomena such as polymorphism and related issues.

7.2 Solid-state properties of drug compounds

7.2.1 Determinant levels of solid-state properties

The properties of solids are the complex result of the molecular features, the intermolecular arrangement, and forces between the molecules as well as physical effects at the micro- and macroscopic level. This situation is summarized in Table 7.1 and will be briefly discussed.

The *molecular structure* of a drug compounds is, in the first place, dictated by its pharmacological activity, which is usually based on the capability of a molecule to interact (usually reversibly) with physiological molecules (i.e. receptor proteins). Thus, drug compounds have specific functional entities (i.e. hydrogen-bond donors and acceptors) and a molecular *body* that ensures a certain three-dimensional arrangement of these groups and their interaction with the binding sites of the biomolecule. Secondly, drug compounds should be able to pass through biological membranes. Therefore, the vast majority of drug compounds: (a) are small molecules (about 90%) with molecular weights below 600 [3] (b) show a certain hydrophilic–lipophilic balance, and (c) are weak bases (about 85%) rather than weak acids or neutral compounds. Between 40 and 50% of all drug compounds are used in a salt form, most often as a hydrochloride. Other molecular features that may finally determine the nature of the solid and have to be considered in the analysis of physicochemical properties are chirality and the conformational flexibility of the molecule. Excipients, on the other hand, comprise simple inorganic salts, small organic molecules, and polymers and represent all extremes of solid-state classes. Analytical techniques that focus on the molecular level are in the primarily spectroscopic techniques (i.e. ultraviolet-visible (UV-Vis), IR, Raman, and nuclear magnetic resonance). All these techniques can be either applied to the liquid (solution) or (sometimes with more or less technical expense) to the solid state.

The *supramolecular arrangement* (crystal lattice) of course depends on the basic molecular features (molecular size, shape, electronic properties). If there is only one stable arrangement of a molecule in the solid state (one crystal form, one packing motif), the solid-state properties of a substance are expected to be less variable. In such cases, inconsistent solid-state properties are mainly an issue of particle physics. However, the majority of drug compounds can exist in different polymorphic forms and/or may form solvates. Any crystalline state may exhibit different degrees of order (long-range order), particularly manifested in large molecules, and may be responsible for continuous changes in the physicochemical properties of a solid. Finally, the state of highest energy, the glassy state, is more or less totally disordered and the physical properties of the solid become isotropic, which means that the solid exhibits equal physical properties in all directions (in contrast to *anisotropic* crystalline solids where the physical properties depend on the direction). The most

Table 7.1 Solid-state property levels of a single substance with some analytical approaches

Level / Issue	Molecular	Supramolecular	Particulate	Multiparticulate (bulk)
Basic issues	Covalent bonding, molecular structure (configuration)	Non-covalent bonding forces multi-molecular arrangement, long-range (crystal lattice) and short-range order	Particle morphology, size, surface	Particle arrangement and packing, interparticular forces, agglomeration, aggregation
Variables and issues to be analysed	Conformation electronic properties	Polymorphism state of long-range order (no order = glass), packing	Crystal habit particle size, shape, and density	Particle-size distribution flow and mechanical properties, bulk and tap density
Dominant analytical techniques	Spectroscopy X-ray diffraction (via the supramolecular state)	X-ray diffraction thermal analysis and microcalorimetry, pycnometry, spectroscopy, solubility	Microscopic techniques micromeritics	Micromeritics powder flow characterization, compaction studies

important experimental way to explore the supramolecular state is the application of X-ray diffraction techniques. Single-crystal X-ray diffractometry yields complete structural information about the three-dimensional arrangement of the atoms (molecules) in the solid, whereas, X-ray diffraction of powdered solids captures only a two-dimensional, but unequivocal, *fingerprint* of the crystal lattice. Complementary structural information can be obtained by spectroscopic techniques. In order to measure the energetic (thermodynamic) quantities of a solid, we need to record enthalpy changes during a phase transition (solid to solid, solid to liquid, solid to gas). This is the domain of thermal analytical techniques and calorimetry. Such energetic quantities can also be obtained via the intrinsic solubility, as well as, vapour pressure measurements. Finally, the density of a solid is an important supramolecular quantity that indicates the packing efficiency of the molecules. The true density can be either calculated when the crystal structure is known or experimentally determined, for example, by helium-pycnometry.

At the *particulate level*, microscopic techniques (light and electron microscopy) must be mentioned first. Microscopy enables the characterization of the single particle in regard to crystal morphology, particle shape, size, and surface roughness, but also other features such as birefringence and other optical properties giving important indications of the crystallinity of a particle. Particle shape, size (distribution), and surface properties are the determinant factors for the properties of *multiparticulate* ensembles (i.e. powders), such as flowability, compaction behaviour, and bulk density. Because the particle and powder properties are critical in almost all processing steps of drugs (as well as of other materials), various techniques and methods are available that are directed at the characterization of the physical and mechanical properties of powders. Those methods that rather refer to the characteristics of the particles are summarized under the term micromeritics, whereas, others can be regarded rather as physicomechanical or physicotechnical methods [4]. However, these techniques are mostly applied after the material (ideally, well-characterized powder of the pure drug) has been blended with excipients and/or other drug components of the formulation. It is far beyond the scope of this short overview of solid-state analytical techniques to discuss all the methods applicable at the multiparticulate level in more detail. A few of these techniques are addressed later.

The division of solid-state properties into these four levels is of course a more didactic approach and it is obvious that the thorough understanding of material properties is always interdisciplinary and requires proper information at all levels.

7.2.2 Types and properties of pharmaceutical solids

Pharmaceutical scientists have to deal with a wide variety of materials (inorganic salts, small organic molecules, peptides, polymers, etc.). Each class of materials shows a characteristic solid-state behaviour and thus the selection of the analytical methods and the analytical strategies should be tuned properly to the kind of material in question. However, the majority of pharmaceutical compounds are small organic molecules and therefore we consider mainly this class of materials here.

A simple classification scheme of solids is given in Fig. 7.1. In order to differentiate between the types of solids, we have to consider the Gibbs phase rule, which is discussed in any physical chemistry textbook. The basic question is whether the solid substance consists of only one chemical entity (component) or more than one. Usually the component is one molecular unit, with only covalent bonded atoms. However, a component can also consist of more constituents if their concentration cannot be varied independently. An example of this is a salt. The hydrochloride salt of a base must be regarded as a one-component system as long as the acid and the base are present in a stoichiometric ratio. A deficiency of hydrochloric acid results in a mixture of the salt and the free base, which behave as two completely different substances (i.e. two different systems). Polymorphic forms, the glassy state, or the melt of the base (or the salt) are considered as different phases within such a system (a phase is defined as the portion of a system that itself is homogeneous in composition but physically distinguishable from other phases). When the base (or salt) is dissolved in a solvent, a new system is obtained; this is also true when a solvent is part of the crystal lattice, as in the case of a solvate. Thus, each solvate represents a different multicomponent system of a compound, whereas, polymorphic forms are different phases. The variables in the solvate are the kind of solvate (hydrate,

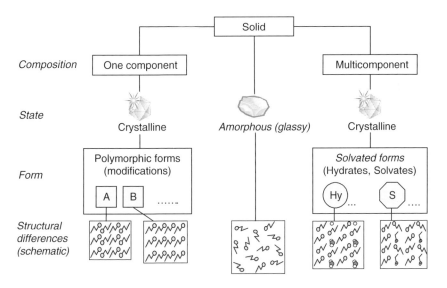

Fig. 7.1 Simple classification of solid-state forms.

ethanol solvate, acetone solvate, etc.) and the amount or molar ratio (e.g. monohydrate, dihyrate, etc.) of the solvent in the crystal lattice, with each combination representing a different solid system since the composition is not the same. However, a solvate may also exist in different polymorphic forms. This case is less abundant but stresses the importance to differentiate between the meaning of *system* and *phase*. For example, aspartame hemihydrate has been reported to exist in two different crystalline states [5] and succinylsulfathiazole monohydrate in three [6]. Within any *system* (constant composition), we may find different polymorphic forms regardless whether the system consists of one or more molecular units.

A substance (e.g. drug compounds) can exist in several polymorphic (one component, ansolvate) and solvated forms but, in most cases, the number of stable crystal forms is limited. One task of the preformulation work is to find and characterize all possible forms, including their interrelations. This can be challenging and laborious depending on the number of existing forms. Figure 7.2 illustrates the possible production ways and interrelations of the most important solid-state types, which are discussed in more detail in the following paragraphs.

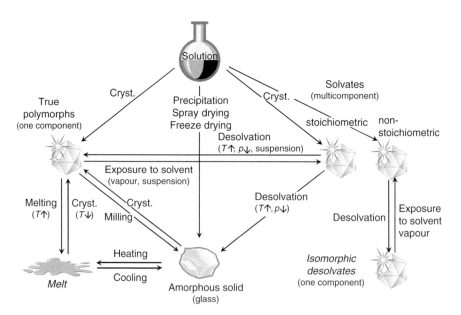

Fig. 7.2 Simple scheme showing the most important types of solid-state forms and the most relevant production processes along with their interrelationships (slightly modified according to Yu *et al.* [114]). Key: Cryst.: crystallization; $T \uparrow (\downarrow)$: increasing (decreasing) temperature; $p \downarrow$: reduced pressure (also reduced partial pressure of the solvent of inclusion).

7.2.2.1 Crystalline solids

Polymorphic forms As already discussed in the previous paragraph, two crystalline solids are regarded as polymorphic forms (or polymorphic modifications) when the difference only concerns the supramolecular arrangement in the crystal but not their chemical composition. In other words, we can discriminate between these species only in the solid state but the solution or melt of two polymorphs is indistinguishable. The consequences of the structural differences are more or less marked differences in the physicochemical properties. One set of analytical methods aims specifically at these physicochemical properties, particularly thermodynamic quantities, others at structural (spectroscopy, diffraction methods), optical (microscopy), or mechanical properties. Table 7.2 summarizes most of the properties that may differ among various solid-state forms and may thus be the object for analytical assays.

One of the most important analytical tasks is to determine the thermodynamic relationship between the different polymorphic forms (i.e. to determine

Table 7.2 Some factors for analytical questions to distinguish and characterize different crystal forms

Structural properties	Crystal structure (space group, unit cell dimension, atomic positions, molecular packing, and disorder)
	Molecular conformation
	Hydrogen bonding
	Order–disorder phenomena
	Packing density
Thermodynamic quantities	Temperatures of phase transitions (melting point and transition temperatures)
	Enthalpies of phase transitions (fusion, solid–solid transformation, sublimation, and solution)
	Heat capacity
	Vapour pressure
	Equilibrium solubility
	Entropy and free energy (thermodynamic stability)
Kinetic properties, stability	Transition kinetics (activation energy)
	Physical stability
	Dissolution rate
	Chemical stability
Surface properties	Surface free energy
	Interfacial tensions
	Surface roughness
Optical properties	Morphological properties
	Birefringence and refraction indices
	Hardness
Mechanical properties	Tensile strength
	Compaction behaviour (elasticity, plasticity, brittleness, and fracture properties)

whether a form is thermodynamically stable or unstable within the temperature region relevant to drug production and storage). There are two possible relationships: monotropism or enantiotropism. In the case of monotropically related pairs of polymorphs, one form is thermodynamically stable in the entire temperature range, whereas, the other form is always unstable. If a thermodynamically unstable form can be produced and handled, we use the term *metastable*, implying that such a form is *kinetically* stabilized. A metastable state has the potential (higher free energy) to transform (in the solid state) to a more stable one (with lower free energy) at any time, but some excess energy is required to overcome the activation barrier and to trigger the transformation process. This can easily happen during various processing steps [7] or storage; therefore, only a metastable form with a high kinetic stability should be considered for drug production. In monotropically related systems, only a transformation from the less stable form to a more stable form is possible in the solid state, but not the reverse. In order to transform a stable form to a metastable form, it is imperative to pass through the liquid (solution, melt) or vapour state (sublimation).

Enantiotropically related crystal forms can be interconverted in the solid state. At a certain temperature, the transition point, the free energy of both forms is equal, which means that their thermodynamic stability is the same. Below this temperature, one form is thermodynamically stable, but above this temperature, the other form becomes the stable form. Thus it is important to know this transition temperature.

In order to quickly estimate the thermodynamic relationship of polymorphic forms, Burger and Ramberger [8, 9] defined certain rules. The *Burger–Ramberger rules* are based on thermodynamic laws and the strength of this concept is that only few experimental data are required to distinguish between monotropism and enantiotropism. Mainly thermal analytical data are required but also the true densities of the polymorphic forms (sometimes their IR spectra) and solubility data can be used. The graphical solution of this approach is a semischematic energy/temperature diagram, which displays the entire thermodynamic situation of a polymorphic system. For further reading, we refer to the original articles of Burger and Ramberger as well as later reviews that are focused on the thermodynamics of polymorphic forms [10, 11]. Important steps and questions in the characterization of a polymorphic drug substance in the preformulation phase are summarized below.

- Perform a polymorph screening (how many forms do exist?) by crystallizing the substance under various conditions from different solvents, the melt, or the vapour phase (sublimation).
- Find the best and fastest method (X-ray powder diffraction, IR or Raman spectroscopy, thermal analysis) to identify each individual form.
- Determine how the forms are thermodynamically related to each other (enantiotropism, monotropism, apply the Burger–Ramberger rules).

- Determine which form is thermodynamically stable at room temperature.
- Determine the kinetic stability of the metastable forms (how fast do they transform to a more stable form at certain conditions?).
- Establish how the polymorphic forms differ in relevant properties (solubility, chemical stability, hygroscopicity, mechanical properties, etc.).
- Examine if it is favourable to use one of the metastable forms (equilibrium solubility should be known) in the solid formulation for poorly soluble compounds.

Solvated crystal forms (pseudopolymorphic forms) Solvates can be regarded as co-crystals between a molecular compound and a solvent, most frequently water (hydrate). One of the main questions concerns the stability of this binary system, which belongs either to that of molecular compounds (also called molecular complexes) or that of solid solutions or inclusion compounds. The terminology of solvates is very confusing and depends on whether we look at these systems from the crystallographic point of view or thermodynamics. Sometimes the solvent is an integral part of the lattice and must be present in order to maintain the chemical stability of the drug in the solid state. Such solvates are usually stable and the removal of the solvent requires rather extreme conditions (e.g. higher temperatures, vacuum). In other solvates, the solvent is rather loosely bound and one has to determine the conditions that are critical for the stability as well as the extent and kind of changes in the crystal lattice. Desolvation may lead to different solid forms and, in most cases, the lattice reorganizes to a new lattice type resulting either in one (or more) polymorphic form or a solvate of lower stoichiometry. Removal of the solvate may also lead to a complete collapse of the lattice (i.e. the glassy state). With other classes of solvates, the lattice type is more or less maintained upon desolvation and the resulting ansolvated phase is isostructural (isomorphic) to the original solvate. Such solvates are called non-stoichiometric as the solvent can take on all values between zero and a multiple of the molar compound ratio, depending on the vapour pressure of the solvent in the environment of the solid. The term *isomorphic desolvate* or *desolvated solvate* describes the fact that the solvent-free crystals of non-stoichiometric solvates show the same molecular packing as the parent solvate. In general, the isomorphic desolvate could be regarded as another polymorph (true polymorph) but the fact that such phases can only be obtained via the solvate and can only be stabilized in the ansolvated state in the absence of solvent molecules (since they readily reuptake the solvent) justifies their treatment as extra class of solids.

It should be stressed, that a successful analysis of solvated crystal forms requires additional methods besides those mentioned with true polymorphs. Such methods are, for example, thermogravimetry or Karl–Fischer titration and moisture sorption/desorption analysis for hydrates. In order to perform studies of solvates at non-ambient conditions (dry atmosphere, certain vapour pressure conditions, low and high temperature) spectroscopic and X-ray diffraction instruments require specific equipment. Also, combinations of several methods can be very advantageous in the characterization of solvates such as coupling of thermogravimetric analysis (TGA) with IR or mass spectrometry. Coupled methods become particularly interesting when more than one solvent is entrapped in a solvate.

Specific analytical objects and questions concerning solvates and hydrates in addition to those listed in Table 7.2 are the following:

- Determine the kind of solvent(s) present in the solvate.
- Determine the stoichiometric ratio of solvent to (drug) compound (if possible at different vapour pressures of the solvent).
- Determine the solvation/desolvation behaviour depending on the vapour pressure of the solvent (hydrates: moisture sorption/desorption isotherm).
- Characterize the thermal desolvation behaviour (kinetics, activation energy, usually by thermogravimetry).
- Determine if it is favourable to use a solvate (usually hydrate) instead of a solvent-free crystal form.
- Determine the conditions at which the solvate (hydrate) looses its solvent, as well as those where any ansolvated form transforms to the solvate.

7.2.2.2 Amorphous solids

The ability of organic compounds to form crystals decreases with increasing molecular size and thus many large drug molecules, proteins, and polymers form glasses rather than crystals. However, small organic molecules also form glasses that are more or less stable. A glass is generally defined as a liquid that has no fluidity. The glassy state is of particular interest for poorly water-soluble drug compounds because this highly energetic state usually shows a higher solubility than any crystalline phase. An amorphous solid represents a metastable state that theoretically may crystallize at any time. Usually the amorphous form is also less chemically stable and more hygroscopic than a crystalline form. Amorphous forms may result from different pharmaceutical processes (e.g. spray drying, milling, lyophilisation, granulation, fast precipitation from solvents, desolvation of solvates) and generation of partially amorphous material, or some decrease in the crystalline order, is very common during the formulation of solid dosage forms. In general, the presence of amorphous material is undesirable in formulation because physicochemical or mechanical properties of solid or semisolid dosage forms may change during storage. In order to stabilize the glassy state and to increase the dissolution rate of poorly water-soluble drugs, solid dispersion techniques are frequently applied in which a glass can be preserved in a water-soluble polymer matrix such as polyvinylpyrrolidone (PVP).

The characterization and detection of amorphous materials require strategies other than those of crystalline solids. Amorphous solids lack the melting point and birefrigence typical of crystalline solids; they show spectral broadening (vibrational, solid-state NMR) and only broad and featureless X-ray powder diffraction patterns because of the missing long-range order. Therefore, analytical assays to determine the amount of amorphous form in solids focus rather on the presence (absence) of the more characteristic crystalline state rather than that of the less characteristic amorphous state. Since the amorphous form of a compound is the state of highest energy among all existing solid-state forms, it is possible, for example, to determine the amount of amorphous content by measuring the enthalpies of solution (solution calorimetry). Other strategies benefit from the higher reactivity of the amorphous state; for instance, the ability to uptake moisture in significantly higher amounts than any of the crystalline counterparts. The most specific descriptor of an amorphous solid is the glass-transition temperature (T_g). This higher-order transition indicates the change from the glassy state (solid, *frozen* or low molecular motion) to the supercooled liquid state (also called rubbery state, high molecular motion) and can be recognized by a distinct change in heat capacity. Unlike the melting point of a crystalline solid, the glass-transition temperature is kinetically controlled and sensitive to the sample history or impurities (including water). It has been shown that T_g values of organic compounds can be roughly predicted from the melting points (T_m) of a compound. The T_g/T_m (in Kelvin) ratios lie in most cases [12] between 0.7 and 0.85 (roughly 2/3 to 4/5). Since the crystallization tendency of the supercooled melt above the T_g is high compared to the glassy state (below T_g), it is important to know this value in order to predict the stability of a glass at different storage temperatures. A number of thermal analytical and calorimetric methods are applicable for the characterization of amorphous solids (e.g. enthalpy relaxation, heat capacity, T_g). For a detailed review about amorphous solids and their characterization, we refer to Yu [13].

7.2.2.3 Regulatory aspects and quality control
Owing to the inherent physical and chemical instabilities of thermodynamically less stable forms, the International Committee on Harmonisation has issued a guidance [14] containing a decision tree for 'investigating the need to set acceptance criteria for polymorphism in drug substances and drug products' (see Figs 7.3 and 7.4). However, note that Part 2 in Fig. 7.3 states that if different polymorphs may be formed no further test or acceptance criterion is needed for the drug substance if the drug safety, performance, or efficacy is not affected. Furthermore, Fig. 7.4 states that no acceptance criteria need to be established for polymorphic changes in a drug product if the change does not affect the safety or efficacy of the product. Thus, although a single form of the drug substance is preferred, it is imperative to establish the physical form of the drug substance and ensure that the performance of the drug product is not compromised.

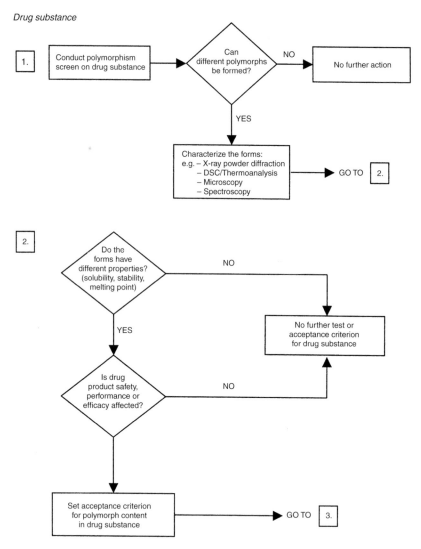

Fig. 7.3 ICH Q6A decision tree that investigates the need to set acceptance criteria for polymorphism in drug substances [14].

Quality control of solid pharmaceuticals is essential to verify the integrity of the drug substance and drug product and such materials require special, yet simple, techniques and methods to analyse these samples. The subsequent sections discuss these methods and techniques and are presented to give an overview of the field but not an in-depth treatise of the subject.

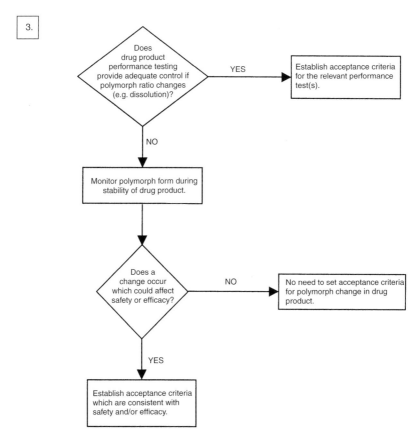

Fig. 7.4 ICH Q6A decision tree that investigates the need to set acceptance criteria for polymorphism in drug products [14].

7.3 Methods of solid-state analysis

In order to characterize solids, we need an assembly of techniques to cover different types of information. These types can be roughly divided into those providing primarily structural information and methods that measure preferentially thermodynamic or particle and bulk properties.

Structural properties	X-ray diffraction methods (powder, single crystal) Spectroscopy (UV, IR, Raman, solid-state NMR)
Thermodynamic properties	Thermal analysis, microcalorimetry, solubility determination, vapour pressure determination Moisture sorption
Particle and bulk properties	Microscopy and micromeritics

7.3.1 X-ray diffraction and crystallography

X-ray diffraction is a formidable tool for the examination of crystalline material. The theory and practice of single-crystal X-ray diffractometry will be covered here only briefly, since there are many reference sources for in-depth information [15–20].

Using radiation with wavelengths in the vicinity of a bond length, X-ray diffractometers can probe the structure of simple ionic compounds (e.g. sodium chloride), molecular compounds (e.g. steroids), and complex assemblies (e.g. hemoglobin). For organic molecules, the typical X-ray source is a hot cathode tube with either a molybdenum anode (Mo $K_\alpha = 0.70926$ Å with an achievable resolution of 0.35 Å) or a copper anode (Cu $K_\alpha = 1.5405$ Å with an achievable resolution of 0.75 Å). The X-rays from these sources diffract off the electrons surrounding the atoms in the sample. Thus, the location of hydrogen atoms, especially acidic hydrogen atoms, is difficult to pinpoint with high accuracy because of the low electron density surrounding these atoms. Conversely, atoms with high electron density, such as bromine or iodine, diffract X-rays very well and thus were once incorporated in molecules as derivatives of the parent molecule to enhance the diffraction patterns and aid in the solution of the crystal structure.

Modern X-ray diffraction equipment is equipped with high-energy X-ray sources from rotating anodes as well as CCD detectors coupled with high-speed computers that can acquire and interpret diffraction data at many angles simultaneously in a relatively short amount of time. Hence, where it once took many months to acquire the diffraction data on a large crystal of excellent quality and then interpret the data using a slide rule to formulate the structure of a given molecule, today it is possible to acquire data within an hour or so on a small crystal (approximately 100 µm in each dimension) and determine the structure in a few minutes on a personal computer. Establishment in 1965, the Cambridge Crystallographic Data Centre (CCDC) oversees a searchable structural database that extracts crystallographic data from over 1400 journals and now contains more than 257 000 entries [21].

Powder X-ray diffraction (PXRD), also known as X-ray powder diffraction (XRPD), is a routine method for both the qualitative and quantitative analysis of solid materials [22, 23]. As the name implies, the samples in PXRD analyses are essentially powders but can also be compacts. When a monochromatic beam of X-rays impinges on a crystalline sample, the angle θ at which the radiation is diffracted is related to the interplanar spacing d within the crystals and is given by the Bragg equation: $n\lambda = 2d \sin\theta$, where λ is the wavelength of the X-ray beam and n is any positive integer. Thus the inherent arrangement of molecules within a crystal dictates the resulting pattern of diffraction peaks, both position and intensity. The data obtained from a powder diffractometer is typically presented as intensity, either as counts per second or relative to the highest peak, vs diffraction angle, typically as $°2\theta$ (degrees two-theta). Most pharmaceutical compounds will diffract X-rays in the range of $4-40°2\theta$ from a copper source ($\lambda = 1.5405$ Å).

The utility of PXRD in pharmaceutical analysis is that each crystal form of a drug will produce a diffraction pattern that can be used as a fingerprint for that form of the drug and thus can be used to screen for polymorphs and solvates during drug discovery, formulation development, and manufacturing. Furthermore, PXRD patterns can be calculated from single-crystal data (see Fig. 7.5) and used as reference standards. Notice that the production sample contains a mixture of polymorphs (the intensity scale has been expanded fivefold to emphasize the minor peaks) – the sample consists mostly of the red prism form (corresponding to the ticks below the PXRD pattern labelled RP) and some yellow prism form (corresponding to the ticks below the PXRD pattern labelled YP), yet there are peaks that do not correspond to any of the known forms indicating the possibility of at least one new, unidentified form (indicated by the arrows). The International Center for Diffraction Data (ICDD) maintains the powder diffraction file (PDF) which contains 148 379 inorganic and organic powder diffraction patterns including calculated patterns from the CCDC [24].

One feature of nearly all organic crystals is that they are anisotropic, that is, the molecules within the crystal are not symmetrically arrayed with respect to the various faces of the crystal. Thus, the PXRD pattern for a single large crystal would appear to be different when X-rays are diffracted from only a single face of the crystal. This situation gives rise to preferred orientation anomalies when dealing with material that is composed of crystals that are shaped like plates or needles. In other words, when a collection of cards is tossed in the air, the end result will afford an assembly of cards in which some will be face up and some will be face down but all of the cards will be oriented randomly with respect to rotation about the plane of the card; however, very few, if any, of the cards will be on an edge. Thus, the assemblage of cards has a preferred orientation in which the plane of each card is parallel to the plane of the collecting surface.

The effects of preferred orientation on PXRD patterns are illustrated in Fig. 7.6 which presents diffraction patterns of a set of theoretical aspirin crystals. Although only one crystal form of aspirin is known, the outward appearance of

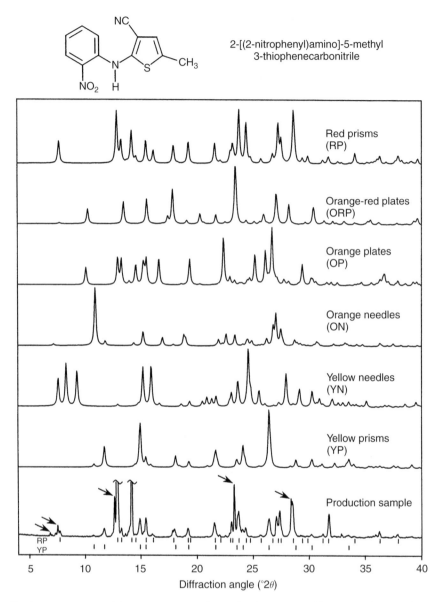

Fig. 7.5 Comparison of the PXRD pattern of a production sample of 2-[(2-nitrophenyl)amino]-5-methyl-3-thiophenecarbonitrile to those of six coloured polymorphs calculated from single-crystal data. The ticks below the PXRD pattern of the production sample (with an expanded intensity scale) correspond to the major peaks of the red and yellow prisms forms. Peaks in the production sample that are not found in the patterns of the six known forms are indicated with arrows.

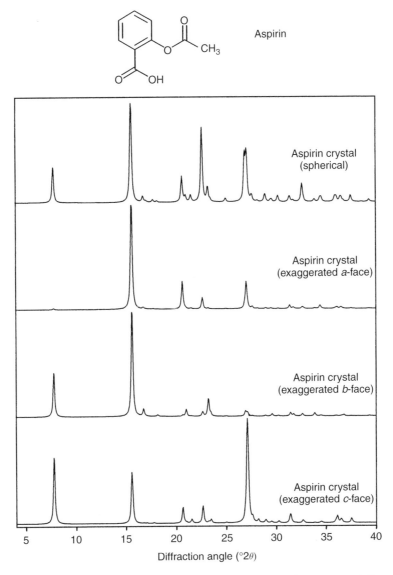

Fig. 7.6 Comparison of calculated PXRD patterns for aspirin crystals having exaggerated dimensions with that of a spherical crystal.

the crystals may be different when grown from different solvents. Thus, the morphology or habit (i.e. the shape) of the resulting crystals may vary depending on the crystallization conditions (see Fig. 7.7) yet the molecules are arrayed in exactly the same manner in each of the crystals. It is therefore important to eliminate or reduce the likelihood of preferred orientation in a sample. This

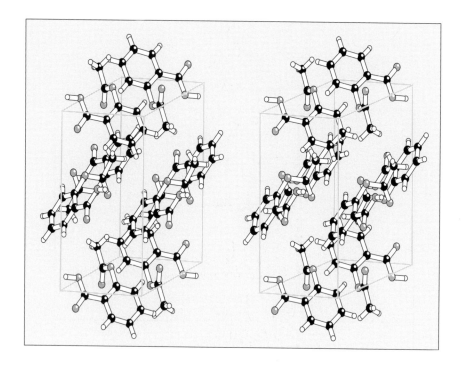

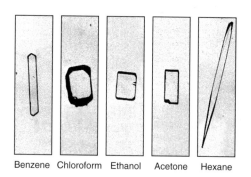

Fig. 7.7 Habits of aspirin grown from five organic solvents.

may be accomplished by grinding the sample to a powder, back filling the powder in a sample holder, spinning or rocking the sample during data acquisition, or a combination of these methods. One caveat about grinding is that it may induce phase transformations (i.e. polymorphic transformation, desolvation, or amorphization) that may lead to artefacts in the resulting data.

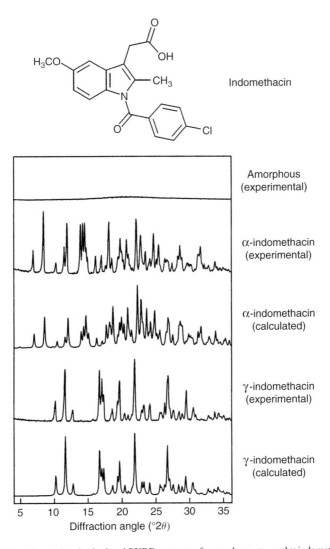

Fig. 7.8 Experimental and calculated PXRD patterns of amorphous, α-, and γ-indomethacin.

Since amorphous material contains no long-range order, an amorphous sample generally gives a featureless, broad diffraction pattern of relatively low intensity, called an amorphous halo (see Fig. 7.8). However, it is possible that an amorphous sample may contain some short-range order (e.g. carboxylic acid dimers or π–π stacking) and thus may exhibit some noticeable features. Some materials may exhibit liquid-crystal-like behaviour and exhibit an amorphous halo with one or more intense, sharp diffraction peaks owing to the spacing between aligned strands of long alkyl groups. Depending on the crystal structure and hence the arrangement of molecules within the crystal, samples that are composed of particles less than 0.1 μm may appear to be *amorphous* by PXRD methods and should be checked for birefringence using a polarizing optical microscope or for the presence of a glass transition or melting point by differential scanning calorimetry (DSC). If a featureless pattern is obtained though the powder consists of crystallites, we talk about *X-ray amorphous samples*.

7.3.2 Spectroscopy

As spectroscopic methods, UV-Vis, IR, Raman, and nuclear magnetic resonance (NMR) spectroscopy are specifically treated by other chapters of this book. The following section attempts to summarize exclusively the use of these techniques in the characterization of phenomena related to the solid state. Each technique is based on a different phenomenon (UV-Vis and IR – energy absorption, Raman – inelastic light scattering, NMR – energy absorption of samples in a magnetic field) and allows a different insight into the molecular and supramolecular structure of a solid [25]. Spectroscopy displays the short-range structure of a molecular solid (electronic environment of molecular functions) and is thus complementary to structural information obtained from X-ray diffractometry (which displays the long-range order structure). Successful solid-state characterization work requires at least one spectroscopic technique. The choice of the most appropriate instrumentation depends on the scope of analytical work (quantitative analysis, polymorph screening, stability testing, etc.) but most solid-state laboratories are equipped with different types of spectrometers in order to benefit from the advantages of each technique.

7.3.2.1 *Ultraviolet-visible diffuse reflectance spectroscopy*
UV-Vis spectroscopy in solution is probably one of the most frequently applied spectroscopic methods in the quantitative analysis of pharmaceuticals (see other chapters of this book). In solid-state analysis, this situation is quite the opposite since most solids are too opaque to permit the use of this technique in the conventional transmission mode. UV-Vis spectroscopy on solids can only be realized via diffuse-reflection techniques connected with mathematical corrections (e.g. Kubelka–Munk function) and lacking the high reproducibility of UV-Vis spectroscopy in solution owing to particle dispersion effects. The number of published papers on the application of UV-Vis spectroscopy to solid pharmaceuticals is very small and these papers include topics such as photostability of dyes and active ingredients in tablets, drug-excipient interactions in drug products, quantitative measurements on discolouration in drug products, and others. For further reading we refer to Brittain [26] and the literature cited therein.

7.3.2.2 Infrared spectroscopy

The IR region of the electromagnetic spectrum extends from ~10 to 20 000 cm^{-1}. Because of instrumental and functional reasons this region is divided into near-, mid-, and far-infrared spectroscopy.

Mid-infrared (MIR) spectroscopy MIR spectroscopy (frequency range: ~2.5–25 nm or 4000–400 cm^{-1}) is a popular technique for identification assays (chemical identity) of drug substances in pharmacopoeias. The fact that different solid-state forms exhibit different MIR spectra represents rather a problem for such purposes. Thus, in the case that the spectrum of a compound, whose identity needs to be determined or confirmed, is different than that of the reference compound, pharmacopoeias suggest either to record the spectra in solution or to recrystallize both the substance and the reference using the same method before recording their solid-state spectra.

IR-active molecular vibrations require a change in the molecular dipole and thus polar groups are more likely to show prominent IR active bands. Hence IR spectroscopy is particularly sensitive to intermolecular dipole interactions, hydrogen bonding, and other kinds of associations, which again are features of many drug molecules owing to the abundance of polar functional groups. Crystal forms of a compound that differ in their intermolecular hydrogen-bonding arrangements are thus always distinguishable by MIR spectroscopy. However, if the different packing arrangements in polymorphs are based on only weak interactions (e.g. van der Waals or London forces), the MIR spectra may be barely distinguishable. This situation may be illustrated by different crystal forms of caffeine and theophylline (Fig. 7.9). The spectra of caffeine Mod. I and II are almost identical, and only minimal differences in a few bands can be recognized at a closer inspection (marked with arrows in Fig. 7.9). Therefore, these two forms have frequently been reported to be undistinguishable by MIR spectroscopy. Owing to the absence of donor groups, the planar caffeine molecules arrange in stacks. In the high-temperature form (Mod. I), the caffeine molecules are highly disordered (orientational disorder) but the stacking is similar to that of the low-temperature form (Mod. II), which can be inferred from the X-ray powder pattern. Caffeine also forms a hydrate with an unusual stoichiometry (0.8 mol water per mol caffeine). The water is rather loosely bound (which can also be concluded from the broad water band at 3500 cm^{-1}), but interactions of the polar water molecules with the caffeine molecule are strong enough to induce the formation of a new crystal lattice. The caffeine molecules in the hydrate are also stacked [27]. The disordered water molecules are arranged in tunnels and interact with the unsubstituted pyridine nitrogen [28]. Thus, the MIR spectrum of the hydrate (Fig. 7.9) shows distinct differences compared to the spectra of the anhydrous forms.

Theophylline differs from caffeine solely by the lack of a methyl group in position 7 of the xanthine ring. Owing to the presence of a hydrogen-bond

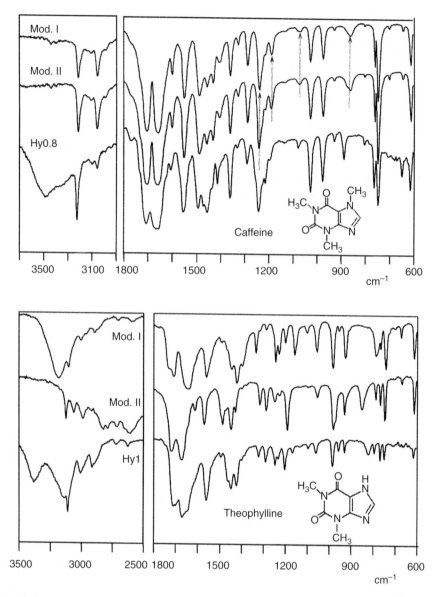

Fig. 7.9 MIR spectra of crystal forms of caffeine and theophylline (Mod. I and Mod. II: polymorphic forms; Hy0.8 and Hy1: hydrates with 0.8 and 1 mol water per drug molecule, respectively).

donating N–H function, the molecule is able to form hydrogen bonds in the ansolvated forms as well as the hydrate. The spectra of these forms are all vastly different even though the molecules in the high-temperature form (Mod. I) are also disordered (therefore, a broad band at about $3200\,cm^{-1}$ is evident). The stronger hydrogen bonding of the water molecules in theophylline monohydrate, compared to caffeine hydrate, is obvious by the shift of the water band to lower wavenumbers.

Figure 7.10 shows the MIR spectra of crystalline and non-crystalline forms of the theophylline derivatives proxyphylline and diprophylline (dyphylline). The chemical structures of the two compounds are the same except for an additional hydroxyl group in diprophylline. These chiral compounds are used as racemate, and since the 1:1 ratio of the enantiomers is maintained in all forms, they must be regarded as polymorphs of a multicomponent system (see previous section). The spectra show clear differences and give a rough insight into structural features of polymorphism without knowing the single crystal structure. The presence of one (proxyphylline) or two (diprophylline) hydroxyl groups is readily discernable in the spectra of the stable forms (Mod. I for each compound) for the O–H stretching vibration (3300–3500 cm^{-1}). Even though Mod. II of proxyphylline is less stable, the hydrogen bond interaction must be stronger than in Mod. I, which is evident from the distinct shift to lower wavenumbers of the O–H stretching vibration. The v_{O-H} band is split in the spectra of proxyphylline Mod. III and diprophylline Mod. II indicating that more than one kind of hydrogen-bonding interactions are present. The absorption bands of the amorphous states show broadening as expected.

These samples demonstrate that MIR spectroscopy provides important structural information about solid-state forms, particularly where different hydrogen-bond associations are present as well as for hydrates and solvates. However, from our experience, MIR spectra of conformational polymorphs [29] with aliphatic chains are often barely distinguishable owing to rather weak interactions between the molecules. In order to identify or quantify such forms by spectrometry, Raman or solid-state NMR spectroscopy stands a better chance of success.

Developments in MIR spectrometers, particularly Fourier-transform (FT) techniques, have enabled the use of a variety of solid sampling techniques which overcome the disadvantages of classic IR-sampling techniques. Classic sampling techniques [30], such as alkali halide pellet preparation (with KBr or KCl) or mineral oil mull preparations, require a mechanical treatment of the sample and may thus induce solid–solid transformations or desolvations.

The diffuse reflectance method (DRIFTS – diffuse-reflectance infrared Fourier-transform spectroscopy) is a popular alternative sampling technique [31] that requires no specific sample preparation or only a dilution with a non-absorbing material (e.g. KBr or KCl). The main disadvantage of DRIFTS is the sensitivity to particle-size effects, which must be considered for quantitative assays.

Attenuated total-reflectance (ATR) IR spectroscopy represents another solid-state sampling technique. Here the samples are placed in close contact to a crystal that is transparent in the IR region. This technique provides only spectral information about the sample surface in direct contact with the ATR crystal and is thus applied in the characterization of coatings rather than in bulk material.

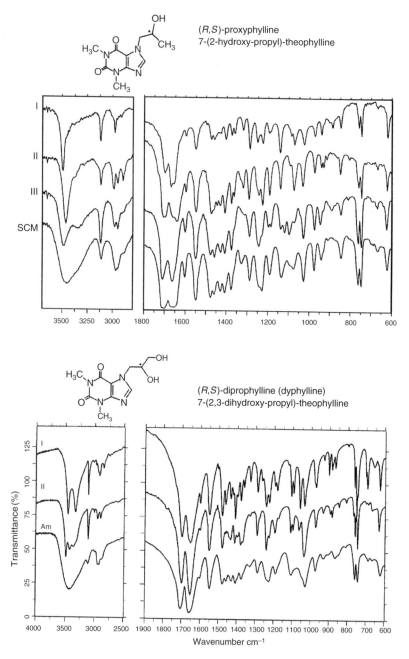

Fig. 7.10 FT-IR spectra of (R,S)-proxyphylline and (R,S)-diprophylline polymorphs and the non-crystalline state; SCM: supercooled melt, Am: glass [51, 107].

Photoacoustic FT-IR spectroscopy (PAS) is based on the absorption of modulated IR radiation by a sample. This non-destructive method involves the absorption of a pulse of light energy by a molecule and the subsequent detection of a pressure wave generated by heat energy released by the molecule upon its return to the ground state [32]. Even though this method is very attractive for the characterization of pharmaceutical solids (e.g. spectra can be obtained from opaque materials such as formulations, possibility of depth profiling, or simple sample preparation), the number of reports in this field is very limited. Studies on polymorphic systems involving PAS have been reported [33, 34].

FT-IR microspectroscopy [35], the coupling of optical microscopy and FT-IR spectroscopy, is one of the most useful IR sampling techniques for the analysis of solid-state problems. The ability to record spectra of samples as small as 20 μm, the option to measure in reflectance or transmittance mode, the possibility to use a hot or cold stage, and to study anisotropy effects using optical polarizers in addition to other features provide an enormous potential for analytical information. Since IR light is absorbed by conventional glass lenses, IR microscopes require special all-reflective optics (Cassegrainian objective). In order to measure in the transmission mode, the samples must be placed on IR-optical plates (e.g. KBr, NaCl, ZnSe, etc.), whereas, silver mirrors are used for the reflection mode. In both modes, the specimen should be rather thin and must usually be flattened by applying some pressure. By using two IR-optical plates as a sandwich, spectra of melt-film preparations can be readily collected in transmission mode. As a rule, a larger number of metastable polymorphic forms are obtained by crystallizing from the supercooled melt than by any other method. The IR spectra of most of the polymorphs may be obtained through such film preparations before any solvent crystallizations have been performed. This is particularly valuable when only a few milligrammes of substance are available for polymorph screenings. Also, the spectra of very unstable or simultaneously crystallizing forms can be easily recorded in this manner [36]. Figure 7.11 shows photomicrographs of such film preparations with the different polymorphic forms of indomethacin and the corresponding IR spectra.

It should be mentioned in advance that FT-IR spectrometer is used increasingly in combination with thermogravimetry to identify the released components (solvents or decomposition products) during the heating process. The applicability of some of the IR-sampling methods for the characterization of solid-state forms is discussed in more detail by Threlfall [37] and Bugay [25] and quantitative analysis has been reviewed by Stephenson *et al.* [38].

Near-infrared (NIR) spectroscopy The NIR region of the electromagnetic spectrum ranges from 800 to 2500 nm (4000–12 500 cm^{-1}). NIR-absorption spectra arise from overtones and combinations of vibrational modes. These vibrations are rather weak in intensity compared to the fundamental bands (e.g. O–H, N–H, or C–H stretching vibrations) and are highly overlapping.

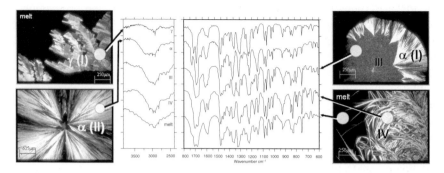

Fig. 7.11 Photomicrographs of the different forms of indomethacin (crystal film, polarized light) and IR spectra recorded with an FT-IR microscope of corresponding crystal films prepared between ZnSe discs.

Owing to recent advances in instrumentation and multivariate data analysis (chemometrics), NIR spectroscopy is an increasingly applied method for qualitative and quantitative assays in the pharmaceutical industry [39]. Multivariate data analysis is required for precise quantitative analysis because of the highly overlapping absorption bands. The classical and most widely applied application of NIR spectroscopy is the quantitative determination of water in materials. NIR spectra of solids can be either measured in transmission or diffuse reflection mode, but the main advantages concern the simple sample preparation and the fact that NIR radiation is not absorbed by ordinary glass. This feature, as well as the developments in fibre optics and computer technology, results in NIR spectroscopy as the method of choice for on-line process monitoring and real-time process control in pharmaceutical industry. The application areas in the solid-state analysis of pharmaceuticals are manifold and cover fluid-bed drying [40] and blend analysis [41–43] as well as characterization and quantification of solid-state forms of drug compounds and solid drug products [44, 25, 38] and other areas. The nuisance factor of diffuse-reflection NIR spectroscopy is the sensitivity to particle-size differences that, for quantitative assays, require adequate calibrations, careful sample preparation, and the application of scatter-correction methods. However, it has been shown [45] that these NIR-dispersion effects can be utilized as an alternative method of particle-size determination to current accepted methodologies.

The development of NIR spectral-imaging systems using IR focal-plane array detectors represents a novel technique that may provide new insights into the structure and function of solid dosage forms [46].

7.3.2.3 Raman spectroscopy

Raman spectroscopy provides information about molecular vibrations that can be used for sample identification and quantitative analysis. The Raman effect is based on the inelastic scattering of laser light by molecules or crystals. Like

IR spectroscopy, the Raman method gives vibrational spectra of the samples in the range of 10–3600 cm^{-1} (mid- to far-IR). These spectra can be used for structural analyses, identification of substances and crystal forms, and quantitative assays [47]. Since Raman spectra show the symmetric vibrational modes, particularly those of homopolar bonds (e.g. C≡C), this technique is complementary to IR spectroscopy, which predominantly displays the Raman-inactive asymmetric modes and vibrations of polar groups.

Raman spectroscopy does not require much sample preparation and may be applied to any optically accessible sample (e.g. aqueous solutions, substances in polymer or glass containers, or inclusions in transparent media). The fact that the relative intensities in Raman spectra are barely affected by sampling parameters qualifies this technique for accurate quantitative analysis of solid components or polymorphs in drug formulations. Since the laser focuses on a very small area, sample homogeneities must be considered. This problem can be minimized by different approaches (see [38] and references cited therein). Another advantage, compared to IR spectroscopy, is that Raman spectra usually exhibit a great number of well resolved and very reproducible adsorption bands. Thus, it is more likely to find spectral regions where characteristic bands of the active ingredient in multicomponent systems, such as tablets and granules, can be distinguished readily without interference from the other components [48].

Raman spectroscopy is ideally qualified for use with an optical microscope [49] enabling the investigation of extremely small amounts of sample down to a size of 1 µm. By the use of confocal microscopes, different depth layers of samples may be observed. In confocal-Raman microscopes, the parallel laser light is focused on a diffraction-limited spot on the sample with a microscope objective of high numerical aperture. Light scattered from this spot in a backward direction is collected and collimated by the same objective and focused through a pinhole onto a photodetector. Because only the focused light can penetrate through the pinhole, light from other depths in the sample is efficiently blocked; out-of-focus regions do not blur the signal. This provides for the possibility to investigate thin films (e.g. tablet coatings) and to obtain three-dimensional microscope images of samples, which is of particular interest for the analysis of drug formulations.

It is important to know that in contrast to other vibrational-spectroscopy techniques, where the FT instruments generally show considerable advantages over dispersive instruments and thus have more or less replaced conventional grating spectrometers completely, dispersive and FT techniques in Raman spectroscopy have different features. The FT technique allows for the use of an NIR laser (1064 nm) that minimizes fluorescence problems and gives high-quality spectra of the mid- and far-IR region in a very short time. Excitation with NIR light results in a much lower Raman intensity that is compensated by the use of an interferometer (Jaquinot advantage). Owing to the high quality of the spectra, FT-NIR–Raman spectroscopy is favourable for the analysis of

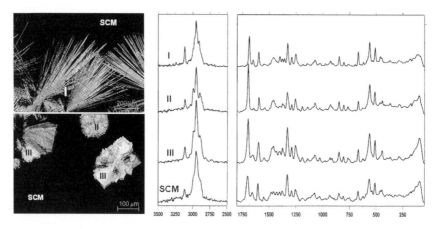

Fig. 7.12 Polarized-light photomicrographs of the different forms of (R,S)-proxyphylline (Mod. I, II, and III, SCM: supercooled melt) in a melt-film preparation between a glass slide and a cover slip as well as the corresponding Raman spectra, recorded with an NIR–FT-IR microscope directly from the film preparations [107].

bulk materials (e.g. raw material characterization, quantitative and qualitative analysis, final product analysis, detection of polymorphic forms, etc.). Dispersive Raman spectrometers use visible lasers (e.g. 758, 633, and 532 nm), which yield higher Raman light intensities and thus higher sensitivities. Because of this and optical reasons [49], the dispersive Raman technology shows some significant advantages in combination with microscopy over FT-NIR–Raman microscopy, such as, better signal-to-noise ratios and the confocal-beam option. For Raman microanalysis, it is thus preferable to use the dispersive, multichannel technique with array-detectors that provide fluorescence rejection.

Raman microscopy is a valuable analytical tool for the characterization of solid-state forms where applications may be extended by combination with a hot stage. This has been demonstrated by Szelagiewicz *et al.* [50] using a dispersive Raman microscope as well as by Griesser *et al.* [107] applying FT-NIR microscopy (see Fig. 7.12). Note that Raman spectra also record the far-IR region below 200 cm^{-1} where the phonon vibrations of the crystal lattice can be observed.

Owing to the many advantages and the technical improvements in instrumentation (e.g. availability of stable diode lasers and high-quantum-efficiency detectors, advances in fibre-optic technology, etc.), Raman spectroscopy is moving out of the shadow of IR spectroscopy. Although these developments arose in the 1990s, the number of applications in the pharmaceutical analysis is still limited but rapidly increasing.

7.3.2.4 Solid-state NMR

Although NMR spectroscopy was used by scientists at the Massachusetts Institute of Technology to study solids more than 50 years ago [52], it took several decades until experimental techniques and advances in instrumentation afforded methods for routine acquisition of spectra from solid samples. In the ensuing years, a wealth of information has been written about solid-state NMR spectroscopy. General information about the fundamental and physical principles of solid-state NMR [106], applications for chemists [53], and reviews on pharmaceutical applications [54, 55] are readily available and for this reason only the very basics will be discussed here.

The relative abundance and the atomic nucleus characteristics of hydrogen atoms that are a boon to solution-state NMR experiments are confounding in the solid state because of dipole–dipole interactions between the abundant protons and the sparse ^{13}C nuclei (natural abundance is 1.11%). Thus, high-power proton decoupling is used to negate these interactions. Because molecules in the solid state are relatively rigid and hence have fixed orientations relative to each other and the applied magnetic field, solid samples must be spun at high speeds (typically 3–15 kHz) and at a particular angle to the applied magnetic field in order to remove these anisotropic effects. This angle is called the magic angle and is 54.7° or 54°44' for the carbon nucleus (the angle between the diagonal and one of the sides of a cube). To enhance the weak signal from the small number of ^{13}C nuclei present in a sample, a technique called cross-polarization (CP) is applied in which the magnetization of the abundant proton nuclei is transferred to the ^{13}C nuclei. The combination of high-power proton decoupling, CP, and magic-angle spinning (MAS) constitutes the basis for typical solid-state NMR experiments. With the advent of newer pulse sequences and higher spinning rates, spinning sidebands can be virtually eliminated from the spectra.

Solid-state NMR is extremely useful in investigating solid modifications of both the drug substance and the solid drug product. Crystalline material typically affords sharp and narrow ^{13}C resonance peaks similar to solution spectra. In addition, because the molecules in one polymorphic form of a drug are arrayed differently than those in another polymorphic form, differences in the solid-state spectra of the polymorphs are quite apparent. For example, Fig. 7.13 clearly shows that there are a number of differences in the solid-state ^{13}C CP/MAS spectra of the two prednisolone polymorphs [56].

Examine the five spectra in Fig. 7.14 of five sulfathiazole polymorphs [57]. Not only are the five spectra distinct, but notice that the spectrum of Form I contains several doublets. The doubling of resonance absorptions in not only Form I but also in Forms II and III (e.g. the resonance absorptions for C9) indicates that these crystal forms contain two molecules in the asymmetric crystallographic unit while Forms IV and V each contain only one molecule in the asymmetric crystallographic unit. That is, the fundamental *building blocks* of Forms I–III

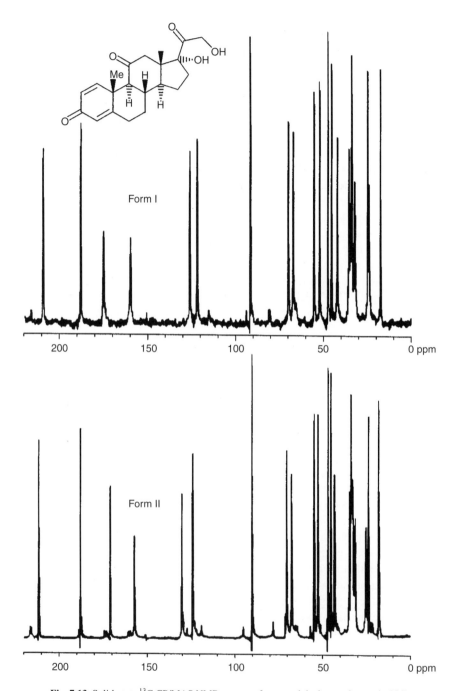

Fig. 7.13 Solid-state ^{13}C CP/MAS NMR spectra of two prednisolone polymorphs [56].

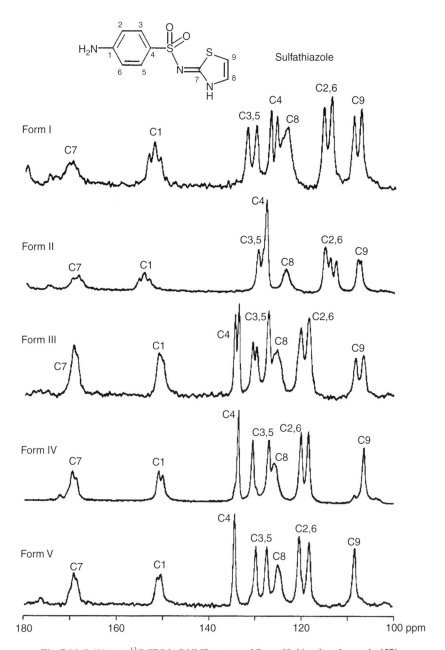

Fig. 7.14 Solid-state ^{13}C CP/MAS NMR spectra of five sulfathiazole polymorphs [57].

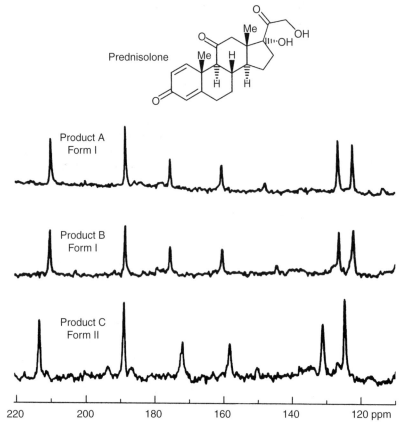

Fig. 7.15 Solid-state ^{13}C CP/MAS NMR spectra of tables from three different manufacturers containing 5 mg of prednisolone in 100-mg tablets. Note that the tablets labelled Product A and Product B contain prednisolone Form I and those labelled Product C contain prednisolone Form II [56].

contain two molecules in different environments while the molecules of Forms IV and V are in uniform environments. These observations are consistent with the crystallographic data.

Solid-state NMR is a non-destructive technique and can be used to look at not only powders but also whole tablets. The spectra of tablets from three different vendors that contain just 5 mg of prednisolone in 100-mg tablets are shown in Fig. 7.15 [56]. Although not shown in Fig. 7.15, excipients do not generally interfere with solid-state NMR analyses because the resonances owing to the most common excipients (carbohydrates, such as lactose, starch, and various forms of cellulose) occur within a narrow range of frequencies (65–80 ppm) while those of the drug substance range from 15 to 240 ppm.

Amorphous material is distinguished by broad peaks in solid-state NMR owing to the numerous conformations and environments of the molecules in the amorphous state (see Fig. 7.16). Using a partial least-squares method, Nyström and co-workers are able to detect less than 0.5% amorphous content in lactose samples [108]. The ^{13}C CP/MAS NMR spectrum of a sample containing both crystalline and amorphous lactose is simply a combination of the sharp crystalline spectrum and the broad amorphous spectrum. A correlation plot (not shown) of the predicted amount of amorphous lactose vs the observed amount of amorphous lactose produced a correlation coefficient of 0.994 over a range of 0–100% crystalline lactose.

As illustrated in the above examples, solid-state NMR is a powerful tool for both qualitative and quantitative analysis of solid pharmaceutical samples. Although the acquisition times for solid-state spectra are generally greater than for solution spectra, solid-state NMR can be used routinely for polymorph screening in conjunction with XRPD when single-crystal data are not obtainable as well as investigating excipients–drug interactions. Furthermore, compounds containing fluorine or phosphorous afford noteworthy ^{19}F or ^{31}P NMR spectra, respectively, that can exhibit chemical shifts over a wide range (−450–800 ppm or −200–230 ppm, respectively) that are useful for qualitative and quantitative analyses.

The features and applications of solid-state NMR can be summarized as follows:

- Identification of crystal forms (true polymorphs and solvates) based on their individual chemical shifts.
- Determination of phase purity (quantitative analysis of mixtures of solid-state forms including amorphous forms).
- Information about structural characteristics in crystal forms such as number of molecules in the asymmetric unit, hydrogen bonding, etc. Complementary method for structure solutions when structure solution from single crystals is not feasible.
- Detection of the solvent present in solvates and its interaction with the lattice (isomorphic desolvates).
- Analysis of molecular mobility in crystals.

7.3.3 Thermal analysis and calorimetry

According to the definition presented by the International Confederation of Thermal Analysis and Calorimetry (ICTAC), thermal analysis is 'a group of techniques in which a physical property of a substance is measured while the substance is subjected to a controlled temperature programme'. Table 7.3 summarizes the most important techniques that are briefly discussed below.

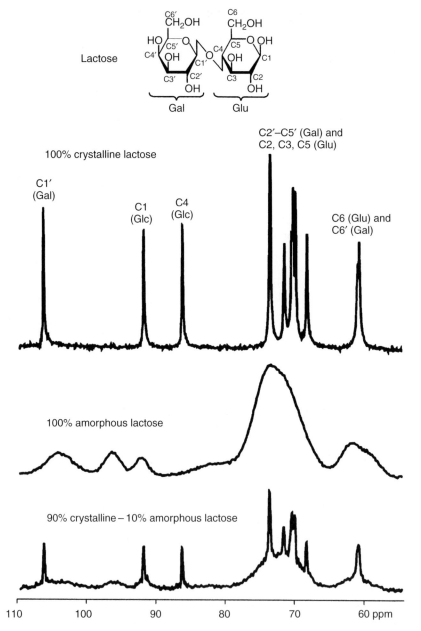

Fig. 7.16 Solid-state ^{13}C CP/MAS NMR spectra of 100% crystalline lactose, 100% amorphous lactose, and a mixture containing 10% amorphous and 90% crystalline lactose. Note the broad peaks for the amorphous material [108].

Table 7.3 Overview of thermal analytical techniques and methods

Technique	Abbreviation	Property	Application examples
Thermomicroscopy (hot-stage microscopy)	TM (HSM)	Optical changes	Phase changes, desolvations, morphological changes, crystal optics
Differential thermal analysis	DTA	Temperature difference	Temperature of phase changes, desolvations, reactions, decompositions
Differential scanning calorimetry	DSC	Energy difference	Enthalpies and temperature of phase changes, heat capacity, desolvations, reactions, decompositions
Thermogravimetry (thermogravimetric analysis)	TG (TGA)	Mass changes	Desolvations, decomposition, oxidation, sublimation
Thermomechanical analysis	TMA	Distance change	Mechanical changes (expansion, contraction, penetration)
Dynamic mechanical analysis	DMA	Frequency	Phase changes, polymer curing
Dielectric thermal analysis		Change in permittivity	Phase changes, changes in polymers
Evolved gas analysis		Thermal conductivity of evolved gas	Decompositions, catalytic and surface reactions

7.3.3.1 Thermomicroscopy (hot-stage microscopy)

Thermomicroscopy (TM) or hot-stage microscopy (HSM) is a thermo-optical technique and has been extensively used in the characterization of solid-state properties of drug compounds [112]. In the early development steps of drug substances (e.g. preformulation, salt selection, etc.) when only a few milligrams of a drug candidate are available, this method is particularly valuable because only minute quantities are required to quickly provide important data, such as melting points, thermal stability, purity (qualitative), as well as the existence of different crystal modifications. This simple and rather inexpensive method requires basically a hot stage attached to a suitable polarized-light microscope.

The hot stages available provide basically different features. The classic and most inexpensive Kofler hot stage [60, 61] is less accurate in temperature ($\pm 0.5°C$) and lacks the feature of linear heating. However, because of the ease of access to the sample, a large number of experimental possibilities exist, which are extremely valuable in hot-stage microscopy. Owing to a small vertical temperature gradient, the sublimation behaviour can be readily observed during heating, which is important for decisions concerning the sample preparation with other thermal analytical techniques such as DSC. Another widely used hot stage is that of by the Mettler company (e.g. Mettler FP82) where the sample (on a glass slide) is heated from both the top and bottom. This sandwich heating provides a good temperature control (±0.1°C) as well as accurate heating programmes and encompass most of the needs in solid-state characterizations. The Linkam company offers a large assortment of hot and cold stages for a wide variety of applications. The specimen is inaccessible and locked within the stage but can be cooled with liquid nitrogen, purged with gases, or optionally evacuated. This type of hot stage also permits observation of freeze drying processes and can be used in combination with IR and Raman microscopes. For documentation purposes and to overcome the disadvantage of subjective results by the investigator, hot-stage microscopes are nowadays commonly equipped with 35 mm or digital photographic cameras or video systems (monitor, camera, recorder, and video text overlay for recoding the temperature signal).

Hot-stage microscopy not only benefits from the features of the hot stage but also the quality and accessories of the microscope. It is obvious that this technique also needs some fundamental knowledge of chemical and optical microscopy. In this context, it should be noted that the advantage of hot stages that combine the features of differential thermal analysis (DTA) and optical microscopy are questionable since both the sensitivity of the DTA signal and the microscopic preparation features suffer much from this combination.

The application field of hot-stage microscopy in pharmaceutical analysis is broad and crosses almost the entire range of materials. Table 7.4 summarizes some applications of thermomicroscopy related to properties of small molecules. Typical photomicrographs of crystal forms that have been crystallized from the supercooled melt in film preparations (between glass slide and cover slip) are shown in Figs 7.11 and 7.12. The forms can be readily distinguished by their different appearances in the polarized light (e.g. shape, pattern, and birefringence colours).

It should be stressed that the valuable features of hot-stage microscopy in the solid-state characterization of pharmaceuticals are much ignored. In order to understand the results obtained by other thermal analytical techniques (e.g. DSC or TGA), it is highly recommended to perform also thermomicroscopic investigations. Visual images are very convincing and contain often much more information than any plotted curve. This may finally save a considerable amount of time and additional experiments, but like most microscopy techniques, TM requires also some experience and patience.

7.3.3.2 Differential thermal analysis and differential scanning calorimetry

The DTA and DSC techniques are both concerned with the measurement of energy changes in a material. The word *differential* indicates that the measurement involves a sample as well as a reference, and the twin-type design is

Table 7.4 Hot-stage microscopy applications [58, 104]

Melting point determination	Melting range (purity)
	Equilibrium melting point
	Instantaneous melting point (for thermally unstable compounds)
	Incongruent melting (polymorphism)
Sublimation	Quick estimate of the vapour pressure
	Micropurification by sublimation
	Sublimation to polymorphic forms (morphology)
	Epitaxy studies
Crystallization (film preparation)	Crystallization of polymorphs from the melt (liquid, supercooled)
	Crystal growth characteristics (melt, solution)
	Crystal growth kinetics
	Morphology of crystals
Transformation phenomena (crystal film, single crystals)	Solid–solid transformations (spontaneous, slow, reversible, irreversible, nucleation and growth characteristics, solvent effects)
Desolvation reactions (hydrates, solvates)	Change to a polycrystalline aggregate upon desolvation (pseudomorphosis) or melting and crystallization phenomena
	Detection of the solvent using a lipophilic oil (bubbles)
	Desolvation kinetics (single crystals)
Special phenomena	Grain growth
	Plastic crystals
	Liquid crystals
Thermal characterization of binary systems including racemates (contact preparation method)	Eutectics
	Peritectics

a characteristic feature of these techniques. The distinction between DTA and DSC lies in the instrumental signal that is proportional to the temperature difference in DTA and to the differential thermal power in DSC (hence *calorimetry*). However, it is controversial where the border between DTA and DSC is drawn and even for a more experienced user the use of these terms is often confusing. It is advantageous to know some of the basics principles of DTA and DSC in view of the manifold instrument designs available in the market.

Differential thermal analysis   DTA is defined as 'a technique in which the temperature difference between a substance and a reference material is measured as function of temperature, while the substance and reference are subjected to a controlled temperature' (ICTAC definition [59]).

The term DTA is used either as a general term for differential thermal techniques (including DSC) or for those techniques that measure temperature differences but are not designed to measure the heat flow. Figure 7.17 shows the

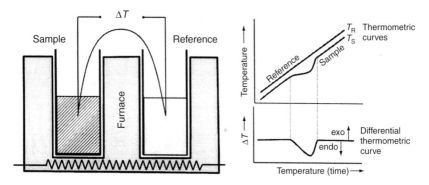

Fig. 7.17 Principle of differential thermal analysis.

principle of a DTA instrument. The temperature change in the sample results in a change of the heat flow from the sample to the furnace, which expresses the heat-flow difference only qualitatively or semi-quantitatively. Note that in DTA curves, the exothermic signal should be represented by an upward peak, whereas an endothermic event should be drawn as downward peak (ICTAC recommendation).

The pure DTA principle exhibits a number of disadvantages and is barely relevant in modern pharmaceutical analysis. However, it is necessary to start with this technique to understand the DSC principle. Table 7.5 summarizes and compares different DTA as well DSC principles. In reality, the theory of DTA and DSC is much more complicated and should be pursued in the specific literature [62–65].

Differential scanning calorimetry As soon as a DTA instrument can be calibrated for quantitative heat measurements we must regard it as *heat-flux differential calorimeter* and by applying a temperature/time programme it becomes a *heat-flux differential scanning calorimeter* (Table 7.5). The first to use this principle was the Boersma type DTA where the measuring system is situated below the sample and reference, and has defined heat conduction paths. Heat-flux DSC instruments use a single large mass furnace that should guarantee that the temperature field of the measuring system is not disturbed by the sample reaction as in a DTA. The measured signal is the temperature difference, which describes the intensity of the exchange and is proportional to the heat-flow rate. Most of the heat-flux DSC instruments used in pharmaceutical thermal analysis are based on the disk-type measuring system as shown in Fig. 7.18. Such systems have high sensitivity, allow high heating (cooling) rates and are designed for small sample volumes. Other heat-flux DSC systems (as shown in Table 7.5, D and E), based on cylinder-type measuring systems usually have large cavities and sample containers, allow only low heating rates,

SOLID-STATE ANALYSIS AND POLYMORPHISM

Table 7.5 Overview of thermal analytical techniques and methods

Scheme	Name	Measuring principle	Description
A	DTA	$T_s - T_R = \Delta T$	Thermocouple is located in the sample; on ΔT
B	Semi-quantitative DTA	$T_s - T_R = \Delta T$	Thermocouple is located outside the sample
C	Boersma type DTA (heat-flux DSC)	$T_S - T_R = \Delta T$	Heat is conducted to the pans
D	Heat-flux DSC	dq_Δ/dt	Thermally coupled sample containers
E	Heat-flux DSC	$dq_S/dt - dq_R/dt = d(\Delta q)dt$	Thermally decoupled sample containers (Calvet principle)
F	Power-compensated DSC	$dQ_S/dt - dQ_R/dt = d(\Delta Q)dt$	S and R are heated from different sources, temperature difference kept close to zero by changing electrical power (P) $\Delta P = d(\Delta Q)dt$

△ thermocouple; ⧠ thermopile; ⌢ heater; S: sample; R: reference; T: temperature; t: time; Q, q: heat.

and show relatively low sensitivity per volume. Such systems are frequently used as reaction calorimeters for a wide variety of applications.

A totally different DSC design is used in power-compensation DSC system. Here, the sample and the reference holders are insulated from each other and have their own individual heaters and sensors as shown in Fig. 7.19. During heating, the same heating power is supplied to both microfurnaces via a control circuit and ideally the temperature of both microfurnaces is identical. When a reaction takes place (e.g. melting process, endotherm) the sample temperature becomes less than that of the reference, which is recorded by the temperature sensor and is immediately compensated by the sample heater. Thus, a power-compensated DSC measures the electrical power that is required to keep both

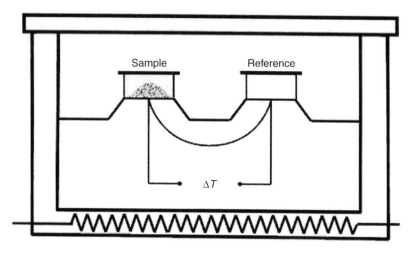

Fig. 7.18 Principle of a heat-flux DSC with a disk-type measuring system.

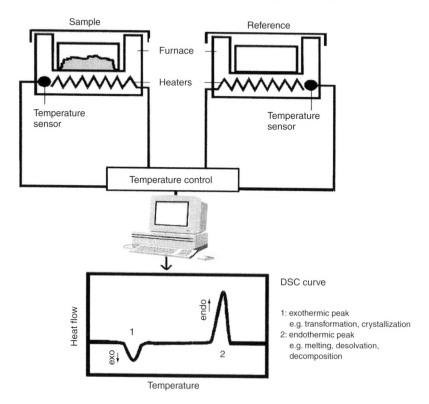

Fig. 7.19 Principle of a power-compensated DSC.

the sample and the reference at exactly the same temperature. The integral over the compensation heating power is proportional to the heat that is released or consumed by the sample. This principle offers a variety of advantages and is frequently denoted as *real DSC* because the heat flow is directly and accurately measured and not calculated as in the case of heat-flux DSC systems. The rapid signal-response time, high resolution, and sensitivity enables the use of smaller sample sizes and higher heating rates. Moreover, the enthalpy calibration is very simple in contrast to heat-flux DSC systems that require, in general, more careful calibration procedures. Calibration of a DSC instrument is performed with standards of known heat capacity (e.g. sapphire) and melting point standards with known enthalpy of fusion (e.g. indium). In DSC curves recorded with a power-compensated DSC, the direction of exothermic and endothermic peaks should be represented in quite the opposite convention as DTA curves (ICTAC recommendation), that is, downward peaks represent exothermic processes and upward peaks represent endothermic processes. Unfortunately, there is no real convention for the presentation of heat-flux DSC curves and depending on the instrument manufacturer one can find both possibilities. However, it is important always to state the direction of the endothermic peaks correctly by using an arrow and adding the abbreviation *endo*.

The DSC analysis is amenable to a wide variety of applications. Properties of pure compounds, such as melting point, boiling point, sublimation temperature, glass-transition temperature, decomposition temperature and energetics, are readily obtainable. In situ production of polymorphs and determination of the phase transitions and their kinetics makes DSC analysis a valuable tool to investigate polymorphs. Studies of physical interactions between drug substance and excipients can be performed quickly and with a minimal amount of material. These and other applications are discussed in a series of publications by Giron [66–69].

Modulated temperature DSC (MTDSC) and dynamic DSC (DDSC) In the modulated DSC operation mode, the temperature oscillates quickly at a selected amplitude and frequency (sine function) in addition to the linear heating rate. This oscillation allows for the separation of the pure thermodynamic signal (reversing signal) and the kinetic signal (non-reversing signal). The theory behind this technique is rather complex and cannot be fully addressed here; thus, the reader is referred to the more specific literature [70, 71]. MTDSC allows the accurate determination of heat capacities in a single run and is particularly applied in the investigation of glass transitions (amorphous forms).

7.3.3.3 Thermogravimetry
Thermogravimetry (TG) (also known as thermogravimetric analysis, TGA) is a technique measuring the variation in mass of a sample as it undergoes temperature scanning in a controlled atmosphere. This variation in mass can be either

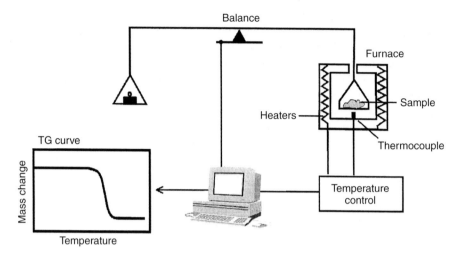

Fig. 7.20 Principle of a thermogravimetric system.

a loss of mass (vapour emission) or a gain of mass (gas fixation). This method readily establishes the stoichiometry of hydrates and solvates or confirms the presence of a volatile component in a sample. The principle of this technique is shown in Fig. 7.20.

Other techniques may be coupled with thermogravimetry to provide very useful data and to confirm suppositions. One such instrument combines DSC and TG (simultaneous DSC/TG) and may facilitate the elucidation of unexpected events obtained from a DSC investigation alone. Other examples of valuable coupling of instruments include TG–FT-IR, TG–Raman, and TG–MS, where the gases that may be produced during a TG investigation are analysed by an FT-IR, Raman, or mass spectrometer, respectively, that have been linked in sequence to the TG instrument [72].

7.3.3.4   Thermomechanical analysis
Thermomechanical analysis (TMA) is a technique measuring the deformation of a sample under a non-oscillating stress when it undergoes temperature scanning in a controlled atmosphere. The stress may be compressive, tensile, or torsional. This technique is widely used in polymer studies and composite materials. TMA allows for the determination of the thermal expansion coefficient, the investigation of phase transitions and penetration performance, measurement of the elastic modulus and compliance, as well as fatigue testing. Rheological measurements, such as viscosity and gelation may be determined using TMA techniques. There are numerous examples of pharmaceutical applications [73–77].

7.3.3.5 Dielectric analysis

Dielectric analysis (DEA), or dielectric spectroscopy, measures the response of a material to a time-varying voltage signal and detects subtle differences in the electrical properties of the sample [78–81]. DEA can detect phase transitions not detected by DSC or dynamic mechanical thermal analysis (DMTA). DEA measures capacitance and conductance versus time, temperature, and frequency. Capacitance, the ability to store an electrical charge, dominates the electrical responses for a rigid material (e.g. a polymer below its glass-transition temperature T_g). Conductance, the ability to transfer charge, is important when a heated material becomes less rigid (e.g. amorphous material, a polymer above T_g, or a melt). Changes in capacitance and conductance provide molecular- or structural-state information of a sample. Parameters monitored by dielectric analysis are ε' (permittivity), which measures the degree of molecular dipole alignment to an applied electrical field, and ε'' (loss factor), which reflects the alignment of dipoles or ions and the energy needed to align them. DEA is particularly useful for mixtures and characterization of dipole and/or ionic phases in mixtures.

7.3.3.6 Microcalorimetry

Microcalorimetry is an extremely sensitive technique that determines the heat emitted or adsorbed by a sample in a variety of processes. Microcalorimetry can be used to characterize pharmaceutical solids to obtain heats of solution, heats of crystallization, heats of reaction, heats of dilution, and heats of adsorption. Isothermal microcalorimetry has been used to investigate drug-excipient compatibility [82]. Pikal and co-workers have used isothermal microcalorimetry to investigate the enthalpy of relaxation in amorphous material [83]. Isothermal microcalorimetry is useful in determining even small amounts of amorphous content in a sample [84]. Solution calorimetry has also been used to quantitate the crystallinity of a sample [85]. Other aspects of isothermal microcalorimetry may be obtained from a review by Buckton [86].

7.3.4 Vapour pressure determination of solids

The vapour pressure is defined as the pressure of a gaseous phase in equilibrium with its condensed phase (solid, liquid, or both at the triple point) at a specified temperature. Owing to the very low vapour pressures of crystalline organic substances (usually far below 10 Pa), such measurements are challenging and thus only a limited number of vapour pressure data of solids are available in contrast to that of liquids. However, the heat of sublimation is a key thermodynamic quantity for organic crystals and can be equated to the lattice energy of a crystalline state. Thus there is considerable theoretical interest in reliable vapour pressure data of organic solids since such data may help computational chemists to improve the algorithms for lattice-energy calculations and the desired goal of predicting crystal structures [109]. As recently shown [51] vapour pressure methods are applicable to determine precise thermodynamic data of different polymorphs. However, the determination of such data does not belong to those methods that can and will routinely be performed in pharmaceutical analysis (for further reading, see [51] and literature cited therein).

7.3.5 Micromeritic measurements

The term *micromeritics* (coined by Godfrey Schmidt by combining μικρόσ *small* and μέροσ *part*; used for the first time by Dalla Valle [113]) encompasses a broad field that includes particle size and particle-size distribution. Other aspects of micromeritics that are important to pharmaceutical scientists are particle shape, specific surface area, porosity, and density. A few of these are briefly discussed in the following.

7.3.5.1 Particle-size analysis

Particle-size analysis is a critical measurement in the pharmaceutical industry. From micron-sized powders for use in dry inhalation impellers to millimetre-sized granules for compression, size does matter. Furthermore, knowledge of the distribution and range of sizes within a sample as well as the shape of the particles is often vital to a successful formulation. The field of particle-size analysis is too vast to cover here and so a cursory outline of the topic will be presented since many books are available on the subject [87–91].

A variety of methods exist for the determination of particle-size ranges and the technique used depends on the sizes of particles in the sample to be analysed (see Table 7.6). Visual inspection of any material should always be part of a routine analysis, especially for evaluating the typical shape of the particles. Mixtures of morphologies are readily detected visually and the observation can be used to select one method preferentially over another since shape may bias the outcome – most methods assume a hard, spherical particle. For example, needles with dimensions of $70\,\mu m \times 70\,\mu m \times 700\,\mu m$ will easily pass through a No. 35 standard sieve (rated as having a $500\,\mu m$ opening as measured between parallel wires) since the diameter of a $500\,\mu m$ square is $707\,\mu m$; however, with enough vibration and agitation, the same needles will pass through a No. 200 standard sieve (rated as having a $74\,\mu m$ opening). Obviously, the conditions of the analysis will determine the data obtained.

For irregularly shaped particles, the *diameter* of a particle may be estimated using one of several methods: Feret diameter, Martin diameter, projected-area diameter, equivalent diameter, or aerodynamic diameter. The two most commonly used diameters, the Feret and Martin diameters, are relative to a randomly chosen reference line, typically the bottom edge of the viewing frame, where the Feret

Table 7.6 Nominal particle-size ranges measured by laboratory sizing methods

Method		Nominal particle-size range (μm)	
		Min.	Max.
Visual			
Microscopic examination	Optical	3	1000
	Transmission electron (TEM)	0.002	1
	Scanning electron (SEM)	0.02	1000
Separation			
	Dry sieving	10	75 000
	Wet sieving	2	500
	Centrifugal classification	0.5	50
	Cascade impaction	0.05	30
Stream scanning			
Zone sensing	Electrical resistance	0.8	500
	Optical	1	500
	Laser diffraction	0.003	2000
Field scanning	Acoustic attenuation	0.01	100
	Photon correlation spectroscopy	0.001	1
Elutriation	Laminar flow	3	75
	Cyclone	8	50
Hydrodynamic chromatography	Packed column	0.03	3
	Capillary tube	0.1	5
Sedimentation			
Gravity	Pipette and hydrometer	1	100
	Photoextinction	0.5	100
	X-ray	0.1	130
	Mass accumulation	0.05	25
Centrifugal	Photoextinction	0.05	100
	X-ray	0.1	5
Surface			
	Gas permeability	0.1	40
	Gas adsorption	0.005	50
Other			
	Brownian motion	0.003	5
	Mercury intrusion	0.02	200

diameter is the maximum width of a particle relative to the reference line and the Martin diameter is the diameter through the centre of mass of the particle (see Fig. 7.21). The projected-area diameter is the diameter of a circle having the same area as the silhouette of the particle (see Fig. 7.21). The equivalent diameter is the area of a sphere having the same volume as the particle. The aerodynamic diameter is the diameter of a spherical particle that would display the same settling velocity as the particle in question. Image analysis software will compute the Feret, Martin, and projected-area diameters as well as aspect ratios of the particles on both static and flowing samples.

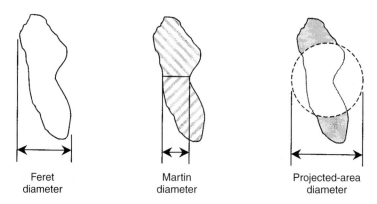

Fig. 7.21 Definition of representative particle-size diameters.

It is rare that particles are of uniform size and shape. Thus, the particle-size distribution of the sample needs to be determined. Typically, the distribution of particle sizes is represented by a frequency distribution curve or histogram where the number or weight of particles having a certain size range is plotted versus the size range or mean particle diameter of the particles. These plots are essential since two samples may have the same average diameter but the particles within the sample may be distributed differently. For pharmaceutical powders, particle-size distribution plots are typically not Gaussian but rather they are skewed. Plotting the frequency vs the logarithm of the particle diameter will often produce a bell-shaped curve called a log-normal distribution.

It is usual to also plot the accumulative percentage of particles either greater than or less than a particular size vs the particle size; the resulting curve is generally sigmoidal. When the accumulative percentage is plotted versus the logarithm of the particle size, a linear relationship is observed, thus characterizing the log-normal distribution curve by the slope and the geometric mean diameter d_g of the sample (i.e. 50% of the cumulative frequency). In addition, numerical d_{10}, d_{50}, and d_{90} values are normally given: d_{10} is the size of particle for which 10% of the sample is smaller, d_{50} is the size of particle for which 50% of the sample is smaller, and d_{90} is the size of particle for which 90% of the sample is smaller.

7.3.5.2 Particle shape

Estimation of the particle shape was mentioned in the previous section on particle-size analysis. Historically, the particle shape was assumed to be spherical to simplify calculations. With the advent of inexpensive high-speed computers and high-resolution imaging systems, image analysis has become a viable option to study the shape and size of particles in greater detail [92]; the systems and software programs currently available are numerous. In conjunction with image analysis, fractal analysis can be used to obtain the shape of particles [93] or even probe the characteristics of the surfaces of particles [94].

7.3.5.3 Specific surface area

An important parameter of bulk powders is the specific surface area S_w, expressed per unit weight. The specific surface area measurement includes the cracks, crevices, nooks, and crannies present in the particles. To include these features in the surface-area measurement, methods have been developed to probe these convoluted surfaces through adsorption by either a gas or a liquid [95–97]. The most widely used surface area measurement technique is the absorption of a monolayer of gas, typically krypton or nitrogen as the adsorbate gas in helium as an inert diluent, using the method developed by Brunauer, Emmett, and Teller [98], known as the BET method. This method utilizes the BET equation

$$\frac{P}{V-(P_0-P)} = \frac{1}{V_mC} + \frac{C-1}{V_mC}\left(\frac{P}{P_0}\right) \qquad (7.1)$$

where P is the partial pressure of the gas, P_0 is the saturation pressure of the gas at the experimental temperature, V is the total volume of the gas at pressure P, V_m is the monolayer volume of gas, and C is a constant approximately equal to, $e^{E_1-E_L/RT}$, where E_1 is the heat of adsorption for the monolayer, E_L is the heat of liquefaction (condensation), R is the gas constant, and T is the temperature. A plot of $P/V(P_0-P)$ vs P/P_0 affords a straight line whose slope is $(C-1)/V_mC$ and intercept is $1/V_mC$, thus providing the two constants V_m and C.

7.3.5.4 Porosity

Aside from highly crystalline material, very few particles have smooth surfaces. The porosity of the particles in a powder can influence the disintegration, dissolution, and bioavailability of the pharmaceutical product. Although gas adsorption may be used to measure pores of less than 10 nm, mercury porosimetry is used to determine much larger pore sizes, including interparticular spacing. An entire issue of *Power Technology* was devoted to mercury porosimetry; one article reviewed the application of this method to pharmaceutical technology [99].

At room temperature, mercury is a non-wetting fluid for most porous materials. As a consequence, higher pressures are needed to force intrusion into small pores (e.g. 400 MPa for 4-nm diameter pores) while lower pressures are used for large pores and interparticular spacing (e.g. vacuum). Vacuum degassing of the sample is utilized to remove moisture from the pores prior to the analysis. Isotherms of the volume of mercury adsorbed versus pressure exhibit a hysteresis between the intrusion and extrusion process; the overall shape of the isotherm affords information about the pore sizes and the size distribution.

7.3.5.5 Density

The ratio of a mass to a unit volume represents a density; however, the actual value of that density is subject to the properties and characteristics of the material being measured. *Molecular density* refers to a perfect crystal and is determined from the molecular weight and the molecular volume; *crystalline density* refers to typical crystals possessing defects and other flaws, and may be calculated, using single-crystal X-ray diffraction, from the number of molecules in the unit cell and the volume of the unit cell, or determined experimentally by flotation in a mixture of two miscible liquids, one of greater density and one of lesser density than the crystal [17]. For powders, several types of densities have been defined. *Bulk density* depends on the size of the particles being measured as well as the particle-size distribution; it is typically determined by pouring the powder into a tared graduated cylinder and noting the volume and mass of the sample. *Tap density* or *drop density* is an extension of the bulk density measurement in which the tared graduated cylinder containing the sample is tapped a set number of times and the resulting volume used in the density calculation. Obviously, the number of taps used and the force of the taps will affect the outcome of the measurement, thus, instruments have been developed to allow for consistency in the measurement. Some investigators may even repeat the tapping procedure until there is no change in the volume. *Particle density* or *granule density* is determined using mercury displacement at atmospheric pressure and should be similar to the true density for non-porous solids. *True density* is the density of the material itself and excludes any pores and voids. The most common method for determining true density is gas (usually helium) pycnometry which is rapid and accurate but utilizes expensive instrumentation. Other methods for determining true density are displacement by a non-dissolving liquid or flotation in a mixture of two miscible liquids, one of greater density than the sample and the other of lesser density. The term *apparent density* has been used to describe granular density, bulk density, and even true density, thus the use of this term is discouraged [95].

7.3.6 Other technologies

7.3.6.1 Atomic force microscopy

Since the development of the atomic force microscope (AFM), or scanning force microscope (SFM), in 1986 [100–102], this relatively new technique has been used to probe the surface of crystals [103, 104] and granules [94] as well as particle interactions [110, 111] and surface defects [105]. As in all other scanning probe microscopes, AFM utilizes a sharp probe moving over the surface of a sample in a raster scan; the movement of the probe on the end of a cantilever in response to the force between the tip and the sample is detected using a laser. The movement of the tip or sample is performed by an extremely precise positioning device made from piezoelectric ceramics, most often in the form of a tube scanner. The scanner is capable of sub-Ångstrom resolution in all Cartesian directions (the z-axis is conventionally perpendicular to the sample). By adhering a nanocrystal on a cantilever or chemically bonding organic substrates onto a probe tip, molecular interactions between the probe and the sample can be evaluated [104, 111].

7.4 Summary

The complex nature of the solid state represents a challenging but also intriguing aspect in the analysis of pharmaceuticals. The key to an efficient and facile characterization of a pharmaceutical compound, as well as to the understanding of the most relevant solid-state properties, is, at first glance, the application of a proper assembly of techniques enabling the assessment of chemical, structural, energetic, and physical aspects of a solid. Furthermore, a good knowledge of the basic solid-state phenomena and the possible impact on the stability, formulation properties, and finally the performance of a drug substance is required in order to develop effective characterization strategies. Since it is practically impossible to explore all solid-state properties and phenomena within the short timescales available in the preformulation phase of drug development, it is important to elaborate analytical programmes and strategies that aim at a minimum in analytical redundancy and a maximum in consistent results. This is only possible when the synergistic effects of different analytical techniques are exploited properly. However, there is no standard recipe for such strategies, which must be continuously adapted to the nature of the compounds under investigation, the available instrumentation (which is always unique, never complete, usually expanding but at least partially always behind the state of the art), the scientific and technical advance in solid state and material research, and many more. The properties of a solid drug compound, particularly the phenomena of polymorphism, can be neither fully understood nor properly realized solely by a confirmed crystallographer, confirmed spectroscopist, microscopist, thermal analyst, and so on, but only in collaboration with each other will they be able to definitely solve most of the solid-state problems that are relevant for a sophisticated, modern drug development.

The complexity of the solid state offers not only an analytical challenge but also interesting aspects and opportunities for the drug development. Thus, a drug can be *engineered* at the supramolecular and particulate level in order to optimize its physical properties. This requires, of course, a substantial knowledge of its solid-state properties and the proper analytical tools.

References

1. Wells, J.I. (1988) *Pharmaceutical Preformulation: The Physicochemical Properties of Drug Substances*, Ellis Horwood, Chichester.
2. Carstensen, J.T. (1998) *Pharmaceutical Preformulation*, CRC Press.
3. Lipinski, C.A., Lombarda, F., Dominy, B.W. & Feeney, P.J. (2001) Experimental and computational approaches to estimate solubility and permeability in drug discovery and development settings, *Adv. Drug Deliv. Rev.*, **46**, 3–26.
4. York, P. (1992) Crystal engineering and particle design for the powder compaction process, *Drug Dev. Ind. Pharm.*, **18**, 677–721.

5. Leung, S.S., Padden, B.E., Munson, E.J. & Grant, D.J.W. (1998) Solid-state characterization of two polymorphs of aspartame hemihydrate, *J. Pharm. Sci.*, **87**, 501–507.
6. Burger, A. & Griesser, U.J. (1989) Polymorphic drug substances of the European Pharmacopeia. Part 4. Identification and characterization of 11 crystal forms of succinylsulfathiazole, *Sci. Pharm.*, **57**, 293–305.
7. Morris, K.R., Griesser, U.J., Eckhardt, C.J. & Stowell, J.G. (2001) Theoretical approaches to physical transformations of active pharmaceutical ingredients during manufacturing processes, *Adv. Drug Deliv. Rev.*, **48**, 91–114.
8. Burger, A. & Ramberger, R. (1979a) On the polymorphism of pharmaceuticals and other molecular crystals I: theory of thermodynamic rules, *Mikrochim. Acta*, **II**, 259–271.
9. Burger, A. & Ramberger, R. (1979b) On the polymorphism of pharmaceuticals and other molecular crystals II: applicability of thermodynamic rules, *Mikrochim. Acta*, **II**, 273–316.
10. Yu, L. (1995) Inferring thermodynamic stability relationship of polymorphs from melting data, *J. Pharm. Sci.*, **84**, 966–974.
11. Grunenberg, A., Henck, J.-O. & Siesler, H.W. (1996) Theoretical and practical application of energy/temperature diagrams as an instrument in preformulation studies of polymorphic drug substances, *Int. J. Pharm.*, **129**, 147–158.
12. Fukuoka, E., Makita, M. & Yamamura, S. (1989) Glassy state of pharmaceuticals. III thermal properties and stability of glassy pharmaceuticals and their binary glass systems, *Chem. Pharm. Bull.*, **37**, 1047–1050. (See also other papers of this series.)
13. Yu, L. (2001) Amorphous pharmaceutical solids: preparation, characterization and stabilization, *Adv. Drug Deliv. Rev.*, **48**, 27–42.
14. Federal Register (2000) *International Conference on Harmonisation; Guidance on Q6A Specifications: Test Procedures and Acceptance Criteria for New Drug Substances and New Drug Products: Chemical Substances*, **65**(251).
15. Dunitz, J.D. (1979) *X-Ray Analysis and the Structure of Organic Molecules*, Cornell University Press, Ithica, NY.
16. Glusker, J.P. & Trueblood, K.N. (1985) *Crystal Structure Analysis: A Primer*, Oxford University Press, New York, NY.
17. Stout, G.H. & Jensen, L.H. (1989) *X-Ray Structure Determination: A Practical Guide*, 2nd edn, John Wiley & Sons, New York, NY.
18. Giacovazzo, C. *et al.* (1992) *Fundamentals of Crystallography* (ed. C. Giacovazzo), International Union of Crystallography Oxford Science Publications, Oxford, UK.
19. Ladd, M.F.C. & Palmer, R.A. (1993) *Structure Determination by X-ray Crystallography*, 3rd edn, Plenum Press, New York, NY.
20. Glusker, J.P., Lewis, M. & Rossi, M. (1994) *Crystal Structure Analysis for Chemists and Biologists*, VCH Publishers, New York, NY.
21. Cambridge Crystallographic Data Centre (2002) *Cambridge Structural Database*, www.ccdc.cam.ac.uk, Cambridge, UK.
22. Bish, D.L. & Post, J.E. (1989) *Modern Powder Diffraction* (eds D.L. Bish & J.E. Post), *Reviews in Mineralogy* (ed. P.H. Ribbe), **20**, Mineralogical Society of America, Washington, DC.
23. Snyder, R.L. & Jenkins, R. (1996) *Introduction to X-Ray Powder Diffractometry*, John Wiley & Sons, New York, NY.
24. International Center for Diffraction Data (2002) *Powder Diffraction File*, www.icdd.com, Newtown Square, PA.
25. Bugay, D.E. (2001) Characterization of the solid state: spectroscopic techniques, *Adv. Drug. Deliv. Rev.*, **48**, 43–65.
26. Brittain, H.G. (1995) Ultraviolet/visible diffuse reflectance spectroscopy, *Drugs Pharm. Sci.*, **70**, 37–58.
27. Sutor, D.J. (1958) The structures of the pyrimidines and purines VII. The crystal structure of caffeine, *Acta Cryst. Sect. C*, **11**, 453–458.

28. Gerdil, R. & Marsh, R.E. (1960) On the arrangement of water molekules in the crystal structure of caffeine, *Acta Cryst. Sect. C*, **13**, 166–167.
29. Bernstein, J. (2002) *Polymorphism in Molecular Crystals*, Oxford University Press, Oxford, UK.
30. Stewart, J.E. (1970) in *Infrared Spectroscopy, Experimental Methods and Techniques*, Marcel Dekker, New York.
31. Fuller, M.P. & Griffiths, P.R. (1978) Diffuse reflectance measurements by infrared Fourier transform spectroscometry, *Anal. Chem.*, **50**, 1906–1910.
32. Vidrine, D.W. (1980) Photoacoustic Fourier transform infrared spectroscopy of solid samples, *Appl. Spectrosc.*, **34**, 314–319.
33. Ashizawa, K. (1989) Polymorphism and crystal structure of 2R,4S,6-fluoro-2-methylspiro [chroman-4,4′-imidazoline]-2′,5-dione (M 79175), *J. Pharm. Sci.*, **78**, 256–260.
34. Berbenni, V., Marini, A., Bruni, G., Maggioni, A. & Cogliati, P. (2002) Thermoanalytical and spectroscopic characterization of solid state dipyridamole, *J. Thermal Anal. Cal.*, **68**, 413–422.
35. Messerschmidt, R.G. & Harthcock, M.A. (eds) (1988) *Infrared Microspectroscopy, Theory and Applications*, Marcel Dekker, Inc., New York.
36. Griesser, U.J. & Burger, A. (1993) The polymorphic drug substances of the European pharmacopoeia, Part 8: thermal analytical and FTIR-microscopic investigations of etofylline crystal forms, *Sci. Pharm.*, **61**, 133–143.
37. Threlfall, T.L. (1995) Analysis of organic polymorphs. a review, *Analyst*, **120**, 2435–2460.
38. Stephenson, G.A., Forbes, R.T. & Reutzel, S.M. (2001) Characterization of the solid state: quantitative issues, *Adv. Drug Deliv. Rev.*, **48**, 67–90.
39. Ciurczak, E.W. (1992) NIR analysis in pharmaceuticals, in *Handbook of Near-Infrared Analysis* (eds D.A. Burns & E.W. Ciurczak), Marcel Dekker, Inc., New York.
40. Morris, K.R., Stowell, J.G., Byrn, S.R., Placette, A.W., Davis, T.D. & Peck, G.E. (2000) Accelerated fluid bed drying using NIR monitoring and phenomenological modeling, *Drug Dev. Ind. Pharm.*, **26**, 985–988.
41. Ufret, C. & Morris, K. (2001) Modeling of powder blending using on-line near-infrared measurements, *Drug Dev. Ind. Pharm.*, **27**, 719–729.
42. Berntsson, O., Danielsson, L.-G., Lagerholm, B. & Folestad, S. (2002) Quantitative in-line monitoring of powder blending by near infrared reflection spectroscopy, *Powder Technology*, **123**, 185–193.
43. El-Hagrasy, A.S., Morris, H.R., D'Amico, F., Lodder, R.A. & Drennen, J.K. (2001) Near-infrared spectroscopy and imaging for the monitoring of powder blend homogeneity, *J. Pharm. Sci.*, **90**, 1298–1307.
44. Otsuka, M., Kato, F. & Matsuda, Y. (2001) Determination of indomethacin polymorphic contents by chemometric near-infrared spectroscopy and conventional powder X-ray diffractometry, *Analyst*, **126**, 1578–1582.
45. O'Neil, A.J., Jee, R.D. & Moffat, A.C. (1998) The application of multiple linear regression to the measurement of the median particle size of drugs and pharmaceutical excipients by near-infrared spectroscopy, *Analyst*, **123**, 2297–2302.
46. Lewis, E.N., Carroll, J.E. & Clarke, F. (2001) A near infrared view of pharmaceutical formulation analysis, *NIR News*, **12**, 16–18.
47. Vankeirsbilck, T., Vercauteren, A., Baeyens, W., *et al.* (2002) Applications of Raman spectroscopy in pharmaceutical analysis, *Trends Anal. Chem.*, **21**, 869–877.
48. Taylor, L.S. & Langkilde, F.W. (2000) Evaluation of solid-state forms present in tablets by Raman spectroscopy, *J. Pharm. Sci.*, **89**, 1342–1353.
49. Turrell, G. & Corset, J. (eds) (1996) *Raman Microscopy, Development and Applications*, Academic Press, London.
50. Szelagiewicz, M., Marcolli, C., Cianferani, S., *et al.* (1999) In situ characterization of polymorphic forms the potential of Raman techniques, *J. Thermal Anal. Cal.*, **57**, 23–43.

51. Griesser, U.J., Szelagiewicz, M., Hofmeier, U.C., Pitt, C. & Cianferani, S. (1999) Vapor pressure and heat of sublimation of crystal polymorphs, *J. Thermal Anal. Cal.*, **57**, 45–60.
52. Purcell, E.M., Torrey, H.C. & Pound, R.V. (1946) Resonance absorption by nuclear magnetic moments in a solid, *Phys. Rev.*, **69**, 37–38.
53. Fyfe, C.A. (1983) *Solid State NMR for Chemists*, CFC Press, Guelph, Ontario.
54. Bugay, D.E. (1993) Solid-state nuclear magnetic resonance spectroscopy: theory and pharmaceutical applications, *Pharm. Res.*, **10**, 317–327.
55. Byrn, S.R., Pfeiffer, R.R. & Stowell, J.G. (1999) Solid-state NMR spectroscopy, in *Solid-State Chemistry of Drugs*, 2nd edn, SSCI, Inc., West Lafayette, IN, 119–137.
56. Saindon, P.J. (1992) *Solid Pharmaceuticals: Analysis and Identification by Solid-state Carbon-13 NMR*, PhD thesis, Purdue University, West Lafayette, IN.
57. Apperley, D.C., Felton, R.A., Harris, R.K., *et al.* (1999) Sulfathiazole polymorphism studied by magic-angle spinning NMR, *J. Pharm. Sci.*, **88**, 1275–1280.
58. Gavezzotti, A. (1998) The crystal packing of organic molecules: challenge and fascination below 1000 Da, *Crystallography Reviews*, **7**, 5–121.
59. Lombardi, G. (1980) For better Thermal Analysis, 2nd edn, published by the International Confederation of Thermal Analysis and Calorimetry (ICTAC).
60. Kuhnert-Brandstätter, M. (1971) *Thermomicroscopy in the Analysis of Pharmaceuticals*, Pergamon Press, Oxford, UK.
61. Kuhnert-Brandstätter, M. (1982) Thermomicroscopy of organic compounds, in *Comprehensive Analytical Chemistry* (ed. G. Svehla), Elsevier, Amsterdam, pp. 239–498.
62. Dollimore, D. (1990) Thermoanalytical instrumentation, in *Analytical Instrumentation Handbook*, (ed. G.W. Ewing) Marcel Dekker, New York, pp. 905–960.
63. Wunderlich, B. (1990) *Thermal Analysis*, Academic Press, Inc., Boston.
64. Haines, P.J. (1995) *Thermal Methods of Analysis. Principles, Applications and Problems*, Chapman & Hall, London.
65. Höhne, G., Hemminger, W. & Flammersheim, H.-J. (1996) *Differential Scanning Calorimetry. An introduction for Practitioners*, Springer, Berlin, Germany.
66. Giron, D. (1986) Applications of thermal analysis in the pharmaceutical industry, *J. Pharm. Biomed. Anal.*, **4**, 755–770.
67. Giron, D. (1990) Thermal analysis in pharmaceutical routine analysis, *Acta Pharm. Jugosl.*, **40**, 95–157.
68. Giron, D. (1995) Thermal analysis and calorimetric methods in the characterization of polymorphs and solvates, *Thermochim. Acta*, **248**, 1–59.
69. Giron, D. (2002) Applications of thermal analysis and coupled techniques in pharmaceutical industry, *J. Therm. Analysis Calorim.*, **68**, 335–357.
70. Reading, M., Luget, A. & Wilson, R. (1994) Modulated differential scanning calorimetry, *Thermochim. Acta*, **238**, 295–307.
71. Gill, P.S., Sauerbrunn, S.R. & Reading, M. (1993) Modulated differential scanning calorimetry, *J. Thermal Anal.*, **40**, 931–939.
72. Rodriguez, C. & Bugay, D.E. (1997) Characterization of pharmaceutical solvates by combined thermogravimetric and infrared analysis, *J. Pharm. Sci.*, **86**, 263–266.
73. Ford, J.L. & Mitchell, K. (1995) Thermal analysis of gels and matrix tablets containing cellulose ethers, *Thermochim. Acta*, **248**, 329–345.
74. Chen, T. & Oakley, D.M. (1995) Thermal analysis of proteins of pharmaceutical interest. *Thermochim. Acta*, **248**, 229–244.
75. Kerč, J. & Srčič, S. (1995) Thermal analysis of glassy pharmaceuticals, *Thermochim. Acta*, **248**, 81–95.
76. Hancock, B.C., Dupuis, Y. & Thibert, R. (1999) Determination of the viscosity of an amorphous drug using thermomechanical analysis (TMA), *Pharm. Res.*, **16**, 672–675.
77. Hancock, B.C., Dalton, C.R. & Clas, S.-D. (2001) Micro-scale measurement of the mechanical properties of compressed pharmaceutical powders. 2: the dynamic moduli of microcrystalline cellulose, *Int. J. Pharm.*, **228**, 139–145.

78. Smith, G., *et al.* (1995) Dielectric relaxation spectroscopy and some applications in the pharmaceutical sciences, *J. Pharm. Sci.*, **84**, 1029–1044.
79. Craig, D.Q.M. (1996) *Dielectric Analysis of Pharmaceutical Systems*, Taylor & Francis, Inc., London, UK.
80. Havriliak, Jr, S. & Havriliak, S.J. (1996) *Dielectric and Mechanical Relaxation in Materials: Analysis, Interpretation, and Application to Polymers*, Hanser Publishers, Munich, Germany.
81. Guma, N.C., Kale, K. & Morris, K.R. (1997) Investigation of film curing stages by dielectric analysis and physical characterization, *J. Pharm. Sci.*, **86**, 329–334.
82. Schmitt, E.A., *et al.* (2001) Rapid, practical and predictive excipient compatibility screening using isothermal microcalorimetry, *Thermochim. Acta*, **380**, 175–183.
83. Liu, J., *et al.* (2002) Dynamics of pharmaceutical amorphous solids: the study of enthalpy relaxation by isothermal microcalorimetry, *J. Pharm. Sci.*, **91**, 1853–1862.
84. Buckton, G., Darcy, P. & Mackellar, A.J. (1995) The use of isothermal microcalorimetry in the study of small degrees of amorphous content in powders, *Int. J. Pharm.*, **117**, 253–256.
85. Gao, D. & Rytting, J.H. (1997) Use of solution calorimetry to determine the extent of crystallinity of drugs and excipients, *Int. J. Pharm.*, **151**, 183–192.
86. Buckton, G. (1995) Applications of isothermal microcalorimetry in the pharmaceutical sciences, *Thermochim. Acta*, **148**, 117–129.
87. Allen, T. (1990) *Particle Size Measurement*, 4th edn, Chapman Hall, London, UK.
88. Stanley-Wood, N.G. & Lines, R.W. (1992) *Particle Size Analysis* (eds N.G. Stanley-Wood & R.W. Lines), The Proceedings of the 25th Anniversary Conference organized by the Particle Characterization Group of the Analytical Division of the Royal Society of Chemistry, September 17–19, 1991, University of Technology, Loughborough, UK. The Royal Society of Chemistry, Cambridge, UK.
89. Washington, C. (1992) *Particle Size Analysis in Pharmaceutics and Other Industries: Theory and Practice*, Ellis Horwood, New York, NY.
90. Gotoh, K., Masuda, H. & Higashitai, K. (1997) *Powder Technology Handbook* (eds K. Gotoh, H. Masuda & K. Higashitani), 2nd edn, Marcel Dekker, Inc., New York, NY.
91. Müller, R.H. & Mehnert, W. (1997) *Particle and Surface Characterisation Methods* (eds R.H. Müller, W. Mehnert & G.E. Hildebrand), MedPharm Scientific Publishers, Stuttgart, Germany.
92. Kaye, B.H. (1999) *Characterization of Powders and Aerosols*, Wiley-VCH, Weinheim, Germany.
93. Carstensen, J.T. (2001) *Micromeritics in Advanced Pharmaceutical Solids* (exec. ed. J. Swarbrick), Drugs and the Pharmaceutical Sciences, **110**, Marcel Dekker, New York, NY, Ch. 5, pp. 81–85.
94. Li, T. & Park, K. (1998) Fractal analysis of pharmaceutical particles by atomic force microscopy, *Pharm. Res.*, **15**, 1222–1232.
95. Martin, A. & Bustamante, P. (1993) *Physical Pharmacy: Physical Chemical Principles in the Pharmaceutical Sciences*, 4th edn, Lippincott Williams & Wilkins, Philadelphia, PA, pp. 436–439.
96. Newman, A.W. (1995) Micromeritics in *Physical Characterization of Pharmaceutical Solids*, (ed. H.G. Brittain), Marcel Dekker, New York, NY, Ch. 9, pp. 254–264.
97. Chikazawa, M. & Takei, T. (1997) Specific Surface Area in *Powder Technology Handbook* (eds K. Gotoh, H. Masuda & K. Higashitani), 2nd edn, Marcel Dekker, Inc., New York, NY, Ch. III. **8**, pp. 337–349.
98. Brunauer, S., Emmett, P.H. & Teller, E. (1938) Adsorption of gases in multimolecular layers, *J. Am. Chem. Soc.*, **60**, 309–319.
99. Dees, P.J. & Polderman, J. (1981) Mercury porosimetry in pharmaceutical technology, *Powder Technol.*, **29**, 187–197.
100. Binnig, G., Quate, C.F. & Gerber, C.H. (1986) Atomic force microscope, *Phys. Rev. Lett.*, **56**, 930–933.
101. Binnig, G., *et al.* (1987a) Atomic resolution with atomic force microscope, *Europhys. Lett.*, **3**, 1281–1286.
102. Binnig, G., *et al.* (1987b) Atomic resolution with atomic force microscope, *Surface Sci.*, **198–190**, 1–6.

103. Danesh, A., et al. (2000) The discrimination of drug polymorphic forms from single crystals using atomic force microscopy, *Pharm. Res.*, **17**, 887–890.
104. Muster, T.H. & Prestidge, C.A. (2002) Face specific surface properties of pharmaceutical crystals, *J. Pharm. Sci.*, **91**, 1432–1444.
105. Li, T., Morris, K.R. & Park, K. (2000) Influence of solvent and crystalline supramolecular structure on the formation of etching patterns on acetaminophen single crystals: a study with atomic force microscopy and computer simulation, *J. Phys. Chem. B*, **104**, 2019–2032.
106. Grimmer, A.-R. & Blümich, B. (1994) Introduction to solid-state NMR, in *NMR Basic Principles and Progress*, **30**, *Solid-State NMR I: Methods* (ed. B. Blümich), Springer-Verlag, Berlin, pp. 1–62.
107. Griesser, U.J., Auer, M.E. & Burger, A. (2000) Microthermal analysis, FTIR- and Raman-microscopy of (*R,S*)-proxyphylline crystal forms, *Microchem. J.* **65**, 283–292.
108. Gustafsson, C., Lennholm, H., Iversen, T. & Nystrom, C. (1998) Comparison of solid-state NMR and isothermal microcalorimetry in the assessment of the amorphous component of lactose, *Int. J. Pharm.* **174**, 243–252.
109. Gavezzotti, A. (1994) Are crystal-structures predictable? *Acc. Chem. Res.*, **27**, 309–314.
110. Bérard, V., Lesniewska, E., Andrès, C., Pertuy, D., Laroche, C. & Pourcelot, Y. (2002) Affinity scale between a carrier and a drug in DPI studied by atomic force microscopy, *Int. J. Pharm.*, **247**, 127–137.
111. Eve, J.K., Patel, N., Luk, S.Y., Ebbens, S.J. & Roberts, C.J. (2002) A study of single drug particle adhesion interactions using atomic force microscopy, *Int. J. Pharm.*, **238**, 17–27.
112. McCrone, W. (1957) *Fusion Methods in Chemical Microscopy. A Textbook and Laboratory Manual*, Interscience Publishers, New York.
113. Dalla Valle, J.M. (1943) *Micromeritics: The Technology of Fine Particles*, Pitman Publishing Corporation, New York, NY.
114. Yu, L., Reutzel, M. & Stephenson, G.A. (1998) Physical characterization of polymorphic drugs: an integrated characterization strategy, *Pharm. Sci. Technol. Today*, **1**, 118–127.

8 Microscopy and imaging in pharmaceutical analysis
Robert A. Carlton

8.1 Introduction

We live in a golden age of science. It is virtually impossible to keep abreast of even the major advances in physics, chemistry, biology, and engineering. Microscopy, even with a long and venerable history, is sharing in these advances. A comparison of the array of microscopy techniques available to us 50 years ago and today is startling. New, but now quite commonplace, techniques include scanning electron microscopy, infrared and Raman microspectroscopy, scanning probe microscopy and confocal microscopy. Even within the discipline of microscopy it is a challenge to keep up with the rapid changes.

The practical applications of the various microscopical techniques have created opportunities for microscopists in industry and, in particular, within pharmaceutical research and development. Microscopy is used extensively, from the earliest stages of drug discovery into late development and even into manufacturing. Pharmaceutical microscopy can be conveniently divided into physico-chemical and biological applications. This chapter will consider exclusively the physico-chemical aspects of microscopy in the pharmaceutical industry. There are three broad areas in which microscopy can play an important role in the development of drugs: solid-state analysis, particle size and morphology studies, and contaminant identification. This chapter presents an overview of how microscopy contributes to each of these three areas. The emphasis will be on practical examples taken from the literature and from the author's experience.

8.2 Solid-state analysis

The physical properties of the drug substance and excipients are important in the manufacture, packaging, shipping, and use of drugs [1]. These physical properties include solubility, dissolution rate, stability, particle size, water absorption, compactibility, and others. The crystalline form of the solid affects each of these properties. The physical form (i.e. polymorphism) of the drug substance can even affect the bioavailability of the drug and its effectiveness as a medicine [2]. Different solid-state forms of a chemical can be patented, thereby protecting the intellectual property of the pharmaceutical company, which can have a marked effect on the sales and profitability of the drug. Finally, regulatory bodies throughout the world are keenly interested in solid-state analysis. Applications to market new drugs are required to have sections detailing the new drug's solid-state properties [3].

Microscopy has a vital role to play in the analysis of polymorphs, solvates, and hydrates. Polarized light and thermal microscopy are particularly well suited for the discovery of new forms and in establishing the relationship among the various forms. The goal of early stage solid-state analysis is the discovery of the different solid-state forms of the drug. At this stage, it is important to know the number of solid-state forms of the drug – or put another way, what is the propensity of the drug to crystallize in different forms. If the drug survives to the next stage, a particular form of the drug must be chosen for development. This decision must optimize the balance between thermodynamic stability, bioavailability, ease-of-processing, and the kinetics of form conversion. Optical and thermal microscopy are excellent tools for determining the thermodynamic stability and for monitoring form conversion. The remainder of this section of the chapter will present applications of microscopy to early and late stage solid-state analysis.

The terminology of pharmaceutical solid-state analysis can be confusing. For this chapter, the following definitions and conventions have been adopted. Chemical polymorphism refers to the ability of crystalline solids to exist in different geometric patterns. A familiar case of polymorphism (more properly allotropy) is the element carbon, which can crystallize as graphite or diamond with radically different physical properties. Imagine having pharmaceutical polymorphs with such extreme differences. It would be critically important to control the crystallization of the desired form. In general, though, pharmaceutical polymorphs do not differ to such a great extent.

The only difference between polymorphs is the arrangement of the atoms or molecules in space. Consequently, polymorphism is peculiar to the solid state and does not exist in the liquid or gaseous states. Many molecules crystallize such that solvent molecules are an integral part of the crystal lattice. These solid forms are termed solvates or hydrates, in the case of water. Organic crystals often crystallize as hydrates. An interesting example of this is β-estradiol which crystallizes as the hemihydrate. In fact, even with non-polar solvents, the hemihydrate will crystallize unless extraordinary efforts are taken to prevent the intrusion of moisture from the air [4, 5].

For this chapter, the word *form* is used to refer to any solid made from the molecule, including polymorphs, solvates, hydrates, and even amorphous solids. In the initial solid-state studies, the precise physical state (polymorph, solvate, etc.) of the drug is generally unknown. In fact, it may require a great deal of time and effort to establish whether or not a solid is a hydrate and how many water molecules exist in the crystal lattice. It is convenient at this stage to categorize all solids as *forms*. Sometimes the word *modification* is employed for the same purpose. Finally, a great deal of emphasis will be placed on discovering the thermodynamically stable polymorph at room conditions (temperature and pressure). By definition, this is the crystal form with the lowest free energy [13].

8.2.1 Early stage solid-state analysis (polymorph discovery)

The goal of early stage solid-state analysis is the determination of the tendency of a compound to crystallize into different forms. It may have many forms like sulfathiazole with at least four polymorphs [6] or methylestradiol with one anhydrous form, two hydrated forms and at least two solvated forms [4]. This information guides the course of the future studies. If only one pharmaceutically significant form exists, then the subsequent studies should be straightforward and relatively rapid. If many forms exist, choosing the optimal form for development may require extensive time and study. It is useful to operate on the principle that all organic compounds crystallize in different forms and that the only questions are *How many?* and *How important?* McCrone [7] put the matter this way:

> Those who study polymorphism are rapidly reaching the conclusion that all compounds, organic and inorganic, can crystallize in different crystal forms or polymorphs. In fact, the more diligently any system is studied the larger the number of polymorphs discovered.

Bettinetti [8] put it succinctly:

> There are no precise rules to predict the polymorphism of a drug, but it has been observed that the phenomenon comes up with a frequency proportional to the time used in its study, and therefore the number of experimental tests conducted.

So, how diligently should we study the polymorphism of a potential new drug? How many experiments should we conduct? Virtually every pharmaceutical company has a different answer to these questions ranging from none to as many as possible. The following discussion is the author's opinion – attempting to strike a balance between extremes.

Polymorph discovery experiments should have two goals. The first goal is to determine the tendency of the prospective drug to form polymorphs. Two of the best means of accomplishing this goal using microscopical techniques are solvent recrystallizations on the polarized light microscope and thermal microscopy. The second goal is to collect a database of information about the solid form to be used for comparisons in later studies. It is commonly observed that the initial crystallization process for new chemical entities favors the formation of metastable forms. A well-documented database of solid-state analytical characteristics is required to detect a form change as crystallization methods are modified. Optical crystallography and thermal microscopy should be included in this database since they are quite sensitive to form changes.

8.2.1.1 Optical crystallography

Optical crystallography was used extensively before the advent of X-ray methods for structure determination and crystallography. Although it is less used today, optical crystallography does have the significant advantages of its ability to distinguish among individual crystals of different polymorphs [9]. Every solid-state form has different optical properties such as extinction angle, refractive indices, birefringence, interference figure, optic sign, optic angle, and the sign of elongation [10]. A useful first step in the solid-state analysis is to determine the optical crystallographic properties of the drug substance as received.

A recent example of the use of optical crystallography involves paracetamol (acetominophen) [11]. Paracetamol is known to have at least two polymorphs [12]. Monoclinic paracetamol (form I) is the thermodynamically stable form at ambient conditions and is the form used in commercial products. It suffers from a manufacturing liability in that it is not suitable for direct compression into tablets. Consequently, it is necessary to mix it with binders which increases the cost of the product and the complexity of manufacturing. Orthorhombic paracetamol (form II) can be directly compressed, but it is very difficult to crystallize from solution, though it can be crystallized from the melt. Nichols and Frampton devised a crystallization protocol that gave a high yield of form II. In the process, Nichols determined the optical crystallography of form II and related the optical orientation to the crystal geometry and to the X-ray crystallography. Table 8.1 presents the morphology and optical crystallography of both forms. Figure 8.1 presents diagrams of typical crystal habits of both forms along with an assignment of the Miller indices. Notice that there are distinct differences in the optical properties of the two polymorphs and, in particular, form I has dispersed extinction (see [10] for an explanation of this phenomenon) whereas form II does not. Nichols and Frampton note that the extinction difference was the basis of their rapid screening method for distinguishing between the two crystal forms in crystallization experiments. In some cases, they were able to determine that traces of form I were present in samples which X-ray diffraction suggested were pure form II. The sensitive nature of optical crystallography is applied quite usefully in such studies of crystallization.

8.2.1.2 Solvent recrystallization experiments

Polymorph discovery by solvent recrystallization is based on the principle that the less stable forms recrystallize before the more stable ones. This is often referred to as Ostwald's law of stages [13]. As the system progresses from the liquid to the solid state, it will go to the state nearest in free energy to that of the liquid state. Frequently a metastable crystal form will appear and then convert to

Table 8.1 This table, for the two polymorphic forms of paracetamol, illustrates the type of data that can be collected using optical crystallography [11]

	Form I	Form II
Crystal morphology		
Habit	Prisms and plates elongated parallel to $\{10\bar{1}\}$. The dominant forms are prisms ($\{011\}$, $\{110\}$) and pinacoids ($\{10\bar{1}\}$, $\{101\}$)	Doubly terminated prisms elongated parallel to c-axis, Some plates occur. The dominant forms are prisms $\{210\}$ and bipyramids $\{211\}$
Cleavage	Parallel to 010 (moderate)	Parallel to 001 (perfect)
Crystal system	Monoclinic	Orthorhombic
Space group	$P2_1/n$	$Pbca$
Beta angle	97.36°	90.00°
Optical data		
Refractive indices ($n_D{}^{25}$):		
$n\alpha$	1.580 (±0.001)	1.491 (±0.001)
$n\beta$	1.643 (±0.001)	1.667 (±0.001)
$n\gamma$	1.704 (±0.001)	1.840 (±0.002)
$n\gamma'$: when lying on $\{10\bar{1}\}$	1.683 (±0.001)	
Birefringence	0.124	0.349
Optic sign	(−)	(−)
Optic axial angle (2V)	86° (calculated)	81° (calculated)
Dispersion of optic axes	$v > r$ (strong, crossed)	$v > r$ (weak)
Acute bisectrix (Bx_a)	α	α
Extinction	Straight for sections perpendicular to the b-axis. Inclined and dispersed when viewed along b-axis; $\gamma \wedge \{10\bar{1}\} = 36.2°$ in white light; 34.0° (656 nm, red); 38.2° (488 nm, blue)	Straight for prisms and symmetrical in rhomb-shaped 002 cleavage fragments
Optical orientation	$X = \alpha = b$	$X = \alpha = c$ $Y = \beta = b$ $Z = \gamma = a$

its nearest free energy state and so on until crystallization of the lowest free energy state. The microscope is particularly well suited to the observation of this process since individual crystals can be observed as the solid is recrystallized.

The diuretic frusemide illustrates how solvent recrystallization experiments can be used to discover new solid-state forms of older drugs. Frusemide was marketed for many years with no reports of polymorphism. Doherty and York [14], however, noted some features of the differential scanning calorimetry (DSC) thermogram of frusemide that suggested the possibility of polymorphs. They performed a number of solvent recrystallization experiments and discovered a new polymorph which they characterized analytically. At nearly the same time, Matsuda and Tatsumi [15] published the results of their experiments with frusemide with the conclusion that there are five polymorphs and two solvates, illustrating the principle that the number of known forms is related to

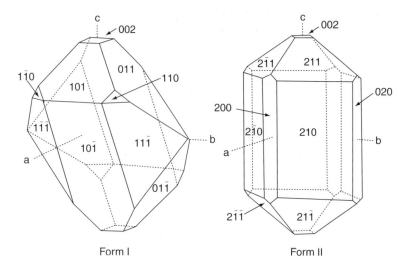

Fig. 8.1 This is a drawing of the crystal faces of the two forms of paracetamol with the Miller indices labeled [11].

the time and effort put into the study. Matsuda and Tatsumi used seven different methods of recrystallization with a wide variety of solvents. Brittain [13] states that, 'Solids should be isolated from solvents having the widest range of dielectric constant that the system will permit, with as many variations in isolation conditions as possible.'

Recrystallization experiments frequently yield crystals having different shapes and morphologies which are not necessarily different polymorphs. For example, Figs 8.2 and 8.3 show crystals of β-estradiol with distinctly different shapes but are, in fact, the same polymorph. The morphology differences are due to different crystallization solvents. It is important then to have some microscopical technique that allows one to distinguish between polymorphs. Optical crystallography, thermal microscopy and microspectroscopy have this ability.

8.2.1.3 *Thermal microscopy*

The thermal microscope is one of the most powerful tools for the study of polymorphism in the analyst's arsenal. Its utility has been amply demonstrated in the work of the Kofler, L. and Kofler, A. [16], Kuhnert-Branstatter [17], and McCrone [7]. A thermal microscope is simply a typical optical microscope (preferably polarizing) that is equipped with a stage that allows heating and cooling of the sample. Modern thermal stages can control temperature to ca. ±0.1°C with a wide variety of heating and cooling rates. It is useful for documentation to connect the thermal microscope to a video recorder as well as an image collecting computer.

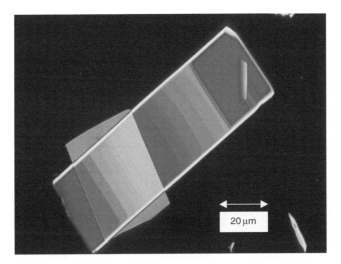

Fig. 8.2 β-estradiol crystallized from ethyl acetate. Polarized light microscopy.

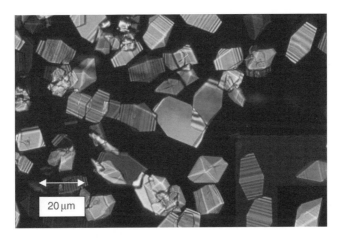

Fig. 8.3 β-estradiol crystallized from isopropyl alcohol. Polarized light microscopy.

The most basic and important data provided by a thermal stage is the melting point. Of course, there are a number of techniques such as DSC that can provide this value. One advantage of thermal microscopy is its sensitivity to small amounts of different forms. Although determination of the melting point seems simple enough, there are, in fact, a number of complications. First, not all of the crystals in the field-of-view will melt simultaneously. Does one choose the temperature at which the first or last crystal melts? McCrone [7] recommends two alternatives. The best, but most time-consuming option is the equilibrium melting point. Experimentally, this value is determined by alternately heating and cooling so that the crystal melts and then recrystallizes. The equilibrium melting point is the midpoint between temperatures for which definite melting and crystallization occurs. McCrone states that done carefully, the two temperatures will differ by approximately 0.2 and 0.3°C. This operation can be somewhat tedious and time-consuming. Especially in early stage drug development it may be better to use the melting temperature of the last crystal in the field of view. The exact value of the melting point will also depend to some extent on the heating rate. A higher rate yields higher melting points and vice versa. The author typically uses a high rate (ca. 20°C/min) for the first test of a new compound and a lower rate (ca. 5°C/min) for subsequent tests. Occasionally, it is useful to use a much lower rate (ca. 1°C/min) in order to study thermal events such as solid–solid form transformations or desolvation in a more thorough fashion.

Another important complication is the effect of residual solvents on the melting point. In general, residual solvents tend to lower and broaden the melting point. In fact, the melting point can be used as a method to determine chemical purity as described by Kuhnert-Branstatter [17]. The broadened melting point can be confused with the presence of polymorphs. Some crystals may melt as much as 5 to 10°C lower due to the influence of solvents. Also, as the available drug substance becomes purer due to improvements in process chemistry, the experimental melting point will probably increase by a few degrees. Consequently, it behoves the microscopist to obtain information about the purity of the compound before conducting any experiments in order to avert unnecessary confusion. Finally, some compounds and even whole classes of compounds have broad melting points. For example, the steroids are known to have broad melting points. A good example is corticosterone form III which is reported to melt between 162 and 168°C [17].

It is possible to directly observe solid-state transformations using thermal microscopy. The transformations include the following:

- Solid–solid polymorph transformations (see Fig. 8.4)
- Solid–liquid–solid polymorph transformations
- Sublimation (see Fig. 8.5)
- Crystallization from the melt
- Desolvation (see Figs 8.6 and 8.7).

At the early stages of drug development, a few thermal tests will suffice to assess the tendency of the new compound to form polymorphs. Fast and slow heating rates, with and without oil should allow one to chart the future course of solid-state studies. (The formation of bubbles when heating in oil is a sensitive method for the detection of solvates and hydrates.) If a number of transformations occur, it is clear that a great deal of time and effort will need

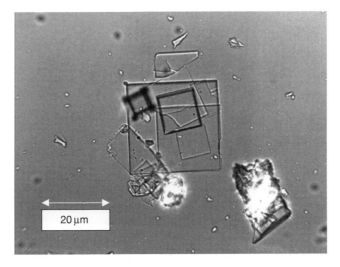

Fig. 8.4 Solid–solid transformation on heating. The transparent particles are growing from the bright ones. Partially crossed polars.

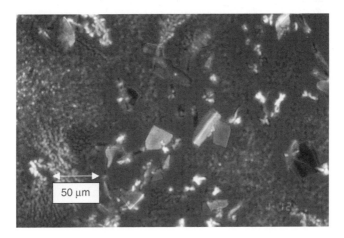

Fig. 8.5 An example of sublimation. This is an image of the upper coverslip and there are at least two different types of crystals growing as the main sample sublimes. Partially crossed polars.

to be allotted to the full polymorphism study since it will be necessary to crystallize pure material representing each solid-state form in sufficient amounts to conduct a wide variety of analytical tests. The microscope occupies a unique niche in solid-state characterization, but it cannot provide all of the answers. To truly understand the solid-state properties of a drug it is necessary to collect the full range of thermodynamic and structural information on every

304   PHARMACEUTICAL ANALYSIS

Fig. 8.6 Desolvation of β-estradiol hemihydrate. This photomicrograph was taken at ~90°C. Partially crossed polarized light.

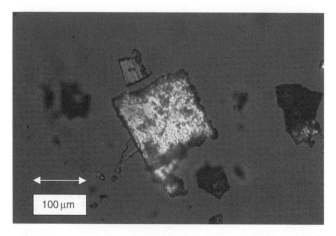

Fig. 8.7 The same crystal as in Fig. 8.6 but the temperature is ~105°C. The mottled appearance of the crystal is an indication of the loss of solvent. Partially crossed polars.

observed form. For a prospective drug with many forms, the real work has just begun.

8.2.2 *Late stage solid-state analysis (polymorph relationships)*

The objective of late stage solid-state analysis is to provide data for the choice of form for development. Microscopy has much to contribute to this analysis but generally plays an adjunct role to other techniques. Microscopy can, however, play a crucial role in some of the thermodynamic investigations leading to the determination of the stable form at ambient conditions and to estimates of the transition temperature for enantiotropic polymorphs.

A thorough and exacting optical crystallographic analysis can be conducted at this time on each form for which such an analysis is possible, meaning on those forms which have well formed and stable crystals. Whereas in the initial optical examination estimates of the principal refractive indices are sufficient, it is appropriate at this point to make an accurate measurement of the indices. Also, the optical crystallographic orientation can be determined and related to the X-ray structural data if available. Nichols [11] work on paracetamol is an excellent illustration of how the optical crystallography can be related to the X-ray structural data. Nichols and Frampton [12] were able to explain the difference in compaction properties between polymorphs I and II based, in part, on the optical crystallography. Form II crystals cleaved perpendicular to the length of the crystals (parallel to the {001} zone), indicating the presence of a well-developed slip plane system which aids plastic deformation and aids compaction. Form I does not break in this fashion.

Krc [18] also provides a good example of the kind of optical crystallographic data that can be collected using flufenamic acid. Flufenamic acid is an analgesic and has at least six polymorphs. Krc characterized as many of these as possible using optical crystallography, X-ray diffraction, and infrared spectroscopy. Two of the forms (I and III) could be grown from solution and so single crystal X-ray crystallography was determined for these. For form II, it was possible to collect sufficient pure material for powder X-ray diffraction.

The other forms were characterized by morphology and by melting point because they could not be isolated as pure material. Krc developed crystal drawings relating the optical properties to the morphology of the crystal. In other words, the optical crystallographic orientation was related to the crystal morphology. Polymorphs I, II, and III can be distinguished using melting point, refractive index and extinction angle (see Table 8.2) among other properties.

One objective of these experiments is an energy–temperature diagram relating all of the forms to each other. Figure 8.8 shows an example of such a diagram for flufenamic acid. This figure indicates that polymorph III is stable at ambient conditions and has an enantiotropic relationship with form I. The enantiotropic transition temperature is 42°C which suggests that it is necessary to monitor solid-state form during processing and during storage since it is conceivable that the temperature would rise above that level.

Table 8.2 A comparison of the various properties of polymorphs of flufenamic acid

Property	Polymorph I	Polymorph II	Polymorph III
Melting point (°C)	134	128	126.5
Refractive index (β)	1.722	1.648	1.616
Extinction angle (°)	51, 56	33, 34	20

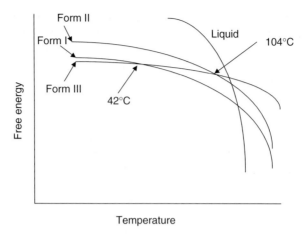

Fig. 8.8 A free energy diagram for flufenamic acid showing three of the seven forms. Forms I and III are enantiotropic with a transition temperature of 42°C. Forms II and III are enantiotropic with a transition temperature of 104°C. Forms II and III are monotropic. (Drawing based on [18].)

A solution phase transformation experiment is a simple and straight forward method of determining the thermodynamically stable polymorph at ambient conditions. The experiment is conducted by mixing crystals of the two polymorphs in a saturated solution. The form that grows is the stable one [19]. This is due to the fact that the stable form has the lower free energy. These experiments can be conducted on a microscope slide or by using a specially designed cell. Nichols and Frampton verified that paracetamol form I was stable compared with form II using saturated benzyl alcohol. Over time the needle-like particles (form II) dissolved leaving the block-like form II crystals. The forms need to be easily identified for this test to work well. A microscope slide experiment has the advantage of being simple and fast, though it does require the use of high boiling point solvents. Haleblian and McCrone [19] have described an apparatus that allows one to use a variety of different solvents. It is a closed cell that can be put on an inverted microscope. Evaporation is minimized allowing one to use volatile solvents.

Solution phase transformation experiments at different temperatures can provide a good estimate of the phase transition temperature for enantiotropic polymorphs. The process is somewhat tedious and time-consuming but can yield a highly accurate value for the transition temperature. The experiment is conducted at successively higher temperatures until the stability relationship reverses as indicated by the growth of the other polymorph. It then is necessary to bracket the transition temperature by repeated experiments. As the transition temperature is approached the experiments generally require more time. It is not unusual for an experiment to run for 24–48 h before it is clear which form is growing. Clearly a closed cell apparatus is required to minimize evaporation. Also it can be necessary to change solvents with increases in temperature. A compound with a low solubility in a solvent at 25°C will have a much increased solubility at 80°C.

8.2.3 Summary

Microscopy has numerous applications in the solid-state analysis of pharmaceuticals. Only a few of these applications have been discussed in this chapter. It is sometimes tempting to attempt to develop a generic plan of attack for solid-state analysis. As useful as this may be to focus experiments, in general, organic crystals are not so predictable. For instance, one compound that the author studied refused to crystallize for much more than a year of study. The decision was made to develop the amorphous form. One evening, after applying for approval for phase 1 clinical studies, a process chemist noticed a thick precipitate in the reaction vessel which turned out to be the crystalline drug form. After that it was nearly impossible to generate the amorphous form and the crystalline form had to be characterized. Solid-state analysis in drug development is challenging, and studies utilizing microscopy can contribute to the understanding of the prospective pharmaceutical.

8.3 Particle size and morphology

The biological and chemical properties of a drug substance can easily overshadow the importance of the bulk physical properties. Yet, these latter properties critically influence the delivery of the drug to the target in the body. Physical properties also influence the often overlooked connection between the manufacturing plant and the human hand and mouth. We must be able to manufacture, package, and transport the drug product – intact and with full purity and efficacy. The choice of drug delivery form (tablet, capsule, etc.) is often determined by the physical properties of the drug. Paracetamol is a good example of this relationship [12]. It is not possible to directly compress form I and so binders must be used for tableting this form. Form II can be directly compressed but has chemical processing difficulties. Since there is a clear connection between the particle and bulk properties, the microscope can play a key role in the investigation of the physical properties as a prospective pharmaceutical. Microscopy is particularly well suited for the analysis of particle shape and morphology.

As was the case with solid-state analysis, it is convenient to make a distinction between early and late stage development. In early stage development, generally there is little drug substance available and crystallization conditions have not been optimized. Particle evaluations using optical and scanning electron microscopy are ideal since the desired information does not generally require full method validation or large populations. At this point, the morphology of the crystals can be critically important. Even if the drug substance will be milled or micronized, knowledge of particle size and shape can be useful. Milling generally results in spherical particles even if the starting material is needle like. Sometimes, however, the needle-like morphology is maintained after milling. This can have an impact on the use of diffraction type particle sizers since they assume sphericity. Often these early batches are used in animal toxicology studies and, if dosed as a suspension, the particle characteristics can impact the results. The author was asked to evaluate a series of suspensions from different lots of material used in an animal study. It was clear from optical microscopy that some suspensions were heavily aggregated. Using only suspensions without aggregates and re-dosing led to more consistent results.

In late stage development, microscopy is generally an adjunct method to faster, more automated methods. In general, particle size methods for the release of drug substance and drug product favors instruments based on laser diffraction. These instruments have advantages of speed, ease of use, good precision, simple sample preparation, and large population statistics. While method validation is never a trivial task, it is relatively straight forward with laser diffraction instruments compared with optical microscopes. Some dosage forms utilize particle size methods specifically designed to test the drug product. An example of this is inhalers used for the treatment of asthma and other respiratory problems. Particle size is generally measured using an Anderson cascade impactor which is designed to simulate some of the properties of the human lung.

Notwithstanding the advantage of other techniques, microscopy has a role in late stage development. Microscopy provides a reality check for other less direct methods of particle size analysis. Virtually every non-microscope based particle size method explicitly makes the assumption that the particles are spherical in shape. Elongated or needle-like particles can lead to large errors in the analysis. Particularly in method development it is useful to examine samples using microscopy. As Allen [20] states in his book on particle size analysis, 'A microscope examination should always be carried out whenever a sample is prepared for particle size analysis. Such an examination allows an estimate of the particle size range of the powder under test and its degree of dispersion.' A microscopical examination can alert the analyst to changes in particle shape due to crystallization and to particle aggregation. It is also possible that another polymorph has crystallized. The first response to inconsistent or unusual results from a laser diffraction instrument should be an examination using the microscope.

For some dosage forms and applications, microscopy with image analysis may be the method of choice. For instance, topical suspensions containing a number of solid excipients make it difficult to directly measure the particle size of the drug substance. Measurement difficulties will be particularly severe, the more stable the suspension, since it will be quite difficult to mechanically

MICROSCOPY AND IMAGING IN PHARMACEUTICAL ANALYSIS 309

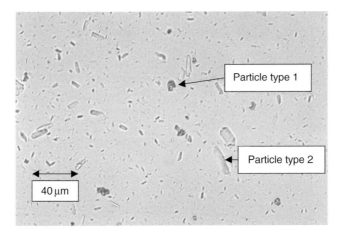

Fig. 8.9 Optical photomicrograph of a topical suspension showing two populations of crystal types. Optical microscopy.

separate the drug substance from the excipients. If the drug substance particles can be distinguished from the excipients on the microscope, then it is possible to develop a particle size method. Figure 8.9 shows an optical micrograph of such a suspension. The various components of the suspension can be distinguished using contrast and refractive index. Even though microscope based methods are a challenge to validate and operate, the author has had success developing such methods and transferring them to manufacturing plants.

This section of the chapter is divided into two parts. The first part discusses the instruments and methods used to evaluate particle morphology. It pays particular attention to nomenclature since the words used in this field are often ambiguous. The second part deals with the details of particle size analysis by microscopy. It pays particular attention to sampling issues and to the use of image analysis.

8.3.1 Particle morphology

Descriptions of particle morphology are intended to categorize samples by shape. It is extremely difficult to make these descriptions quantitative and so it is helpful to illustrate these descriptions with photomicrographs. The scanning electron microscope (SEM) is an ideal instrument for investigations of particle morphology. Figures 8.10 and 8.11 show two polymorphs of the same compound. To state that one is fibrous and the other blocky does not convey nearly as much information as the associated photomicrographs. Even for milled or micronized material, photomicrographs can be used to assess the success of the particle size reduction process. Figure 8.12 is a SEM photomicrograph of a

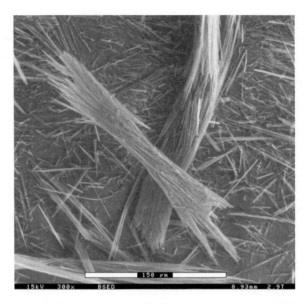

Fig. 8.10 One polymorph of a new chemical entity illustrating the morphology differences. Scanning electron microscopy.

Fig. 8.11 Another polymorph of the same new chemical entity as in Fig. 8.10 illustrating morphology differences. Scanning electron microscopy.

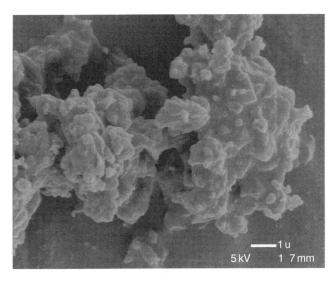

Fig. 8.12 Triamcinolone acetonide, micronized. Scanning electron microscopy.

sample of triamcinolone acetonide that has been micronized. It is clear that substantial aggregation has occurred and that each aggregate is composed of much smaller particles. Particle size measurements on this type of particle will be difficult since it appears that the aggregates have both loosely and tightly bound particles. The measured particle size may depend upon the degree of mechanical stress applied during measurement and sample preparation.

A serious problem with morphological descriptions is the subjective nature of the descriptions. The ambiguous nomenclature used in the field is also a hindrance to clear, meaningful descriptions. For instance, how is one to understand a statement that the particle is a needle and then to distinguish this from a fiber, a rod, columnar, or an acicular particle? Aldrich and Smith [21] have published an admirable nomenclature scheme for categorizing particle shapes using axial dimensions (see Fig. 8.13). For example, a needle is distinguished from a rod by the short dimensions (labeled x and y, z is the long dimension). Using Aldrich and Smith's scheme the difference lies in the aspect ratio. A needle's aspect ratio is greater than 10:1 whereas a rod has an aspect ratio between 10:1 and 10:2. This is a rather fine distinction and somewhat arbitrary, but it does remove much of the ambiguity in such descriptions. Whatever scheme one adopts, it is important that the microscopist clearly defines and communicates the meaning of the terms used in morphological descriptions.

The microscope is often best used as a problem-solving tool. As such, the ideal situation occurs when the requestor of the information (engineer, analyst, or pharmaceutical scientist) can be present during the microscopical examination. Photomicrographs only present a few selected fields-of-view and these

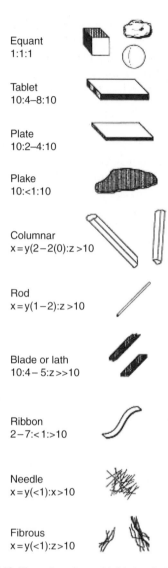

Fig. 8.13 Illustration of crystal habit descriptions [21].

selections are made by the microscopist in an attempt to represent the whole. The requestor, on the other hand, may be as much interested in the atypical regions as the typical ones and may even be most influenced by what isn't there. A good illustration of this is in the case of tablet coatings. Coatings have a variety of purposes including protecting the core tablet during shipping and use, modifying drug release, improving appearance, and masking taste. The author worked with a pharmaceutical scientist who was having problems with tablet coating.

In a SEM examination, the more representative fields of view were not particularly helpful. In this case, however, the engineer was present during the examination and immediately recognized the source of his coating problem based on some relatively rare views. Since the one surface was relatively rare, the microscopist might be tempted to exclude it as being non-representative when making photomicrographs.

8.3.2 Particle size analysis/image analysis

A clear goal of a particle examination is the attempt to relate some property of the individual particles to some bulk physical property of the powder. For example, the dissolution rate of drug substance in the body can often be related to the size of the drug substance particles [22]. A microscopical examination has the maximum utility if it can relate some measured property of the particles to some physical property of the bulk. Establishing this sort of relationship nearly always requires a numerical analysis of size. While it is possible to measure or estimate particle size using only a microscope and microscopist, the operation is tedious and prone to operator error. The use of a digital image analyzer can, in some cases, completely automate the analysis, and in other cases, automate some of the more tedious and error-prone operations of the work.

The steps involved in image analysis are as follows:

1. Sample preparation
2. Image generation
3. Image collection/digitization
4. Image processing
5. Calibration
6. Image analysis
7. Statistical analysis and data presentation.

The following paragraphs discuss each of these steps.

Sampling is by far the most important part of particle size analysis by microscopy (and probably all particle size techniques). A kilogram of drug substance will contain many millions of particles. Since, at most, the particle size analysis samples a few thousand particles, the measured particles must be selected with care. Allen [20] presents an extensive discussion of bulk sampling issues relevant to all particle size analysis, irrespective of the particular technique. Our interest, though, is primarily directed toward sampling as it relates to the specimen used for particle size analysis by microscopy. We will assume that the 50 mg or so of sample delivered to the laboratory is truly representative of the bulk powder.

The size range of the particles to be measured and the objective of the analysis determine the type of microscope to be used. The optical microscope is best applied when the particle sizes are between 3 and 150 μm. The SEM is the better instrument for particles less than 3 μm. The TEM can be applied for very small particles (less than ca. 20 nm). If the method is intended for transfer to a manufacturing facility then the optical microscope is probably the best choice due to cost. The size limits for analysis can be deceiving. For instance, the resolution limit for optical microscopes can be as low as 0.2 μm. In other words, particles that are small can be detected and resolved with an optical microscope. Studies have shown, however, that the diffraction halo surrounding the particle will increase the apparent size of the particle up to approximately 3 μm [20]. The most serious size increase is for particles less than 1 μm. Each microscope has its own limitations that must be considered during method development. Most of the remaining discussion uses optical microscopy to illustrate how particle size methods can be developed using image analysis.

The first issue that we face in optical microscopy is the selection of a fluid to disperse the particles. Dry mounts are almost never satisfactory due to particle aggregation and to the very high particle contrast of the particles (due to the high refractive index difference between the particle and air). In the author's experience, silicone oils are the best first choice for mounting media. Ideally, the fluid should disperse the particles without dissolution. Further, the refractive index of the liquid compared with that of the particles should be such that there is moderate particle contrast. Figure 8.9 shows two particle populations with different contrast values based on the different refractive indices of the particles. It is possible to distinguish between the two particle populations based on this difference. Finally, the mounting medium should have low volatility so that evaporation is minimized during analysis. Silicone oils that satisfy these requirements can almost always be obtained. Particle dispersal and sample preparation issues are somewhat more difficult in the SEM. While simply sprinkling some of the particles onto a stub will suffice for morphology examinations, this method is not sufficient for accurate particle size determinations. For some samples, filtration works quite well. The sample can be mixed into a liquid in which it is not soluble and then filtered using nucleopore filters. Often this will disperse the sample so that individual particles can be measured.

The next sample preparation issue is how many particles need to be measured from how many fields-of-view from how many slide preparations. This is a complicated subject requiring much more space to discuss than is convenient in this chapter. Allen [20] has a good discussion of these issues. In brief, though, the sampling population depends upon the shape of the particle size distribution and the range of particle sizes. A sharp, narrow, Gaussian distribution may only require 100 or so particles whereas a broad, skewed distribution may require tens of thousands. Also, the number of particles also depends upon whether the distribution is by number or by weight. Generally, a weight distribution requires a larger population than one by number. Allen states that for a 2% error in a number distribution, at least 625 particles must be measured.

Once the total number of particles has been determined, it is necessary to determine the number of slides (or SEM stubs) and fields-of-view from each slide. It is the author's experience that there is greater variability between slides than within one slide. Consequently, it is better to prepare more slides and measure fewer fields-of-view on each slide. If each field-of-view contains at least ten particles, then 20 fields-of-view from five slides will yield approximately 1000 particles. This is a typical population for particle size analysis by microscopy. Of course, establishing the sampling methodology is a part of the development and validation of an accurate and rugged method. In many cases, where speed is crucial and the need for accuracy less demanding, the microscopist may opt for a smaller population. As a side note, it is absolutely necessary that the fields-of-view for measurement be chosen randomly. The microscopist should not directly influence this choice. It seems inevitable that the analyst will choose fields containing at least a few easily observed particles. This almost always leads to bias in the distribution – frequently toward larger particles. Many optical microscope stages can be equipped with devices that allow one to advance the slide in defined increments. This is an ideal way of ensuring random sampling. Many modern SEMs are equipped with motorized stages and programs for moving the stage in a raster pattern.

The next step in the analysis is a consideration of image generation. Of course, one has to use a microscope in sample preparation, but the images may not have been optimized for analysis. For example, during the early stages of the method development, the microscopist may have used secondary electron detection in the SEM. This signal may be appropriate for resolution but not ideal for measurement since the edges of the particles tend to be much lighter than the interior. Particles imaged using backscatter detectors, on the other hand, generally have a more uniform brightness which eases image processing and analysis. This is also the appropriate point in method development to choose the magnification. For narrow particle size distributions, the choice of magnification is simple since one magnification will generally resolve the entire range of particles. For broad distributions, the choice is more complicated. Low magnification ensures that the largest particles are counted but at the cost of poor accuracy in measuring the small particles. High magnifications have the opposite problem. The choice of magnification can best be guided by the purpose of the investigation. If the method is designed to monitor aggregation, for instance, then a low magnification may be ideal. If the method is designed to assess the effect of milling on individual particles, then a higher magnification may be desirable. Some researchers have published schemes for using multiple magnifications and combining the results. These schemes are difficult to apply and the user should be especially careful when validating the method.

For image analysis, the image must be collected and then converted into a digital image computer file. For analog cameras, these two steps are separate and generally the computer itself has a board for converting the analog signal into a digital file. Digital cameras combine both steps in the camera itself. Digital cameras have excellent resolution but the collection time, for optical microscopes, is long. This can be problematic for slides with particle movement. For instance, preparations in water will begin to evaporate immediately. The particles will move as the water evaporates from the edges of the coverslip. This movement can be too fast for the digital camera and in this case it is preferable to use analog ones. Most modern SEMs directly produce digital images.

Image processing involves modifying the image for subsequent measurement. A wide variety of digital filters are available on modern image analyzers which allow one to increase particle contrast, remove noise, and sharpen edges. Also, the image can be segmented into a black and white image. Ideally, this converts all of the particles into white and everything else is black. A number of binary operations are available to fill particles, if not all of it is white, to close perimeters, and to remove noise. Image processing can be useful (and fun) but should be minimized since it can be both time-consuming and error prone in automatic analysis. It is better to expend some effort in preparing an image that requires little modification before measurement.

Finally, we are ready for the main course – particle measurement. The geometrical properties of each feature are measured and stored. The most fundamental measurements are area which is determined by the number of pixels (picture elements) in the feature and feature perimeter. It is necessary to calibrate the system in order to translate pixels into real values such as micrometers. Generally, this is accomplished using sample features of a known size. For instance, in optical microscopy it is common to use a micrometer slide containing a grating of known dimensions. For accurate work, it is possible to obtain certified micrometer slides that have been calibrated by one of the national laboratories such as the National Physical Laboratory of Great Britain or the National Institute for Standards and Technology in the United States. It is important to calibrate using as much of the field-of-view of the microscope as possible. For optical microscopes, even objectives that have planar corrections still have some curvature. This results in different sizes at the edge than at the center of the field-of-view. SEMs suffer from a different but related problem. Consequently, the best calibration procedure uses as much of the microscope field as possible. Also, it may be important to limit the particles that are measured to some central region of the field-of-view. Many modern image analyzers allow the microscopist to choose the size of the measurement region.

There is a confusing array of measurement options in modern digital image analyzers. Besides area and perimeter, it is possible to obtain measurements of geometrical features such as equivalent circular diameter, aspect ratio, shape factors, Feret diameters, Martin diameters, to name just a few. In most instruments, the operator has the flexibility to design his or her own geometrical property using the built-in measurements in some mathematical combination defined by the microscopist. This chapter is much too short to allow for an extensive discussion of the various measurement properties. Frankly, it is not particularly easy to learn the significance of the measurement options either from the literature or from the instrument manufacturers. Consequently, it may be useful to collect many of these properties in method development and then test the correlation of these values with bulk properties. This can make the choice of measurement parameter less arbitrary.

It is necessary to develop statistical parameters to describe the measurement property population. The statistical parameters of greatest utility are the following: mean value, standard deviation, median value, 10th percentile (M10), and 90th percentile (M90). Typically modern image analyzers provide all of these parameters and many more. The choice of parameter to be used to describe the particle population may be dictated to some extent by the dosage form or the end use of the data. Regulatory agencies, such as the FDA, often look for control of the mean, M10, and M90. At a minimum, these parameters, along with the standard deviation of the mean, should be reported.

The most common distribution used in image analysis is the number distribution. This relates the number of particles, generally as a percentage of the total, to a measured geometrical property such as the equivalent circular diameter. A number distribution may be appropriate for the analysis of aerosols but may not be ideal for comparisons with other particle sizing instruments or even with bulk properties. These comparisons may be better with a weight distribution. To formulate this distribution it is necessary to have an estimate of the density of the particles. Conversion from a number to weight distribution can be problematic due to the sensitivity of the latter to large particles. The mass of a single 60 µm diameter particle is equal to the mass of approximately 10 000 particles with a diameter of 3 µm. It is necessary to ensure that a sufficient number of large particles have been measured to accurately characterize the weight distribution.

Figures 8.14–8.16 illustrate this process with a suspension. In this case, the objective was to evaluate the effect of various processing parameters on the degree of particle aggregation. Consequently, the individual particles were ignored. In the first figure, the polarized light microscope image shows a population of two particle types based on brightness. The particles of interest are the large aggregated black particles. In the next figure, the image has been segmented and the image inverted. What was black is now white. Notice that the white particles in the original image have been eliminated. In the last figure most of the individual particles have been removed and the image is now ready for measurement.

8.3.3 Summary

This section has presented a few of the methods and applications of particle size analysis using microscopy and image analysis. The technique has gained

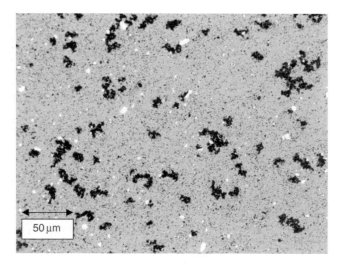

Fig. 8.14 Photomicrograph of topical suspension. The aggregated black particles are the particles of interest. Polarized light microscopy.

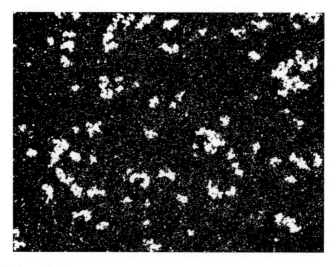

Fig. 8.15 The image has been segmented and inverted. The white particles in the original have been eliminated and the particles of interest are white.

greatly from advances in computer technology. Image processing operations that required substantial time and computing power in the near past can be done rapidly with personal computers. Interestingly, applications and studies using this new power have lagged somewhat behind the advances in hardware and software. There are undoubtedly many novel uses of image analysis in the particle size analysis of pharmaceuticals.

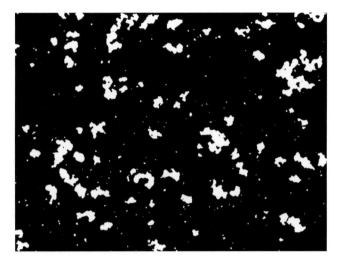

Fig. 8.16 The image has been smoothed and the smaller particles have been removed. This image is now ready for measurement.

8.4 Contaminant identification

Microscopy is ideal for the examination of particulate contaminants in drug substance and drug product. It is an area in which microscopy can have a major impact in drug development. Contamination in drug substance and drug product can delay clinical trials or even force them to be rescheduled, which can have an enormous impact on the timing of regulatory filings and the eventual marketing of the drug. It is crucial then, to accurately and rapidly identify particulate contaminants and their source. Polarized light microscopy, microspectroscopy (IR and Raman), and X-ray spectrometry using the SEM are excellent tools for this kind of analysis. A brief description of the process that the author utilizes in contaminant analysis will be followed by a number of examples. Since companies are generally not anxious to publish information concerning contamination, published examples are rare.

The process is invariably the same and begins with a visual examination. There is a natural tendency to immediately leap to the high-powered, high-priced instruments but this tendency should be resisted. It is important to know something about the appearance of the particles and their matrix before beginning the more detailed examination. In one case, the author was able to make the identification with a visual examination and a magnet. The black, somewhat metallic appearing particles, moved under the influence of a magnet. In fact, the magnet was covered in particles after being inserted into a sample of the powder. Energy dispersive X-ray spectrometry (EDS) was only used to confirm the presence of iron filings and the process engineer could locate the source of the contamination and solve the problem with just this information. The visual examination is followed by examination under a stereomicroscope. The enhanced magnification helps to classify the particle as *animal*, *mineral*, or *vegetable* and is also used to remove particles for further examination.

The author has found the following principle, learned at the McCrone Research Institute, to be an invaluable guide to contaminant analysis. An analytical test should answer a specific question or the test should not be conducted. For instance, the polarizing light microscope can be used to answer the question of the crystallinity of many samples. If the contaminant particles display birefringence between crossed polars, then they are likely to be crystalline. This categorizes the particle and helps point the direction to the next test. The tendency to use every arrow in the quiver should be resisted. Making sense of a mass of data collected from many different instruments makes the task more difficult than it needs to be. The first task then is to categorize the particles. If the particle appears to be metallic in the stereomicroscope and is opaque in transmitted optical microscopy, then it is appropriate to use EDS next in order to confirm that it is metallic and to identify the metal. If it appears polymeric on the optical microscope, then the Fourier transform infrared (FTIR) microscope may be the next obvious tool. If it appears to be crystalline, then thermal microscopy and FTIR microscopy are probably the next tests to be conducted.

The thoroughness of the investigation will depend upon the situation. If the contaminant is in a pre-clinical, early chemical process, then a complete identification may not be necessary. Clinical batches will almost always require a properly documented analysis. The author uses an arbitrary rule that three consistent results from three different analytical techniques are required to constitute proof. For instance, if the melting point, IR spectrum and EDS are consistent, then this is sufficient evidence for identification of the contaminant. As in a forensic examination, however, it is advisable to phrase the conclusion such that the 'analytical results are consistent with the identification of the contaminant as...' This is due to the difficulty in establishing absolute proof.

One example of the sometimes subtle nature of contaminants involves an early stage compound that was non-hygroscopic. Yet, the water vapor absorption of the various process samples varied from 0.1 to 2% (up to 90% relative humidity). The early batches were uniformly low whereas the later ones were inconsistent, some were low and some high. There did not appear to be any consistent pattern. None of the analytical examinations showed any contaminants or problems. The melting points were consistent as were the infrared spectra. By SEM, however, there were a number of small spots which appeared to have a different contrast than the larger particles. And, in one obvious case, a large cubic crystal was growing on one of the drug particles (see Fig. 8.17). An EDS qualitative examination revealed the presence of

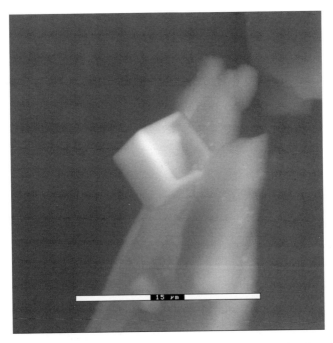

Fig. 8.17 A salt crystal growing on a larger crystal of a new chemical entity. Scanning electron microscopy.

sodium and chlorine. The cubic crystal was salt (this was confirmed by elemental analysis) which explained the water absorption values, but how did it get there and why on such an inconsistent basis? The answer lay in a consideration of the multi-step synthesis. Early in the synthesis, a sodium salt was used and late in the synthesis a chlorine-containing chemical was used. The difference between one process batch and another was the degree of washing using water. For the high water absorption samples, there was little or no washing with water and high salt contamination. For the low water absorption samples, the opposite was true. The solution lay in consistent washing with water.

8.5 Conclusion

This chapter has presented a survey of some of the ways that microscopy can aid in the development of pharmaceuticals. Microscopy can make important contributions to the solid-state characterization of the drug substance, to particle size analysis, and to contaminant identification. It is often joked, though, that the most important instrument in microscopy lies just above the eyepieces – a skilled, well-trained microscopist. It is no joke that the techniques described in this chapter do require training and experience. These subjects are not generally taught at universities and so the microscopist must learn from short courses, at technical meetings, by experience and through a study of the literature. There is no substitute for gaining experience by duplicating experiments described in the literature. Much of the best in microscopy is interpretive and descriptive. Like all good analytical techniques, microscopy has maximum impact when applied by a well-trained experienced microscopist.

It is a wonderful time to be an industrial scientist and, in particular, a pharmaceutical microscopist. The world of organic solid-state chemistry is full of intriguing and interesting compounds. Every new chemical entity that the author has studied has had some odd secret to be teased out. Microscopy can play an important role in the application of science and technology to the discovery and development of new drugs.

References

1. Brittain, H., Bogdonawich, S., Bugay, D., DeVincentis, J., Lewen, G. & Newman, A. (1991) Physical characterization of pharmaceutical solids, *Pharm. Res.*, **8**(8), 963–973.
2. Brittain, H. (1994) Perspective on polymorphism, *Pharm. Technol.*, August, 50–52.
3. Byrn, S., Pfeiffer, R., Ganey, M., Hoiberg, C. & Poochikian, G. (1995) Pharmaceutical solids: a strategic approach to regulatory considerations, *Pharm. Res.*, **12**(7), 945–954.
4. Kuhnert-Brandstatter, M. & Winkler, H. (1976) Thermoanalytic and IR-spectroscopic studies on various forms of drugs from the estradiol and androstane group. Part 1, *Sci. Pharm.*, **44**, 177–190.
5. Salole, E.G. (1987) The physicochemical properties of oestradiol, *J. Pharm. Biomed. Anal.*, **5**(7), 635–648.
6. Anwar, J., Tarling, S. & Barnes, P. (1989) Polymorphism of sulfathiazole, *J. Pharm. Sci.*, **78**(4), 337–342.
7. McCrone, W. (1957) *Fusion Methods in Chemical Microscopy, Interscience*, New York, NY.
8. Bettinetti, G. (1988) Analysis of the polymorphism of drugs, *Il Farmaco*, **43**(3), 71–99.
9. Threlfall, T. (1995) Analysis of organic polymorphs, a review, *Analyst*, **120**, 2435–2460.
10. Hartshorne, N.H. & Stuart, A. (1970) *Crystals and the Polarizing Microscope*, 4th edn, American Elsevier, New York, USA.
11. Nichols, G. (1998) Optical properties of polymorphic forms I and II of paracetamol, *Microscope*, **46**(3), 117–122.
12. Nichols, G. & Frampton, C. (1998) Physicochemical characterization of the orthorhombic polymorph of paracetamol crystallized from solution, *J. Pharm. Sci.*, **87**(6), 684–693.
13. Brittain, M. (ed.) (1999) Polymorphism in pharmaceutical solids, Marcel Dekker, New York, NY.
14. Doherty, C. & York, P. (1988) Frusemide crystal forms; solid state and physicochemical analyses, *Int. J. Pharm.*, **47**, 141–155.
15. Matsuda, Y. & Tatsumi, E. (1990) Physicochemical characterization of frusemide modifications, *Int. J. Pharm.*, **60**, 11–26.
16. Kofler, L. & Kofler, A. (1954) *Thermomikromethoden*, Wagner, Innsbruck.
17. Kuhnert-Brandstatter, M. (1971) *Thermomicroscopy in the Analysis of Pharmaceuticals*, Pergamon Press, Oxford.
18. Krc, J. (1977) Crystallographic properties of flufenamic acid, *Microscope*, **25**, 31–45.
19. Haleblian, J. & McCrone, W. (1969) Pharmaceutical applications of polymorphism, *J. Pharm. Sci.*, **58**(8), 911–929.

20. Allen, T. (1997) *Particle Size Measurement*, 5th edn, Chapman and Hall, London.
21. Aldrich, D.S. & Smith, M.A. (1995) Pharmaceutical applications of infrared microspectroscopy, in *Practical Guide to Infrared Microspectroscopy* (ed. H.J. Humeki), Marcel Dekkar, Inc., New York, 323–375.
22. Florence, A.T. & Attwood, D. (1988) *Physicochemical Principles of Pharmacy*, 2nd edn, Chapman and Hall, New York, USA.

9 Process analysis in the pharmaceutical industry

Martin Warman and Steve Hammond

9.1 Introduction

Process analysis is defined simply as a measurement taken, and a result generated in time for the data to be used to impact on the process. This means process analysis is not exclusive to techniques that require that a sample is taken. In 1987, Callis [1] defined five categories of process analysis, defined as off-line, at-line, on-line, in-line and non-invasive, as is shown in Fig. 9.1. The simplest and the most widespread example of process analysis is when samples are taken from the process stream and analysed *off-line* in a remote laboratory. This process is slow, samples are taken with low frequency, and there are often substantial delays between sample submission and analysis. However, the off-line method does allow for analysis by expert analysts using many of the techniques described in previous chapters of this volume.

The drawbacks to these types of process analysis mean that they are of limited use as in-process control tools. The first improvement is simply the movement of the testing from the laboratory to the manufacturing area. This is

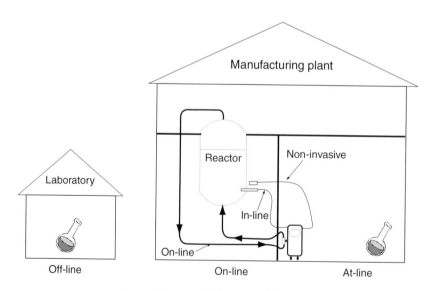

Fig. 9.1 Process analysis as defined by Callis.

defined as *at-line* analysis. As in the case of off-line analysis, the sample is removed from the process stream but, as the analysis is carried out in the area, results are generated much more quickly. However, as the analysis may be performed by the plant personnel, rather than by trained analysts, the instrumentation must be relatively easy to use and sample preparation must be kept to a minimum.

The remaining three categories form the basis of what is now commonly described as *process analysis*. In each case, no sample is taken, and analysis is carried out without any manual intervention. The subtle differences between on-line, in-line and non-invasive are in reality academic – they are often thought of simply as on-line analysis. For the purposes of this chapter, any analysis carried out without the need to withdraw a sample is defined as *on-line* analysis.

Having now defined process analysis, we now need to consider the underlying philosophy behind its use. Traditionally, pharmaceutical manufacturing has been driven by the need to produce a product of guaranteed quality. This has led to a universal system of validated methodologies, with every step of the process being shown to be repeatable, always generating the same result. The need for guaranteed quality has not diminished, rather it has increased, however manufacturing in this way means that safeguards need to be installed. Processes are always validated for worst-case scenario, which by definition means building in inefficiencies. Variability in raw materials, processing equipment and process control needs to be factored in.

Even if these factors are taken into account in the validated process, occasionally pharmaceutical manufacturing processes result in a product that is regarded as an *exception*. Investigation or re-working of these exceptions has the biggest impact on productivity in the pharmaceutical industry.

Process analysis can be used to provide extra insight into the nature of the processes and products, and as such should be part of a process control philosophy that will reduce variation. They even have the ability to produce a paradigm shift in the principles of pharmaceutical manufacturing, with validated measurements taken at *Critical Control Points* (CCP), controlling processes in such a way that product quality becomes guaranteed by measurement, and processes which are adjusted to optimise throughput.

This is at the heart of initiatives to generate total control philosophies or even parametric release, providing a plant with a greater degree of assurance that they will get it *right the first time*.

9.2 Pharmaceutical manufacturing

In principle, pharmaceutical manufacturing is a two-stage process. Firstly, the active pharmaceutical ingredient (API) is made (either by chemical synthesis, or biological process, or a combination of both). This process is sometimes described as primary production or alternatively – as used in this chapter, drug substance manufacture. The API is then used in the formation of the final dosage form in secondary or drug product manufacture. Both of these stages have CCP.

9.2.1 Drug substance manufacture

If we look at the manufacture of an API we find many processes: the initial determination of raw material quality, the dissolution/mixing of the initial reaction mixture, the reaction itself, recovery of the API followed by its drying, milling and blending and passing over to drug product manufacture. Even this is not the end of the story; pharmaceutical manufacture is not normally carried out in dedicated, single-use equipment trains. Production is normally carried out in a series of batches, which together give product campaigns. Once the campaign is over the equipment is cleaned, verified, and changed over to a new campaign of a totally different product.

If we are to propose the conversion to a process controlled by measurement, the possibilities for process analytical technologies (PAT) are huge. Let us consider where the CCPs occur.

9.2.1.1 Raw material testing

Raw materials are normally accepted as being of useable quality based on two pieces of evidence. One is a certificate of analysis (COA) supplied by the supplier. This COA will often contain only analytical tests that are considered to have an impact on process robustness. This is probably one of the biggest variables causing lack of process robustness and therefore variability in both drug substance and drug product manufacture.

What is needed is an analytical technique that is non-specific, not just testing properties that are known to be significant but checking variability from the *norm*, for all possible parameters. For this reason, the primary technique used in raw material testing is near infrared spectroscopy (NIR) [2, 3]. NIR is a vibrational spectroscopy with absorbencies due to changes in the stretching and bending of molecular bonds. Differences in NIR spectra often result from differences in hydrogen bonding. Most of the raw materials used in drug substance manufacture have these bonds. NIR also fits the profile of a simple use technique with no need to prepare the sample, making it ideal for at-line or on-line applications.

Both types are used commonly in the pharmaceutical industry. In an at-line assay, samples are taken and analysed either by transmission/transflectance NIR if liquids, or reflectance if solids. The spectra generated are then used to build identification libraries. These libraries must include the normal variation found in a *good* raw material distribution. Using these libraries, correlation or conformity

Fig. 9.2 Automated NIR analysis in a raw material receiving area.

of a new batch or even each drum of each raw material can be verified. This type of qualitative analysis is covered in current monographs, e.g. United States Pharmacopeia (USP), and guidance on validatable methods given. Although the typical speed of analysis is less than one minute per sample, the throughput of samples can be further increased using autosampler-based systems (Fig. 9.2).

These systems have shown themselves to be capable of identifying raw materials which are variable enough to cause downstream processing problems even though the more sensitive but more specific traditional types of analyses have found the materials to be acceptable. Often this approach is taken further and a prediction model developed whereby a qualitative model, using conformity or correlation algorithms checks the *identity* (ID) then a quantitative model uses the same data to predict a value of actual physical parameters such as moisture, concentration, and even particle size [4].

The movement of these types of measurement from an at-line to on-line technique is relatively easy. There are some downsides, on-line NIR systems tend to be fibre optic based, and current construction materials used in the fibres (normally they are made of NIR grade silica) absorb light above 2100 nm and so restrict the wavelength range, losing some of the combination region. Furthermore, disposable vials may be used with at-line systems meaning no cleaning is necessary. For on-line use, the probes used need to be cleaned to avoid cross contamination. Fourier transform-near infrared (FT-NIR) systems have been developed with multiple measurement points, with a choice of reflectance of transmission probes, incorporating clean in place (CIP) systems. Post analysis, the probe is simply placed in a clean station and automatically cleaned ready for next use (Fig. 9.3).

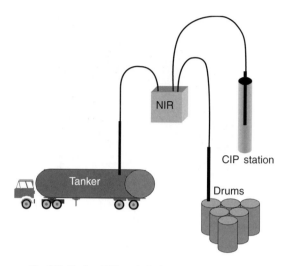

Fig. 9.3 On-line NIR analysis for raw material testing.

These systems are available with Ex rating for using in hazardous areas, although fibres up to 100 m long mean the actual spectrometer and control PC can be housed away from the area, possibly in a non-zone rated area.

9.2.1.2 Reaction mixture
Prior to initiation of a reaction, starting materials are often pre-mixed in a separate dissolution vessel, and only when they are fully dissolved and mixed, are they transferred to the reaction vessel. The homogeneity of this mixture is critical to the even initiation of the reaction. Traditionally, off-line analysis could be used to test the homogeneity of this mixture. However, the variance between repeated on-line NIR spectra shows when mixing is complete.

The importance of using variance between consecutive data points will be shown again later, but the principle is the same no matter where it is used in the manufacturing process. If consecutive spectra of a moving sample are similar, the conclusion is: the variance is small. If the variance is large then the mixture is not homogeneous (Fig. 9.4).

If the variance at an absorbence for the component of interest is plotted by calculating a moving block mean or standard deviation at that absorbence against time, a real time plot of uniformity is achieved, and when the variance is at a minimum the uniformity is at a maximum.

9.2.1.3 Reaction monitoring
The reactions that produce drug substance are relatively easy to automate. However, automation by itself does not guarantee a robust process. It is still necessary to rework, and validation requires consideration of a worst-case

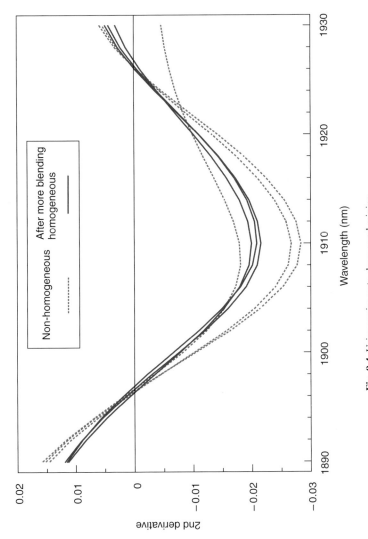

Fig. 9.4 Using variance to show good mixing.

scenario and is therefore inefficient. However reaction monitoring is not straightforward. There is no technique that can monitor every activity that occurs during a typical drug substance reaction. If the vessel or equipment train is only used for a single purpose, a specific tailored solution can be developed. There are many good examples of dedicated reaction monitoring, for example, the use of NIR for hydrogenation [5] (Fig. 9.5).

In this example, mid-infrared (mid-IR) could equally be used. There are many other examples of where a specific spectroscopy could be applied to a specific monitoring problem, in addition to NIR and mid-IR, there are examples using UV, Raman and fluorescence. However it is very rare that a specific question is posed. In the majority of cases, the ability to monitor one step in what could be a multi-step reaction, does not assure reliable performance of the entire process.

For reaction monitoring to become widely accepted, a system is needed that can monitor all processes. So let us consider the advantages and disadvantages of each of the possible spectroscopic techniques, starting with ultra violet (UV) spectroscopy. UV spectra are caused by transitions between electronic energy levels. These are high-energy transitions, which means that UV is excellent at monitoring low-level concentrations. However, several components in a typical reaction mixture will have a UV chromaphore, and a typical UV spectrum is not very information rich. If we are to use UV for reaction monitoring we need to maximise the amount of information available. Figure 9.6 shows the difference between a typical UV spectrum of an API in solution and what can be achieved by increasing the spectral resolution.

This increased spectral information can be maximised still further by applying the same chemometric principles used in NIR. If a derivative is applied to the data, we can resolve overlapping absorptions to differentiate between components with essentially similar UV spectra. This does not however address fundamental issues with probe design. UV spectroscopy, like NIR uses transmission probes

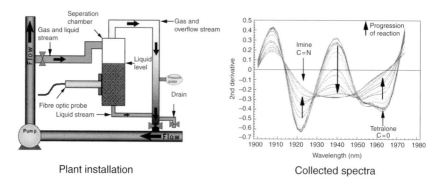

Fig. 9.5 Using NIR to monitor a bus loop hydrogenation.

PROCESS ANALYSIS IN THE PHARMACEUTICAL INDUSTRY   331

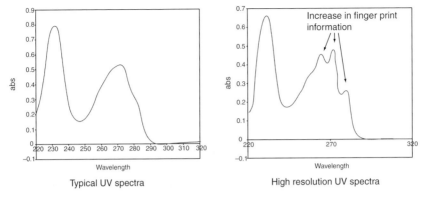

Fig. 9.6 Increased information with increased UV spectral resolution.

for liquids and reflectance probes for solids. If we want a solution that can monitor any reaction step, the probes have to be compatible with all the possible steps. Reflectance probes do not work in clear solutions and transmission probes are susceptible to bubbles and slurries (either from catalysts or during crystallisation steps). A rarely used but very good alternative is attenuated total reflectance (ATR). In ATR light is bounced along a crystal and where the interface occurs with the sample matrix it penetrates into that matrix. The pathlength is equal to the penetration depth (approximately equivalent to the wavelength) multiplied by the number of bounces. In this way, a typical NIR-ATR will have 10 bounces and use a sapphire crystal for robustness, giving an approximate pathlength of 2 μm. If this seems very small, the fact that the absorbencies in a typical reaction mixture are so large that often transmission pathlengths of 1 mm return so little light that the absorbence is off-scale. These short pathlengths are actually advantageous because UV can be used to target small changes in trace level components making it useful in monitoring process-related impurities (PRI).

Adjacent to the UV and visible is NIR, which has been shown to play a role in reaction monitoring. NIR spectra are very information rich, they offer the best single spectroscopy solution to monitoring all the changes that will occur in the reaction matrix. However NIR spectra are very small, the combination and overtone absorbencies are only a fraction of the mid-IR fundamentals (Fig. 9.7). This means typically the sensitivity of NIR is approximately 0.1%. It means using NIR for accurate determination of reaction end point is very difficult. Figures 9.8 and 9.9 show typical NIR data from reaction monitoring and a principle component analysis plot showing the reaction progress.

These plots sum up the issue with using NIR for reaction monitoring; it is excellent at tracking changes, making it good for monitoring reaction kinetics, but not sensitive enough to accurately to detect reaction end point, and has

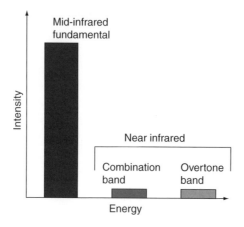

Fig. 9.7 Relative intensities of mid-IR and NIR absorbencies.

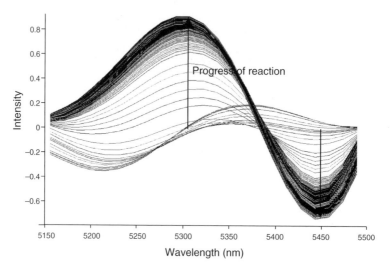

Fig. 9.8 Using NIR for reaction monitoring.

little chance in monitoring PRIs. In addition NIR probes like those for UV, are typically either transmission or reflectance, with the same drawbacks. As with UV an alternative is needed. One option comes in the form of a non-invasive measuring head manufactured by Carl Zeiss, the OMK. This is a reflectance head made up of 15 individual fibres angled so as to have a collection point at a fixed distance in front of the head. The sample is illuminated via a collimating lens. In this way NIR absorbence spectra can even be obtained from a clear liquid (Fig. 9.10).

PROCESS ANALYSIS IN THE PHARMACEUTICAL INDUSTRY 333

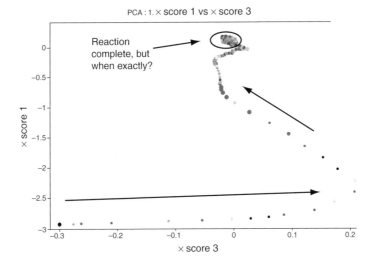

Fig. 9.9 PCA plot from NIR data.

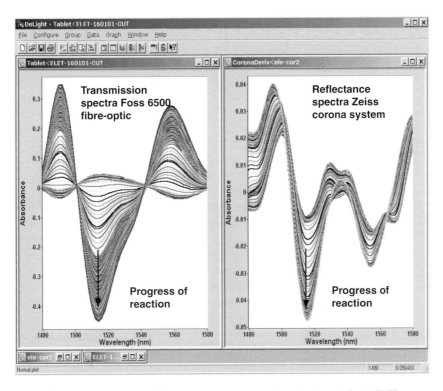

Fig. 9.10 Comparison of NIR spectra from a standard transflectance probe vs OMK.

These spectra were obtained through the oil-filled heating jacket of a laboratory scale automated reactor. Although undoubtedly a useful tool, this head is not a solution to general reaction monitoring. It copes well with the presence of bubbles and is excellent for monitoring slurries during crystallisation, but if the reaction mixture is black, as is often the case in the presence of catalyst, no reflectance is measured.

The only solution capable of monitoring all reaction steps is again ATR. However unlike UV, typical pathlengths for transmission NIR are not limiting. A normal NIR transmission probe uses 2 mm pathlength; going over to ATR, with 10 bounces we only have 20 µm pathlength, i.e. 1/100 of the, already poor, sensitivity. Advances in NIR detectors give some of this sensitivity back and NIR-ATR can achieve one-tenth the sensitivity of transmission measurements. This reduction in sensitivity is acceptable since NIR should be targeted at reaction kinetics and other matrix level monitoring, i.e. whether the reaction is following the expected signature.

Adjacent to NIR is mid-IR. Mid-IR is without doubt one of the fundamental building blocks of present day reaction monitoring [6]. Mid-IR absorbencies are strong, giving good sensitivity, commonly down to 10 ppm concentration, and are bond specific. The most common configuration of instrumentation is Fourier transform IR (FT-IR) utilising light guides or optical conduits to deliver light to a multiple bounce ATR crystal. The effective pathlength is good (between 20 and 120 µm for a 6-bounce ATR crystal). A variety of detectors are available from room temperature DTGS, through liquid nitrogen cooled MCT and even MCT with Stirling cooling to remove the need to use liquid nitrogen. In general DTGS detectors are less sensitive (at approximately one-tenth the sensitivity of MCT), giving a sensitivity of about 100 ppm, whereas the Stirling-cooled MCT has the drawback of cost (typically \$15 000 more than a standard MCT). Also various ATR crystals are available, with diamond, silicon and zirconium being routinely used, plus some more exotic materials such as zinc selenide and zinc sulphide.

This now touches on the two main drawbacks of using mid-IR. Each of the materials absorbs mid-IR energy to a degree. The amount of absorption has a big influence on the usability. If we consider just the two most common options, diamond and silicon, silicon has a very good transmission (low absorbence) of mid-IR light with no specific absorption bands giving a working range from 2500 to 20 000 nm (4000–500 cm^{-1}). It is however relatively soft, meaning it can be scratched by abrasives in the reaction mixture and is not fully pH compatible; this varies from vendor to vendor but typically silicon cannot be used above pH 10. Diamond, however is very hard, very robust and can be used in any pharmaceutical reaction. The problem is diamond absorbs mid-IR in several wavelength regions giving a reduction in light throughput from 3800 to 5500 nm (2600–1800 cm^{-1}), at complete attenuation at 4450–4550 nm (2250–2200 cm^{-1}). This is significant as nitrile groups absorb in this region.

This means that the ATR crystal material needs to be selected appropriately according to the matrix being measured. There are examples whether an appropriate combination of chemical and physical resistance, and sensitivity at the required wavelength is not available. We estimate that approximately 70% of all pharmaceutical reactions can be monitored by selecting the best combination of probes.

There is another issue however when using mid-IR in the plant. Unlike the laboratory, pharmaceutical plants are not temperature controlled and vibration-free environments. As already indicated, the mid-IR light is delivered via light guide. These light guides are tubes down which the light is usually focused, so as to allow correct alignment of the instrument and the ATR, they include adjustable bends or *knuckles*. These knuckles incorporate a mirror so that light is collected and then focused onto the next knuckle. These light guides need to be accurately set up; often there is a utility in the vendor software that simply measures the return energy, so this is not an issue. But any movement/vibration of the light guide alters the alignment of the mirrors changing the intensity of the returning signal, causing an attenuation change. This means even with the incorporation of vibration suppression in the light guides, the capabilities of mid-IR for plant use are reduced. These attenuation changes effectively increase noise, and reduce sensitivity. In many cases a full plant installation using DTGS (reducing sensitivity) then light guides (reducing sensitivity) has a limit of quantitation of 0.1% about that of NIR.

This situation can be improved. For plant use light guides can be replaced with fibre optics. Chalcogenide (AsSeTe) can be used to make fibres which have good mid-IR throughput, not as high as a good, purged, moisture-free light guide but easily sufficient. This means no attenuation occurs due to vibration, so the system noise is less and sensitivity is higher. There is a pay back – chalcogenide has specific mid-IR absorption bands, specifically at 4800–5200 nm (2100–1900 cm^{-1}), and a diamond and chalcogenide combination has a *hole* from 4450 to 5200 nm (2250–1900 cm^{-1}). Although this combination is the only truly plant-hardened pairing, this means that the mid-IR will be able to monitor far less than the 70% of all reaction steps previously accepted.

Accepting these limitations mid-IR is a powerful tool. It is capable of monitoring reaction kinetics, appearance and disappearance of intermediates, prediction end point, and even in some cases the presence of PRIs. Figure 9.11 shows the type of data that can be generated and the way mid-IR can be used to track multiple changes as they occur with a reaction step.

Figure 9.11 shows a typical waterfall plot of spectra against time, specific absorbencies for starting material (1247 cm^{-1}), intermediate (1143 cm^{-1}), product (1312 cm^{-1}) and side reaction (1710 cm^{-1}) can be identified and models built monitoring those changes. One potential new use of mid-IR is its ability to measure super-saturation and thereby aid the control of crystallisation processes.

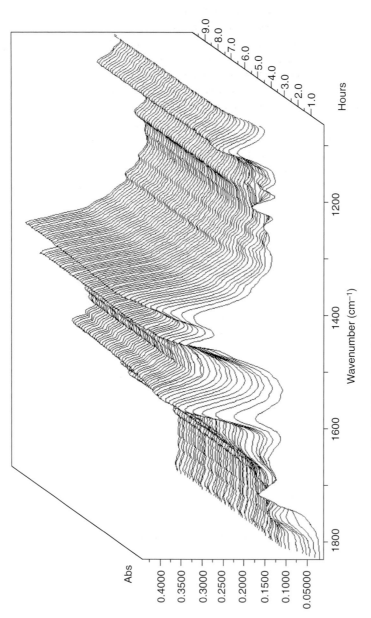

Fig. 9.11 Typical reaction data using a diamond ATR probe.

Raman spectroscopy completes our assessment of vibrational methods [7]. Raman spectra are generated by the inelastic scattering of light. An introduction to Raman theory and instrumentation is provided elsewhere in this volume. One of the main advantages of Raman is that the technology is already plant hardened. The instrument can be housed in a non-hazardous area and fibres laid out into the plant. Also the two standard configurations of a probe are compatible with the process. The first, and the preferred option, is the immersion probes. Typically these are set with a focal point approximately 200 μm within the sample. This means that they are not badly affected by particulates or bubbles. They can also be purchased, constructed from C276 Hastelloy and with process optics. The alternative, which appears preferable, is a non-invasive probe. These obtain data through a sight glass and require no product contact but are less sensitive. Raman systems, like NIR, are easily multiplexed [8] allowing multiple measurement points from a single instrument. Although still not well established, Raman has the potential of being able to monitor reaction kinetics, appearance and removal of intermediates, reaction end point and PRIs. Raman is also very good at monitoring polymorphism.

In summary, the use of standard spectroscopic techniques for process analysis is well established, however each individual technique has drawbacks. The way forward is probably the combination of several of these techniques, for example high resolution UV for targeted information about trace-level impurities, NIR for non-specific information about how the reaction is proceeding and kinetics, mid-IR for end point, intermediates, additional kinetic, and super-saturation, and Raman for end point, intermediates, additional kinetic and information about polymorphism.

Having established the role of spectroscopic techniques, we should not forget chromatography. Development of optimised separation techniques is considered elsewhere in this volume. Can these techniques be moved on-line? The answer is: sometimes. If we take, for example, process gas chromatography (GC), the analysis of a vapour phase is relatively easy. Direct injection of very simple mixtures is also possible, but as soon as the mixture does not follow the same rules of compatibility that apply off-line, the sample has to be pre-treated. This pre-treatment could be simple quenching, filtration, or dilution, through to complex pre-column derivatisation. Anything is possible; however each step applied that is tuned to a specific reaction reduces its usefulness as a general-purpose tool. Add to this the need to carry out the usual off-line system suitability tests such as checking reproducibility and specificity prior to each run, means these systems are still relatively labour intensive. The same arguments apply to high performance liquid chromatography (HPLC), plus the added complication of needing to maintain mobile phase reservoirs. The best process chromatography option is a combination of the two. Supercritical fluid chromatography (SFC) has the advantages of GC, very high resolving power and universal detection (FID or UV) but the ability to handle non-volatile samples, and also samples with a range of polarities. In general process chromatography is restricted to single-use plants where a highly tuned solution can provide a specific measurement question.

9.2.1.4 Crystallisation monitoring

The final step of most reactions is the formation of the API solid, usually via a crystallisation step. We have already seen that mid-IR can be used to measure supersaturation. There are also considerable developments in the use of active acoustic emission spectroscopy to monitor equilibrium changes associated with nucleation. However, these systems are not yet ready for mainstream manufacturing.

The particle size of the API can have a significant impact on downstream processablity. If crystals are too big they must be milled, but if crystals are too small or if there are too many *fine* particles, the product will need to be reworked, as downstream processing may be difficult.

There are techniques available that allow accurate monitoring of the actual crystallisation process. The first and most widespread is focused beam reflectance measurement (FBRM). In FBRM, a spinning laser tracks across the surface of any particle at its focal point. The reflectance from the particle returns to a detector and the duration of this reflectance is known as a chord length (Fig. 9.12).

Changes in the measured chord length give an estimate of trends in particle size. Figure 9.13 shows the use of FBRM to monitor, the number of particles over 100 μm, the number of particles between 50 and 100 μm, the number of particles below 50 μm, and within that segment, the number below 10 μm.

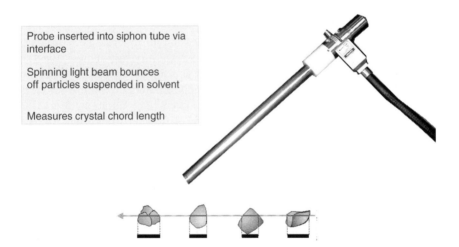

Fig. 9.12 Using FBRM to generate particulate chord measurement.

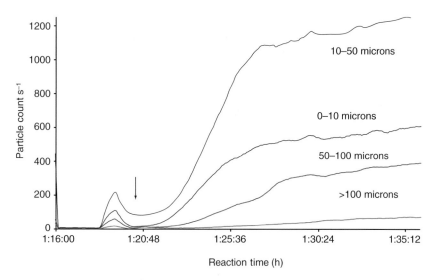

Fig. 9.13 FBRM plots of particle size against time.

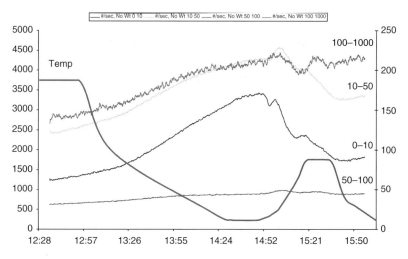

Fig. 9.14 Effects of temperature on particle size distribution.

The importance of this information as a process analysis tool is clearly shown if we look at how these distributions can be altered. Figure 9.14 shows these same distribution ranges and how they change under the effects of temperature. As temperature decreases the number of fine particles (0–10 μm and 10–50 μm) increases, however warming the mixture then re-cooling, generates a final distribution with far fewer fine particles.

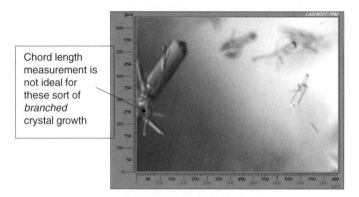

Fig. 9.15 Video microscopy image of a typical crystallisation process.

This means particle size distribution within a final crystallisation can be adjusted to fit the requirements of the formulation process. Even though FBRM is a very good process-monitoring tool, it does have drawbacks. It does not provide a direct measure of particle size. Far better would be video microscopy. There are several systems available that carry out reflectance video microscopy coupled to image analysis software to generate accurate particle size information. All of these have a relatively narrow focal plane, and although the ability to collect images is available, the problem is the development of software capable of recognising the particle when it is in focus (out of focus particles are not reported accurately) and then generating particle size data from the in-focus images. This software is currently available off-line and it is only a matter of time before it is available as an on-line monitoring tool. As an example Fig. 9.15 shows a typical video image obtained during crystallisation.

9.2.1.5 Dryer monitoring

Once the crystallised API is isolated the next step is to dry it. No matter how good the isolation step, there are always trace levels of the solution used to wash the API from the process. This could be organic solvent such as methanol or acetone, or could simply be water. NIR is now recognised as the standard way to monitor the remaining volatiles in the API [9]. Most API drying takes place in either pan or filter dryers. In each case a correctly positioned, appropriately designed NIR probe is capable of monitoring the process. This requires a probe with a highly polished sapphire window capable of withstanding the high vacuum conditions in the dryer, with the window designed to prevent product sticking. These probes are available as part of complete dryer systems from most NIR vendors. The probe is inserted as far into the dryer as possible so that it does not foul the agitator blade as it turns, and so that it does not get covered by a plug of *wet* material. On all the common pan dryers, good probe position is possible.

NIR spectra are collected at regular intervals, and a trend plot generated of changes at the unique absorbence of the solvent being tracked. Even if the NIR is simply used to judge when to take a single sample for off-line assay, the benefits are huge. In a typical pharmaceutical process, this drying step can be anywhere from 6 to 60 h. Simply removing the need to draw samples during this time reduces the drying cycle by approximately 25%. Reducing the temperature and breaking the vacuum for sampling can take 2 h. Figure 9.16 shows a NIR trend plot of drying and demonstrates how drawing samples affects the drying profile. Each of the red boxes represents a pause in drying while a sample is taken.

In other examples a true picture of the drying process can be generated, for example Fig. 9.17 shows the effects of agitation. It clearly shows how a screw agitator which moves up and down within the dryer brings *wet* sample back to the probe showing that localised drying is occurring at the base of the dryer.

Although this particular application is well established, it only works if the probe is seeing a sample representative of the process. If plugging of wet sample at the probe tip occurs, then fresh sample will not be seen. On such occasions, systems are available to insert/extract the probe between agitator rotations or air pressure can be used to blow the tip clean. Both should be avoided if at all possible.

9.2.1.6 *Monitoring the milling process*

As previously established, particle characterisation has a big impact on downstream processability. Milling is carried out by passing the dried API through a simple pin or air jet mill. If the output from these mills can be monitored, the mill settings can be adjusted and particle characteristics of the mill output changed.

Historically the off-line techniques for particle sizing are so slow that results would not be generated in time for the process to be adjusted. Even if fast enough, there are many questions associated with using the current *standard* techniques for particle sizing. Sieving techniques are not capable of measuring the particle size of fines. Laser scattering assumes perfect spherical particles. Far better process control data are generated using vision-based particle size systems, but not fast enough to be of use. There is no off-line way of generating data to control the milling process.

The largest part of the analysis time is the time taken for sample preparation, but we already have a sample finely distributed in air (most milling systems use air conveying to transfer product from the mill to the product hopper, before filling into drums) and provides a means to move an off-line system on-line? Laser scattering on-line is commercially available but requires further sub-sampling which is product specific and liable to blockage. On-line vision analysis is only just becoming available. Rather than using reflectance microscopy they are based on transmission microscopy, where the sample is illuminated from one side of the transfer pipe and data collected on the opposite side. Images are

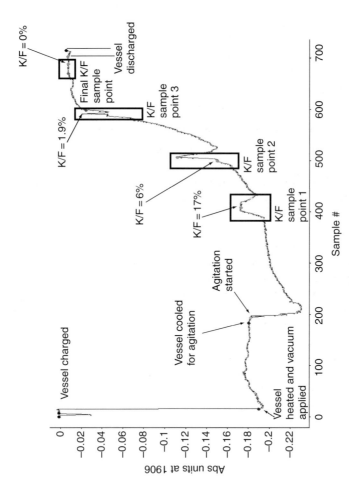

Fig. 9.16 Typical dryer profile showing disturbance caused by sampling.

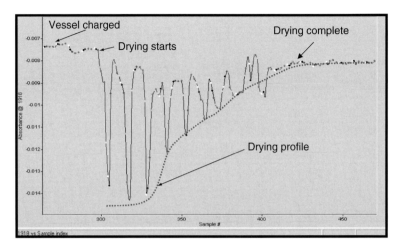

Fig. 9.17 Typical dryer profile showing disturbance caused by agitation.

then processed and size or volume distributions calculated in real time. Using these distributions and comparing to acceptance limits, real time adjustment of the milling process will be available.

9.2.1.7 Cleaning monitoring
As previously described the pharmaceutical manufacturing process consists of campaigns of production, before cleaning and changeover to a different process. The importance of changeover cannot be stressed too highly. Typically a campaign lasting 14–21 days is followed by a changeover lasting for several days. Changeover typically accounts for 30% of the manufacturing cycle, yet is nearly always ignored when looking for increased productivity.

As with the reaction step, changeover is a validated process, with validated cleaning processes, followed by off-line monitoring of a final rinsate solution, to determine that no more product is being removed. As with any validated process, the cleaning process is validated for worst case and is very inefficient. This can easily be demonstrated using on-line UV measurements. There are now commercially available systems specifically designed for monitoring the cleaning solutions, measuring product removal in real time. These systems are inserted into the cleaning circulation line (Fig. 9.18).

Typically a series of cleaning solutions, chosen for their ability to dissolve product, are circulated for a pre-defined period. The solution is then discarded, replaced and the next solution circulated. There is not usually any assessment of product removal until the final step is complete. However by monitoring the UV absorbence of the product we see that quite often the individual steps are not complete or they last too long and sometimes are totally unnecessary. Figure 9.19 shows some typical product removal curves.

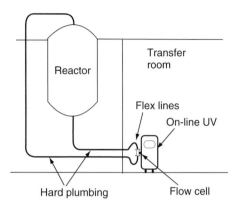

Fig. 9.18 Schematic of on-line rinsate monitoring installation.

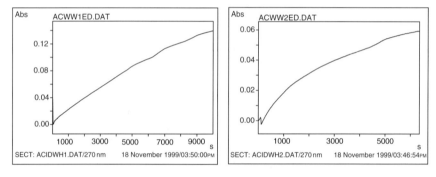

Fig. 9.19 Typical product removal curves monitored by on-line UV.

In both the above examples product was still being removed and the cycle should have been extended. Instead the solution, which was still removing product, was discharged and replaced. In addition to this being wasteful in terms of solvent, discharging, refilling, reheating a cleaning solvent typically takes 4–8 h, all wasted manufacturing capacity. Even if the UV is only used to generate new validated regimes, it saves large amounts of process capacity. Controlling the cleaning process on the basis of an on-line measurement technique will free up to 50% of the changeover period, equivalent to 15% of the overall manufacturing cycle, giving 15% extra manufacturing capacity.

9.2.2 Drug product manufacturing

Although the API is designed to provide efficacy, without a properly controlled manufacturing process the API will not be formulated and delivered correctly. Drug product manufacture can be complicated and varied, but most processes have three components.

9.2.2.1 Raw material monitoring

As with drug substance manufacture, the correct characterisation of raw materials plays a huge part in robust drug product processes and probably represents the primary CCP in drug product manufacture. Testing of these raw materials has traditionally been carried out off-line using highly sensitive but very specific techniques. The transfer of raw material testing to at-line using non-specific NIR testing allows a qualitative approach [3]. Qualitative testing can be performed in two ways. The first is known as conformity testing. Samples of known performance are used to build reference libraries. The standard deviation within that reference library is calculated. The test samples need to fall within an acceptable number of deviations from the reference set, if not it is unacceptable (Fig. 9.20).

If libraries are built containing several materials, a series of conformity tests can be carried out. This approach allows the testing of not just quality but also identification (Fig. 9.21). Figure 9.22 shows a real application of this approach with three types of lactose being shown to have distinctly different libraries, related to particle size.

The second approach is one of correlation. A spectral match is calculated by comparing the slope generated from comparative data points on the sample and reference spectra. The cosine between the slopes is calculated and a spectral match index or correlation determined (Fig. 9.23).

What this actually means is the nearer the spectral match value is to 1.000, the better the correlation. Correlations near 1.000 are generated from spectra that are similar but offset. The lower the index the poorer the correlation, through to −1.000 when the spectra are mirror images (Fig. 9.24).

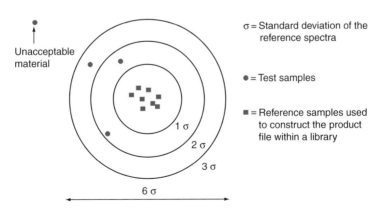

Fig. 9.20 Conformity approach to raw material testing.

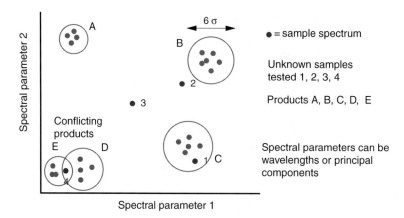

Fig. 9.21 Conformity approach to identity testing.

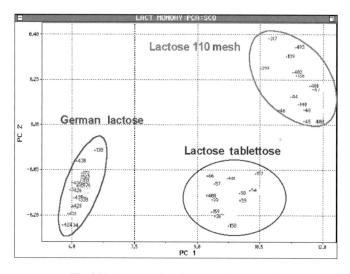

Fig. 9.22 Example of conformity testing approach.

The main characteristics causing lack of process robustness in drug product manufacture are again particle size and shape. NIR is capable of giving information on variability in particle size and shape. The reason for this is simply the fact that we are using diffuse reflectance NIR, and changes in particle size and shape cause changes in depth of penetration of NIR light (Fig. 9.25). This change in pathlength is wavelength dependent and causes an alteration in the slope of the spectra (Fig. 9.26).

Even this type of testing does not always provide enough information to maintain a robust process. NIR gives an indication of gross particle characteristics

PROCESS ANALYSIS IN THE PHARMACEUTICAL INDUSTRY 347

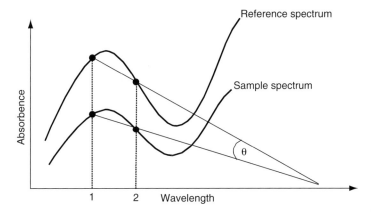

Fig. 9.23 Calculating spectral match index.

Fig. 9.24 Interpreting the meaning of spectral match index data.

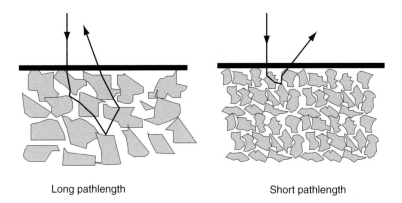

Fig. 9.25 Effects of particle size on NIR pathlength.

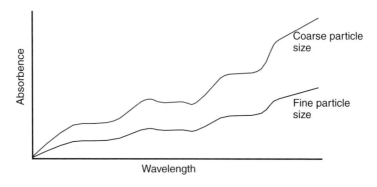

Fig. 9.26 Slope bias of NIR spectra caused by different particle size.

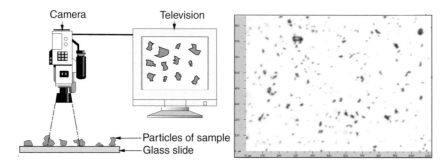

Fig. 9.27 Vision-based particle size analysis.

when often manufacturing issues are caused by subtle changes in number of fines (shown by number distributions) or shape. The only techniques capable of generating these numbers are vision-based particle size systems. These systems take digital images of samples using long focal length microscopy (Fig. 9.27).

Optimum process robustness can only be achieved by full control of acceptable raw material characterisation, both chemical and physical. The best way to do that is using at-line NIR backed up with off-line vision-based particle sizing.

9.2.2.2 *Formulation monitoring*

Formulation, the formation of the mixture that goes to form the final dosage, can have many steps. For the purpose of this chapter we shall consider these steps individually. They would normally be used in combination to generate the final dosage form.

The first step to consider is blending. Blending is carried out (perhaps several times) prior to direct compression, granulation and roller compaction. Traditionally, blending is another validated process, with evaluation (usually API concentration) being carried out on samples drawn from the blender post-mixing. So the success of the whole blending process is based on as little as ten small samples drawn from the blender after blending is complete. There is no chance for genuine process analysis. The use of NIR changes that. NIR allows the measurement of every component found in a typical pharmaceutical blend [3].

The easiest implementation is to use NIR to monitor content uniformity of not only API but also the other excipients at-line. Modern NIR spectrometers are so robust and the software sufficiently easy to use that this sort of analysis, with virtually no sample preparation, can be carried out by plant operators. NIR spectra could be acquired in diffuse reflectance by presenting a sample of the blend to an NIR spectrometer. More information is generated by taking a unit dose weight of blend, manually pressing this into a wafer, then generating a transmission spectrum through the wafer (Fig. 9.28).

Rather than simply getting surface measurements, transmission measurements give content uniformity of the whole wafer. The spectra can then be used to monitor the variance within the blend for all components by measuring the standard deviation between samples taken from various points in the blend.

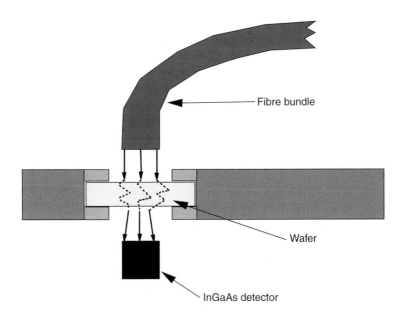

Fig. 9.28 NIR transmission through a blend wafer.

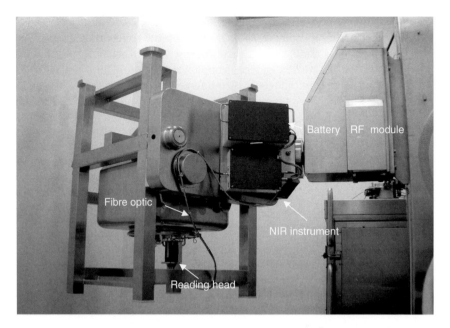

Fig. 9.29 Non-invasive blend monitoring system.

This does not tell the whole story. If we acquired NIR spectra during the process we can monitor blending as a kinetic process [10]. In order to do that a system has been developed utilising a non-contact measuring head attached to a diode array based spectrometer, that is powdered by battery, and communicates via radio frequency. This system *looks* through a sample interface, a window installed on to the blender lid (Fig. 9.29).

By activating the acquisition only when the blender is inverted, and sample is covering the window, a fixed amount of blend (the system has been developed such that the volume of each acquisition is 200–300 mg) is monitored. As the sample changes at each inversion, the blend process can simply be monitored by either tracking the changes in the unique absorbence of each component (Fig. 9.30) or by measuring variance between consecutive spectra at those same unique peaks (Fig. 9.31).

Whichever way the data are calculated the blending process is complete long before the blending is stopped. However the variance plot will also detect de-blending, because the variance between consecutive spectra would increase.

NIR is not the only technology available for blend monitoring, light-induced fluorescence (LIF) [11] has also been used as has frequency domain photon migration (FDPM) [12, 13]. A second type of formulation process is wet granulation. In wet granulation a dry excipient blend is dosed with API and wetted under shear to form particles of known size, each of which contain the

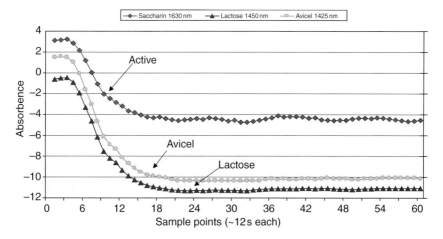

Fig. 9.30 Changes in unique absorbencies during blending process.

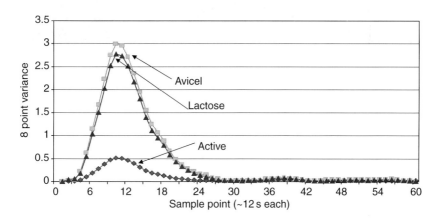

Fig. 9.31 Variance in unique absorbencies during blending process.

correct amount of API and excipient. Granulation is carried out in an enclosed system. Particle size and composition may be monitored by NIR and particle size distribution by FBRM, but these techniques require the insertion of a probe and are therefore disruptive. An alternative is passive acoustic emission spectroscopy (AES). AES simply listens to the sounds of the process, the impact of particles on equipment and on each other. These sounds change as granulation progresses (Fig. 9.32).

Each step of the granulation can be heard using AES, and a signature of the process established. A comparison of a lot to the *normal* can be made using class analysis and the process adjusted to bring the granulation back into normal variance.

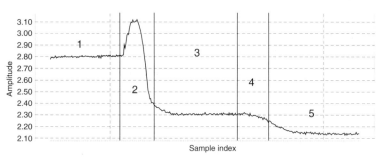

Fig. 9.32 Using AES to monitor the granulation process.

The wet granulate needs to be dried, usually in fluidised bed dryers (FBD). FBD consist of a bowl containing product through which hot air is blown at a throughput such that the product bed fluidises. As with drug substance pan dryers, NIR is the technique of choice. However the diffuse reflectance NIR probes currently used [14] in drug substance pan dryers are not ideal. The combination of the small sampling area of a traditional probe (as little as 2 mg) with the fast moving sample (which is interspersed with the drying air) means that the spectral noise is much higher resulting in loss of sensitivity. A much more common solution is the monitoring of effluent gas [15]. NIR gas cells (up to 6 m long) have successfully been used for this application. This is still not ideal, without monitoring the actual bed, subtle changes, such as differences in water *type*, e.g. free vs bound, are not measured. The situation has changed with the development of large sample area measuring heads (similar to those used to monitor the blending process). The large sample area means that the effects of density variations which occur within the fluidised bed are reduced, and when attached to fast scanning diode array instrument allowing up to 50 averages (or co-adds) to be taken within a second, result in reduced spectral noise and good sensitivity. In addition these sample heads are *focused*, having an acquisition point at a set distance in front of the head. This means with the correct head, selection spectra can be acquired through the glass viewing windows inserted into the FBD bowl. The large sample area also means that the small residue that may occur on the window has little contribution to the actual spectra. Using this system even extreme drying conditions, including high throughput FBD can be monitored (Fig. 9.33), where filter cleaning is detected in the process profiles.

The real benefit of using a technique that monitors the bed can be seen if we look at the NIR spectra of a fluid bed drying of a product with free and hydrated water. Figure 9.34 shows spectral changes in the second overtone water region whilst drying a product of this type. If we simply monitored the changes in the free water or even total water the correct end point would not be identified.

9.2.2.3 Tablet cores

The next CCP is when the dry blend or granulation is pressed into a tablet core. If the granulation and blending steps are properly controlled this is usually relatively straightforward. However, segregation can still occur if the transfer from blending into the charge hopper of the tablet press is not properly designed. NIR has been used [16], with multiple probes positioned around the transfer. Issues in this area can be *engineered out*. Post tableting, automated tablet-core test stations are a common sight on the output chutes. These auto-test stations carry out functions such as removing fines, metal detection (to prevent metal fragments from the machine making it through into the final product), weight, hardness and thickness. These are PAT applications although not always recognised as such.

The weight, hardness, thickness is measured to try and generate information about core consistency. An off-line HPLC test for potency carried out on ten samples selected as being representative of variability across the whole batch. The analysis of these samples is far too slow to be regarded as PAT. However

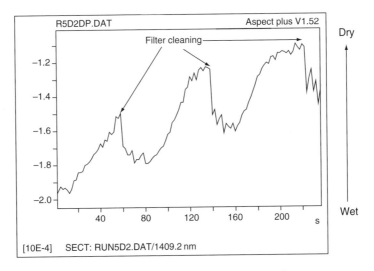

Fig. 9.33 Using NIR to monitor the fluid bed drying.

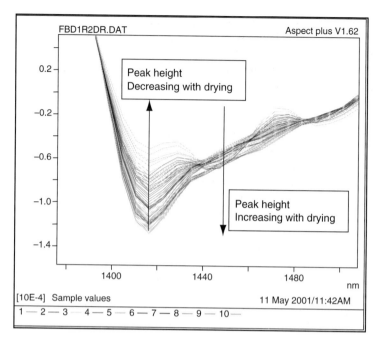

Fig. 9.34 Monitoring changes in free and hydrated water during fluid bed drying.

if this off-line testing can be simply replaced by at-line transmission measurements using NIR. The same instrumentation is used as for measuring blend wafers and the same type of variance measurement made. As this technique is fast, approximately 30 s per measurement, many more samples could be taken. At-line NIR is becoming important with the introduction of Product Quality Research Institute (PQRI) guidelines of acceptability of tablet cores. New guidelines indicate the need to increase the current testing regime to 20 sampling pints across the batch. Initially three samples will be analysed and variance calculated; if acceptable the sample point is OK if not a further four samples are analysed, and compared to a different set of acceptance criteria. This means at least 60 and possibly 140 samples will need to be monitored across the batch. NIR will be useful in complying with these new guidelines. However if NIR can be used at-line, why not move the analysis on-line. Incorporating NIR into an auto-test station has already been achieved and is commercially available. This can be incorporated simply to comply with the guidelines or could be used for much higher intensity of monitoring. Even including core positioning, on-line NIR is capable of monitoring a sample every 90 s. As most batches last 6 h or more, this sampling frequency of 40 per hour means sampling regimes twice those including in PQRI are possible. With the correct calibration model, outliers can be identified quickly so that the press can be stopped to permit correction of the problem and the cores brought back into specification. The weight, hardness and thickness parameters are currently used in this way.

In future this may be taken to the next step of 100% core testing and outlier rejection. At this point the process is not being simply controlled but the quality of every core is being measured by PAT.

9.2.2.4 Tablet coating
Often the cores are coated before release. Traditionally, measurement of coating effectiveness is only inferred from dissolution testing, the coating is dissolved off and then concentration of this solution measured wither by HPLC or by direct UV. Both these are destructive and do not show the actual thickness that was originally in place. The penetrating power of reflectance NIR mean it can be used to monitor either the increase in spectral contribution from the coating material or even the reduction in spectral contribution of the core spectra [17]. Both systems have been successfully used. This is another area where the large sample area of the non-invasive measuring heads used for monitoring FBD and blending is useful. Within the 30 mm diameter sample area many tablets can be placed. It is possible to carry out a measurement of up to ten tablets, and generate an average coating value for all 10 in less than 1 s. Using either of the two NIR systems at-line analysis of tablet coating is possible.

9.2.2.5 Packaging
The packing of the final dosage form to produce the finished product is often overlooked as part of the manufacturing process. This impression is reinforced by the degree of automation. Packaging lines, that run at speeds so fast that they are hard to follow by eye, infer total efficiency. Incorrect labelling, missing inserts and missing products are some of the most common reasons for product complaint, and highlight this as a CCP. The raw materials used in packaging are subjected to similar requirements to those going into actual drug production. NIR can again be used for identification, as in drug product raw material identification. By using transflectance NIR not just to identity but also to detect contamination, both surface and internal can be highlighted.

The content of sealed vials and bottles can be qualified using transmission NIR, and non-invasive Raman and NIR cameras have been used to identify the correct filling of blister packs.

9.3 Conclusions

Process analysis is establishing itself as a fundamental component in Pharmaceutical Production. The FDA have now acknowledged the importance of these powerful measurement technologies, and have begun the development of guidelines on the use of PAT.

References

1. Callis, J., Illman, D. & Kowalski, B. (1987) *Anal. Chem.*, **59**, 624A.
2. Plugge, W. & VanDerViles, C. (1993) *J. Pharm. Biomed. Anal.*, **11**, 435.
3. Axon, T., Brown, R., Maris, S., Hammond, S. & Ting, F. (1998) *J. Near Infrared Spectrosc.*, **6**, A13.
4. Pasikatan, M., *et al.* (2001) *J. Near Infrared Spectrosc.*, **9**, 153.
5. Ward, W., *et al.* (1998) *Appl. Spectrosc.*, **52**, 17.
6. Rein, A.J. (1989) *Am. Lab.*, **21**(3), 126.
7. Roberts, M.J. (1991) *Process Control and Qual.*, **1**, 281.
8. Marteau, P., *et al.* (1994) *Analysis*, **22**, 32.
9. Norris, T., Aldridge, P. & Sekulic, S. (1997) *Analyst*, **122**, 549.
10. Sekulic, S., *et al.* (1996) *Anal. Chem.*, **68**(3).
11. Lai, C., *et al. AIChE Journal*, **47**, 2618.
12. Shinde, R., *et al. J. Pharm. Sci.*, **88**, 956.
13. Pan, T. & Sevick-Muraca, E. (2002) *Anal. Chem.*, **74**, 4228.
14. Rantanen, J., *et al.* (1998) *Powder Technol.*, **99**, 163.
15. Coffey, C., Predoehl, A. & Walker, D. (1998) *Appl. Spectrosc.*, **52**, 717.
16. Lowery, M., *et al.* (1999) *9th International Conference of Near Infrared Spectroscopy*.
17. Anderson, M., *et al.* (2000) *Anal. Chem.*, **72**, 2099.

Index

absolute configuration 212
accelerated solvent extraction 43
ACCORD–HMBC 119, 123
accurate mass 191
acetazolamide 214
acetominophen 298
acetonitrile 45, 88, 130, 191
acetylene 216
acid 51, 90, 93, 94, 97, 130, 138, 140, 173, 184, 241, 245, 259, 305
acid/base dissociation constant 241
acidic 44, 45, 57, 87, 88, 93, 96, 97, 102, 148, 173, 185, 254
acoustic emission spectroscopy (AES) 351
acquisition 112
active pharmaceutical ingredient (API) 1, 4, 11, 13, 14, 15, 16, 19, 20, 23, 30, 31, 32, 34, 44, 48, 51, 61, 106, 325, 326, 330, 338, 340, 341, 344, 349, 350, 351
aldehydes 203
alkyne 126
aluminas 58
amide 186, 203, 225, 232
amine 46
amino 46
amino acid 88
amorphous 217, 220, 222, 223, 224, 225, 227, 228, 240, 250, 251, 259, 260, 263, 273, 281, 283, 296, 307
amperometry 55
amylose 85
angular velocity 180
anion 185
anisotropic 126, 242, 255, 269
antibiotic 14, 27, 44, 86, 88, 89, 92, 96, 97, 98, 102, 107
antihistamine 33
aromatic 44
array detectors 55
aspartame 246
assay 95
'at-line' analysis 325

atmospheric pressure chemical ionisation (APCI) 168
atomic emission detector (AED) 53
atomic force microscope (AFM) 288
attenuated total reflectance (ATR) 204, 213, 230, 263, 331, 334, 335
automation 54, 216, 235, 328, 355
autosampler 138

back pressure 46
Barr Ruling 26
baseline 43
basic 13, 44, 47, 58, 61, 87, 96, 97, 100, 102, 115, 119, 150, 241, 242, 243, 244, 245, 269, 275, 289, 301
Beers law 225
benzene 95
benzodiazepines 90
bioanalysis 48, 79, 87, 95, 96, 99, 102
bioavailability 15, 19, 198, 217, 226, 240, 287, 295, 296
biochemistry 119
biological fluid 37, 95
blending 326, 349, 350–3, 355
Bragg equation 255
brucine 113, 119, 122, 123
buffer 56, 57, 130, 193
bulk density 244, 288
Burger–Ramberger rules 248

caffeine 261
calibration 2, 7, 8, 9, 12, 19, 22, 23, 139, 175, 192, 197, 198, 224, 225, 281, 316, 354
Cambridge Crystallographic Data Centre (CCDC) 254
capillary electrochromatography (CE) 31, 34, 35, 44, 54, 55, 57, 58, 60, 61, 98, 99, 100, 101, 102, 138, 145, 165, 168
capillary HPLC 39, 137, 173
capillary zone electrophoresis 55, 138
carbamate 232
carbamazepine 229

carbonyl 93, 107, 118, 122, 142, 150, 203, 215, 216, 218, 219, 220, 225, 232
cascade impactor 308
cation 48
CCD 58, 254
CCP 325, 326, 345, 353, 355
CEC 55, 57, 58, 100, 101, 138, 165
cefazolin 225
cefepime 224
cellulose 58, 85, 86, 272
certificate of analysis (COA) 326
21 CFR 24
chalcogenide 335
channel hydrate 221
chemical ionisation 106, 167, 170, 173, 184
chemical shift 108, 110, 115, 119, 122, 123, 125, 131, 149, 150
chemometric 32, 66, 67, 330
chiral resolution 79
chirality 74, 211
CIGAR–HMBC 119
cimetidine 152, 153, 218
clean in place (CIP) 327
column 32, 37, 38, 42, 45, 46, 47, 48, 49, 51, 52, 53, 54, 56, 58, 61, 62, 65, 66, 88, 89, 90, 92, 95, 96, 97, 101, 102, 128, 134, 137, 151, 165, 172, 173, 185, 186, 193, 234, 337
column switching 63
combinatorial chemistry 137, 138, 199, 213
compaction behaviour 244
computer-assisted interpretation 125
computer assisted structure determination 125
conformation 10, 119, 148, 153, 217, 236
conformity 326, 327, 345
conjugation 218
contaminant 212
corticosterone 302
cortisone acetate 217
COSY 107, 108, 112, 115, 151
coupling constants 112, 123
CP-MAS 150, 152, 154, 269, 273
critical control points 324
cross polarisation 108
crystal lattice 220, 221, 242, 245, 249, 261, 268, 296
crystalline form 216, 250, 295, 307
crystallographic 20, 218, 249, 254, 269, 298, 305
crystallography 254, 298, 300, 305
cyclodextrins 55, 57, 86, 88, 96, 97, 99, 100, 145
cysteine 89, 152

data system 166, 167, 193, 199, 200
database 125, 186, 201, 254, 297, 298
deconvolution 177, 186
decoupling 149, 150, 151, 269
degradation 23, 34, 106, 107, 128, 137, 147, 184, 196
density 57, 126, 148, 172, 211, 240, 244, 254, 284, 287, 288, 317, 352
DEPT 108, 117, 126, 150
derivatisation 34, 43, 45, 53, 80, 95, 100, 337
derivatised silicas 58
detection 4, 31, 34, 37, 38, 48, 50, 51, 52, 53, 54, 55, 58, 59, 66, 80, 95, 96, 100, 101, 102, 112, 113, 117, 118, 128, 130, 131, 137, 138, 139, 144, 147, 150, 152, 165, 180, 182, 194, 195, 201, 225, 226, 234, 235, 251, 265, 268, 302, 315, 337, 353
deuterated solvents 57, 113, 127, 137
diastereomeric 80, 94
diastereotopic 125, 126
dielectric analysis (DEA) 283
dielectric spectroscopy 283
differential scanning calorimetry (DSC) 217, 221, 228, 229, 231, 260, 275, 276, 277, 278, 279, 281, 282, 283, 299, 301
differential thermal analysis (DTA) 276, 277, 278, 281
diffuse reflectance 205, 224, 234, 236, 263, 346, 349, 352
diffusion 108, 113
diffusion constant 113, 114, 115, 145
dimeric 145
diode lasers 228, 268
1,2-diphenyl-1,2-diaminoethane 92
dipolar coupling 149, 150, 151, 152, 153
dipole–dipole interaction 149
diprophylline 263
dirithromycin 219
discovery research 32
dispersive instrument 208
dissolution 27, 42, 148, 217, 239, 240, 241, 250, 287, 295, 313, 314, 326, 327, 355
DOSY 108, 113, 115
DPFGSE 108, 113
DRAWS 152
DRIFTS 205, 219, 226, 228, 263
drop density 288
drug product 4, 11, 12, 13, 14, 15, 19, 20, 23, 34, 52, 79, 105, 216, 226, 228, 240, 241, 251, 252, 269, 307, 308, 319, 326, 344, 346, 355

drug substance 1, 17, 33, 43, 52, 54, 79, 87, 95, 99, 105, 125, 128, 144, 146, 153, 154, 184, 210, 217, 218, 220, 221, 224, 226, 228, 234, 236, 241, 248, 251, 252, 269, 272, 281, 289, 295, 298, 302, 307, 308, 309, 313, 319, 321, 326, 328, 330, 345, 352
dryer monitoring 340
DTGS 334, 335
dynamic range 175, 176, 183

efficiency 46, 47, 50, 54, 61, 85, 86, 94, 97, 98, 101, 209, 231, 244, 268, 355
electromagnetic spectrum 203, 261, 265
electron beam 168, 171
electron ionisation 167, 169, 177, 193
electron multiplier 175, 182
electroosmotic flow 56
electrospray 168, 172, 173, 175, 177, 184, 185, 186, 188, 192, 193, 199
enantiomer 74, 76, 78, 79, 80, 86, 87, 90, 92, 94, 95, 96, 97, 99, 101, 102, 145, 212, 229, 263
enantioselectivity 86, 88, 90, 92, 94, 96, 97, 98, 102
enantiotropism 248
energy dispersive X-ray spectrometry (EDS) 319, 320
energy/temperature diagram 248
eniluracil 215
enol 211, 218
environmental analysis 52
enzyme 74
excipients 242
excitation sculpting 113, 127, 135

FAB-MS 107
factor analysis 225, 229
factorial designs 66
Feret diameter 284, 316
fibre optics 205, 209, 210, 232, 335
flow cell 59, 128, 129, 131, 134, 135, 137, 138, 234
flow rate 54, 62, 131, 278
flow rates 44, 46, 48, 59, 98, 129, 172, 173
flowability 244
flufenamic acid 305
fluidised bed dryers (FBD) 353, 355
fluorescence 55, 95, 208, 209, 226, 267, 268, 330
focused beam reflectance measurement (FBRM) 338, 339, 340, 351

Food and Drug Administration (FDA) 7, 9, 10, 11, 12, 13, 18, 20, 21, 24, 25, 26, 27, 28, 29, 30, 34, 69, 317, 355
Fourier transform 204, 208, 334
fragmentation 166, 169, 170, 175, 181, 188, 189, 191, 201
free energy 240, 248, 297, 298, 306
frusemide 299
FT-IR 53, 60, 205, 208, 228, 232, 234, 236, 265, 282, 320, 334
FT-Raman 208, 209, 226, 229
functional group 39, 43, 44, 93, 106, 107, 126, 139, 203, 204, 207, 208, 214, 217, 221, 261

gas chromatography (GC) 31, 34, 35, 41, 43, 44, 49, 50, 51, 52, 53, 54, 58, 59, 60, 61, 62, 68, 69, 71, 72, 99, 143, 145, 169, 175, 193, 194, 196, 234, 337
gas liquid chromatography (GLC) 50, 165
gas solid chromatography (GSC) 50
GC/MS 53, 234
gel capsules 216
Gibbs phase rule 245
Gold Sheet 28, 30
Good Laboratory Practice (GLP) 5, 7, 9, 10, 18, 19, 29
Good Manufacturing Practice (GMP) 4, 5, 7, 10, 11, 13, 14, 15, 16, 17, 18, 19, 20, 22, 23, 24, 29, 30, 54
gradient enhancement 117
granulation 250, 349, 350, 352, 353
granule density 288
gravimetric 147
Grignard 232
griseofulvin 221

Headspace analysis 52
HETCOR 107, 108
heterocyclic 120
heteronuclear coupling 123
heteronuclear interaction 149
heteronuclei 118, 120
Heuristic evolving latent projections 66
high performance liquid chromatography (HPLC) 31, 34, 35, 39, 44, 45, 46, 47, 48, 51, 53, 55, 57, 58, 59, 60, 61, 62, 63, 65, 66, 68, 69, 71, 98, 105, 128, 129, 130, 137, 138, 139, 140, 143, 145, 147, 165, 168, 172, 173, 185, 192, 193, 198, 337, 353, 355

high performance silicas 58
HMBC 108, 117, 118, 119, 123, 127, 128
HMQC 108, 115, 118
homonuclear couplings 119
hot-stage microscopy 275
HSQC 108, 109, 117, 118, 127, 128
hybrid instruments 181
hydrate 216, 220, 221
hydrogen bonding 122, 203, 211, 212, 215, 216, 217, 218, 220, 221, 242, 261, 262, 263, 273, 326
hydrophobic 86
hyphenation 55, 59, 105, 137, 138

ibuprofen 228
ICRMS 177
ID 327
identification 26, 34, 35, 105, 106, 128, 131, 135, 137, 138, 184, 186, 192, 195, 197, 201, 205, 208, 216, 217, 218, 224, 226, 227, 231, 234, 235, 261, 266, 267, 295, 319, 320, 321, 326, 345, 355
IMPEACH–MBC 119, 123
impurity 35, 37, 50, 51, 60, 79, 87, 95, 99, 100, 101, 102, 106, 119, 125, 128, 131, 135, 139, 143, 145, 147, 192, 196, 201, 208
INADEQUATE 151
induced shifts 99
infrared 127, 148, 165, 203, 204, 205, 207, 208, 209, 211, 212, 214, 218, 219, 220, 221, 224, 225, 228, 229, 231, 232, 234, 236, 242, 248, 261, 263, 265, 266, 267, 268, 295, 305, 320, 326, 330
infrared spectroscopy 234–5
injection systems 59
integration 127, 140, 141, 145, 147
interaction 33, 50, 74, 80, 86, 90, 92, 93, 97, 98, 101, 148, 149, 150, 151, 165, 169, 184, 191, 199, 221, 235, 242, 263, 273
intermediate precision 146, 147
internal standard 139, 140, 142, 145, 180, 196, 197, 198, 225
International Center for Diffraction Data 255
International Conference on Harmonisation (ICH) 5, 12, 13, 15, 17, 18, 30, 37, 101, 106, 146
inverse gas chromatography 53
ion cyclotron resonance 174, 177
ion exchange 60
ion trap 175, 188
ionic strength 44, 55, 87, 88

IR 107, 203, 204, 205, 207, 208, 212, 214, 215, 216, 217, 218, 219, 220, 221, 222, 224, 225, 226, 228, 229, 231, 232, 234, 236, 241, 248, 250, 254, 260, 261, 263, 265, 267, 268, 276, 282, 319, 320, 330, 331, 334, 335, 337, 338
IR microscopy 229
ISO 9000 2, 5, 6, 7, 8, 9, 29
isomerism 203, 229
isothermal microcalorimetry 283
isotropic 242

keto 211, 218
ketones 203
Kofler 275

lactone 219
lactose 272, 273, 345
laser 205, 208, 209, 216, 224, 225, 232, 266, 267, 288, 308, 338
lattice vibrations 220, 222, 229
ligand exchange 99
limit of detection 195, 226
linewidth 137, 141, 149
liquid chromatography 234
liquid–liquid extraction (LLE) 52
logP 241
long range couplings 118, 119
Lufenuron 229
lyophilisation 250

magic angle spinning (MAS) 108, 149, 150, 152, 269
MALDI 177, 201
Martin diameter 284, 316
mass analyser 165, 173, 181, 182, 166 167
mass chromatogram 193, 195, 197
mass spectrometry (MS) 55, 58, 96, 106, 107, 124, 127, 165, 168, 174, 191, 195, 199, 201, 210, 211, 234, 250, 282
matrix 37, 38, 42, 43, 151, 198, 204, 205, 228, 234, 250, 319, 331, 334, 335
MCA 13, 15, 30
MCT 334
mercury porosimetry 287
metabolite 78, 79, 95, 96, 106, 115, 128, 129, 137, 138, 201
method development 31, 32, 35, 38, 39, 41, 42, 43, 44, 48, 54, 55, 58, 59, 61, 62, 63, 65, 66, 79, 88, 89, 90, 93, 98, 100, 101, 146, 308, 314, 315, 317
method validation 145

micellar electrokinetic chromatography (MEKC) 55, 57
microcalorimetry 254, 283
microemulsion electrokinetic chromatography (MEEKC) 55, 57
micropellicular particles 45
microscopy 242, 254, 288, 295, 296, 304, 307, 308, 319, 321, 322
microspectroscopy 265, 295, 300, 319
Miller indices 298
milling 250, 308, 315, 326, 341, 343
MIR 261, 263
mixture analysis 106
mobile phase 44, 45, 48, 49, 50, 66, 80, 87, 88, 90, 94, 96, 98, 102, 173, 337
moisture sorption 254
molecular crystals 240
molecular formula 127, 128, 194
molecular structure 43, 106, 127, 203, 208, 211, 214, 222, 236, 241, 242
molecular vibrations 203
molecular weight 106, 107, 118, 127, 137, 165, 166, 172, 183, 184, 185, 186, 188, 288
monographs 217, 241, 328
monotropism 248
mull 204, 229, 263
multivariate 225, 266
Mutual Recognition Agreement (MRA) 15, 16, 21, 29, 30

n-hexane 86, 97
NAMAS 8, 9, 29
nanospray 173, 186
narrow bore HPLC 129
natural product 123, 124, 137
near infrared spectroscopy (NIR) 265, 266, 267, 268, 324, 327, 330, 331, 332, 334, 335, 337, 340, 341, 345, 346, 347, 349, 350, 351, 352, 354, 355
neural networks 225
neutral molecules 88, 166, 169, 173
nOe 107, 109, 113, 119
NOESY 107, 109, 112, 113, 115, 128, 129, 135, 152
NSAID 90
nuclear magnetic resonance spectroscopy (NMR) 34, 55, 60, 99, 105, 106, 107, 108, 109, 112, 113, 114, 115, 117, 118, 119, 122, 123, 125, 127, 128, 129, 130, 131, 134, 135, 136, 137, 138, 139, 140, 141, 142, 143, 145, 146, 147, 148, 149, 150, 152, 153, 154, 165, 180, 183, 210, 211, 218, 234, 251, 254, 260, 263, 269, 272, 273, 289
Nujol 204, 225, 229

occupational exposure limit (OEL) 51
octadecylsilane 236
off-line monitoring 343
off-resonance 141
optical microscope 205, 209, 260, 267, 300, 313, 315, 320
optical microscopy 265, 276, 308, 314, 316, 320
organic modifier 44, 58, 135
Ostwald's law of stages 298
out-of-specification (OOS) 26, 27, 28, 30
out-of-specification laboratory results 26

packaging 295, 355
packaging materials 12, 14, 33, 34
packing symmetry 149
pan dryer 340
paracetamol 229, 298, 305, 306
parallel analysis 199
partial least squares 225
particle density 288
particle size 205, 224, 241, 284, 286, 295, 308, 309, 311, 313, 314, 315, 317, 321, 338, 339, 340, 345, 346, 347, 348, 351
partition coefficient 241
partitioning 48, 49, 57
patent 193
patents 90
peak broadening 129
peak pick 135
peak purity 131, 193
peak shape 44, 46
peak tailing 46, 97
peak width 54, 129, 141
PEEK 47
penicillin 3, 4, 12, 196
pentaerythritol tetrastearate 229
peptide 48, 172, 186, 201, 244
perfusion particles 45
pH 44, 46, 47, 55, 56, 57, 58, 87, 333
pharmacokinetics 95, 196
pharmacopoeias 216, 240, 261
phase correction 142
phase cycling 151
phase I/II 18, 19
phase III 20
phenyl acetate 131
phosphine oxides 90
phospholipids 119
phosphonates 90
Pirkle-type 86, 90, 92, 97, 98
pKa 47, 56, 241
polar embedded groups 47
polarisation transfer 108

polyamides 58
polymer 47, 55, 92, 97, 214, 228, 250, 267, 282, 283
polymorph 148, 153, 154, 157, 216, 218, 248, 249, 260, 265, 273, 296, 297, 299, 300, 302, 305, 306, 308
polymorphism 216, 217, 218, 231, 240, 241, 251, 263, 289, 295, 296, 297, 299, 300, 303, 337
polysaccharide 86, 87, 88, 92, 97, 102
porosity 287
positive ion 169, 171, 172, 185
positive/negative switching 171
povidone 226
powder diffraction 215, 218, 222, 225, 229, 231, 248, 251, 254, 255, 273
powder diffraction file 255
powder X-ray diffraction (PXRD) 254, 255, 260
pre-approval inspections 21, 29
pre-clinical 17, 32, 319
pre-concentration 37, 52, 55
precision 37, 42, 142, 143, 145, 146, 147, 192, 196, 197, 198, 228, 240, 308
precursor ion 175, 181, 182, 188, 189, 191, 192, 195
prednisolone 269, 272
preferred orientation 255
presaturation 128, 135
pressure 18, 47, 54, 58, 59, 98, 100, 138, 168, 169, 170, 172, 173, 188, 190, 193, 195, 204, 205, 231, 244, 249, 250, 254, 265, 283, 284, 287, 288, 297, 341
primary production 326
principal component analysis 225, 331
process analysis 324, 325, 337, 339, 349, 355
process analytical technologies (PAT) 326, 353, 355
process monitoring 195
process related impurities (PRI) 331, 332
processability 217, 341
product ion 181, 182
product ions 169
propan-2-ol 86, 90, 97
protein 48, 74, 86, 105, 152, 172, 177, 186, 199, 201, 212, 242, 250
proteomics 186
proton coil 118
proton resonances 115
proton transfer 169, 170
protonated solvent 113, 128
protonated species 169, 172

proxyphylline 263
pseudopolymorphs 217
pulse interval 140, 141
pulse sequence 120, 124, 151
pulsed field gradient 108
pump 48, 74, 131
pycnometry 244, 288

quadrupole ion traps 175
qualitative analysis 268, 327
quality 1, 2, 3, 4, 5, 6, 7, 8, 9, 10, 11, 12, 13, 14, 15, 16, 17, 18, 19, 22, 23, 24, 28, 29, 30, 32, 38, 55, 106, 110, 112, 113, 118, 129, 130, 142, 145, 147, 150, 151, 152, 198, 210, 216, 228, 229, 234, 235, 240, 251, 252, 254, 267, 276, 325, 326, 345, 355
quality assurance 2, 4, 5, 13, 28
quality control 1, 3, 5, 13, 16, 21, 30
quality system 4, 5, 28
quantitative analysis 205, 210, 254, 260, 265, 266, 267, 273
quantitative NMR 105, 139, 141
quaternary ammonium salt 185

racemate 78, 212, 263
racemic 74, 78, 79, 95, 102, 212, 229
racemisation 76, 95
radiochemical 55
radiofrequency 150, 174
Raman 60, 148, 165, 203, 205, 207, 208, 209, 211, 212, 213, 214, 215, 216, 217, 218, 219, 220, 221, 222, 224, 225, 226, 228, 229, 230, 231, 232, 234, 235, 236, 242, 248, 254, 260, 263, 266, 267, 268, 276, 282, 295, 319, 330, 337, 355
Raman microscopes 267
Raman optical activity (ROA) 212
raw material 1, 14, 20, 34, 50, 268, 325, 326, 327, 345, 348, 355
Rayleigh scatter 206
reaction monitoring 330, 331, 334
recovery 140, 147, 196, 198, 326
REDOR 152
regioisomers 51, 123
relative humidity 221, 320
relative standard deviation 141, 143, 146
relaxation 119, 123, 130, 140, 142, 145, 146, 147, 251, 283
repeatability 147
reproducibility 142, 146, 147, 196, 204, 260, 337

residual solvent 51, 52, 54, 302
resolution 38, 44, 45, 46, 53, 57, 58, 60, 61, 62, 66, 74, 86, 87, 88, 89, 95, 97, 99, 100, 101, 109, 110, 117, 129, 130, 131, 134, 141, 145, 150, 151, 152, 154, 175, 192, 228, 254, 281, 286, 288, 314, 315, 316, 330, 337
resolving power 60, 165, 174, 175, 177, 180, 182, 191, 337
response factor 51
retention characteristics 44, 45, 66
reverse phase 47
ritonavir 217
robustness 146, 147, 216, 326, 331, 346, 348
ROESY 109, 112
run time 37, 38, 61, 62, 63

salt 87, 212, 242, 245, 275, 321
sample introduction system 167, 168
sample preparation 31, 42, 43, 45, 52, 53, 54, 61, 96, 102, 118, 139, 147, 204, 205, 209, 211, 213, 216, 218, 219, 220, 226, 236, 263, 264, 266, 267, 275, 308, 311, 314, 315, 325, 341, 349
satellite 118, 143
scanning electron microscope (SEM) 295, 308, 309, 313, 314, 315, 316, 319, 320
scanning force microscope (SFM) 288
scanning probe microscopes 288
secondary manufacture 220
selected ion monitoring 183, 197
selection rules 203, 204, 207, 214
selectivity 34, 47, 55, 56, 57, 59, 94
sensitivity 52, 108
sidebands 141, 143, 145, 150, 269
signal averaging 109
signal-to-noise 109, 110, 117, 130, 131, 134, 140, 141, 143, 145, 147, 148, 268
silanol 46, 47
silica 39, 45, 46, 47, 54, 56, 90, 92, 97, 100, 234, 235, 327
silicon 90, 334
simulated moving bed (SMB) 39, 97, 98, 102
size exclusion chromatography 60
Sodium Dodecyl Sulphate (SDS) 57
solid phase extraction 43, 52, 138
solid phase microextraction 52
solid-state form 216, 217, 218, 224, 226, 228, 229, 247, 251, 261, 263, 265, 266, 268, 295, 296, 298, 299
solid state NMR 148, 303, 305

solid-state forms 217
solubility 46, 96, 98, 148, 217, 240, 241, 244, 248, 249, 250, 254, 295, 307
solution calorimetry 218, 283
solvates 216, 220, 249
solvent suppression 113, 114, 128, 130, 135
specificity 101, 139, 146, 148, 195, 337
spectral match index 345, 346
spin-locking 150
spironolactone 219
spray drying 250
stability 12, 17, 19, 20, 23, 34, 48, 50, 97, 146, 175, 184, 192, 196, 199, 217, 220, 240, 248, 249, 251, 260, 275, 289, 295, 296, 306
π–π stacking 259
standard deviation 140, 317, 327, 345, 349
starch 272
starting material 1, 15, 34, 51, 231, 308, 327, 335
stationary phase 39, 45, 46, 47, 49, 50, 58, 80, 90
stearic acid 226
stereoisomers 126
stereoselective 79
steric effects 126, 202
Stokes 206
Stokes scatter 206
stopped flow 129, 130, 131, 134
structure confirmation 106, 126
structure determination 105, 106, 298
structure elucidation 53, 105, 106, 107, 108, 109, 112, 117, 119, 122, 123, 126, 127, 139
sublimation 248, 275, 281, 283, 302
sucrose 110
sulfonamide 219
sulphoxides 90
superconducting magnet see nuclear magnetic resonance spectroscopy (NMR)
supercritical fluid chromatography (SFC) 31, 32, 41, 52, 58, 59, 60, 97, 98, 102, 138, 337
supercritical fluid extraction 43, 52
Supercritical Fluid Extraction (SFE) 43
surface energy 53
surfactants 57
system suitability 148

T_1 112, 140, 145, 150, 154
tablet coating 355
tap density 288
tautomerism 122, 142, 211, 219
teratogenic 74

thalidomide 5, 74, 80, 95
theophylline 261, 262, 263
thermal desorption 52
thermal methods 218
thermogravimetic analysis (TGA) 231, 250, 276, 281
thermogravimetry 218, 228, 249, 250, 265, 282
thermomechanical analysis (TMA) 282
thermomicroscopy 275
thermospray 172, 184, 188
thin layer chromatography (TLC) 31, 32, 34, 41, 58, 59, 60, 71, 73, 99, 234, 235
thiol 211
time of flight spectrometer 181
tip angle 140, 145
TOCSY 110, 129
torsional constraints 152
total ion chromatogram (TIC) 194, 195
Townsend discharge 169
toxicity 226
toxicology 1, 17, 18, 183, 308
trifluoroacetic 90, 97, 173
triple quadrupole 181, 182, 188, 190
true density 244, 288

USP 27, 140, 142, 327
UV 43, 50, 51, 54, 59, 80, 95, 98, 100, 107, 131, 134, 135, 137, 139, 142, 143, 165, 192, 193, 194, 254, 260, 330, 331, 332, 334, 337, 344, 355

validation 2, 17, 20, 22, 23, 42, 61, 105, 144, 145, 225, 308, 315, 328
van der Waals 261
vibrational optical activity (VOA) 212
vibrational spectroscopy 203, 210, 211, 216, 218, 228, 326
video microscopy 340

warfarin 48, 90
WATERGATE 113, 128, 135
WET 113, 128, 135
Wittig reactions 119

xanthine 261
X-ray diffraction 212, 244, 250, 254, 288
X-ray powder diffraction (XRPD) 148, 153, 225, 230, 254

zwitterionic 44